Lecture Notes in Mathematics 1965

Editors:
J.-M. Morel, Cachan
F. Takens, Groningen
B. Teissier, Paris

Charles Chidume

Geometric Properties
of Banach Spaces
and Nonlinear Iterations

 Springer

Charles Chidume
Abdus Salam International Centre for
Theoretical Physics
Mathematics Section
Strada Costiera 11
34014 Trieste
Italy
chidume@ictp.it

ISBN: 978-1-84882-189-7   e-ISBN: 978-1-84882-190-3
DOI: 10.1007/978-1-84882-190-3

Lecture Notes in Mathematics ISSN print edition: 0075-8434
                    ISSN electronic edition: 1617-9692

Library of Congress Control Number: 2008938194

Mathematics Subject Classification (2000): 47XX, 46XX, 45XX, 49XX, 65XX, 68XX

9 8 7 6 5 4 3 2 1

Springer Science+Business Media, LLC

springer.com

To my beloved family:
Ifeoma (wife),
and children:
Chu Chu; Ada; KK and Okey (Okido).

Preface

The contents of this monograph fall within the general area of nonlinear functional analysis and applications. We focus on an important topic within this area: *geometric properties of Banach spaces and nonlinear iterations,* a topic of intensive research efforts, especially within the past 30 years, or so.

In this theory, some geometric properties of Banach spaces play a crucial role. In the first part of the monograph, we expose these geometric properties most of which are well known. As is well known, among all infinite dimensional Banach spaces, Hilbert spaces have the *nicest* geometric properties. The availability of the inner product, the fact that the *proximity map or nearest point map* of a real Hilbert space H onto a closed convex subset K of H is Lipschitzian with constant 1, and the following two identities

$$||x + y||^2 = ||x||^2 + 2\langle x, y \rangle + ||y||^2, \qquad (*)$$

$$||\lambda x + (1 - \lambda)y||^2 = \lambda ||x||^2 + (1 - \lambda)||y||^2 - \lambda(1 - \lambda)||x - y||^2, \qquad (**)$$

which hold for all $x, y \in H$, are some of the geometric properties that characterize inner product spaces and also make certain problems posed in Hilbert spaces more manageable than those in general Banach spaces. However, as has been rightly observed by M. Hazewinkel, "... many, and probably most, mathematical objects and models do not naturally live in Hilbert spaces". Consequently, to extend some of the Hilbert space techniques to more general Banach spaces, analogues of the identities $(*)$ and $(**)$ have to be developed. For this development, the duality map which has become a most important tool in nonlinear functional analysis plays a central role. In 1976, Bynum [61] obtained the following analogue of $(*)$ for l_p spaces, $1 < p < \infty$:

$$||x + y||^2 \leq (p - 1)||x||^2 + ||y||^2 + 2\langle x, j(y) \rangle, \ 2 \leq p < \infty,$$

$$(p - 1)||x + y||^2 \leq ||x||^2 + ||y||^2 + 2\langle x, j(y) \rangle, \ 1 < p \leq 2.$$

Analogues of $(**)$ were also obtained by Bynum. In 1979, Reich [408] obtained an analogue of $(*)$ in uniformly smooth Banach spaces. Other analogues of $(*)$ and $(**)$ obtained in 1991 and later can be found, for example, in Xu [509] and in Xu and Roach [525].

In Chapters 1 and 2, basic well-known facts on geometric properties of Banach spaces which are used in the monograph are presented. The materials here (and much more) can be found in any of the excellent books on this topic (e.g., Diestel [206]; Lindenstrauss and Tzafriri [312]). The *duality map* which is central in our work is presented in Chapter 3. Here, we have also computed explicitly the duality maps in some concrete Banach spaces. In Chapters 4 and 5, we sketch the proofs of the analogues of the identities $(*)$ and $(**)$ obtained in 1991 and later. In the last section of Chapter 5, we present characterizations of real uniformly smooth Banach spaces and Banach spaces with uniformly Gâteaux differentiable norms by means of continuity properties of the normalized duality maps. Applications of the geometric properties of Banach spaces presented in Chapters 1 to 5 to iterative algorithms for solutions of nonlinear equations, an intensive and extensive area of research work (Berinde [28] contains 1575 entries in the reference list on this topic) begin in Chapter 6. To motivate some of the reasons for our choices of the classes of nonlinear operators studied in this monograph, we begin with the following.

Let K be a nonempty subset of a real normed linear space E and let $T : K \to K$ be a map. A point $x \in K$ is said to be *a fixed point* of T if $Tx = x$. Now, consider the differential equation $\frac{du}{dt} + Au(t) = 0$ which describes an evolution system where A is an *accretive map* from a Banach space E to itself. In Hilbert spaces, accretive operators are called *monotone*. At equilibrium state, $\frac{du}{dt} = 0$, and so a solution of $Au = 0$ describes the equilibrium or stable state of the system. This is very desirable in many applications in, for example, ecology, economics, physics, to name a few. Consequently, considerable research efforts have been devoted to methods of solving the equation $Au = 0$ when A is accretive. Since generally A is nonlinear, there is no closed form solution of this equation. The standard technique is to introduce an operator T defined by $T := I - A$ where I is the identity map on E. Such a T is called *a pseudo-contraction* (or is called *pseudo-contractive*). It is then clear that any zero of A is a fixed point of T. As a result of this, the study of fixed point theory for pseudo-contractive maps has attracted the interest of numerous scientists and has become a flourishing area of research, especially within the past 30 years or so, for numerous mathematicians. A very important subclass of the class of pseudo-contractive mappings is that of *nonexpansive* mappings, where $T : K \to K$ is called nonexpansive if $||Tx - Ty|| \leq ||x - y||$ holds for arbitrary $x, y \in K$.

Apart from being an obvious generalization of the contraction mappings, nonexpansive maps are important, as has been observed by Bruck [59], mainly for the following two reasons:

- Nonexpansive maps are intimately connected with the monotonicity methods developed since the early 1960's and constitute one of the first classes of nonlinear mappings for which fixed point theorems were obtained by using the fine geometric properties of the underlying Banach spaces instead of compactness properties.
- Nonexpansive mappings appear in applications as the transition operators for initial value problems of differential inclusions of the form $0 \in \frac{du}{dt} + T(t)u$, where the operators $\{T(t)\}$ are, in general, set-valued and are *accretive* or *dissipative* and *minimally continuous*.

If K is a closed nonempty subset of a Banach space and $T : K \to K$ is nonexpansive, it is known that T may not have a fixed point (unlike the case if T is a strict contraction), and even when it has, the sequence $\{x_n\}$ defined by $x_{n+1} = Tx_n, n \geq 1$ (the so-called *Picard sequence*) may fail to converge to such a fixed point. This can be seen by considering an anti-clockwise rotation of the unit disc of $\mathbb{R}^2$ about the origin through an angle of say, $\frac{\pi}{4}$. This map is nonexpansive with the origin as the unique fixed point, but the Picard sequence fails to converge with any starting point $x_0 \neq 0$. Krasnosel'skii [291], however, showed that in this example, if the Picard iteration formula is replaced by the following formula,

$$x_0 \in K, \quad x_{n+1} = \frac{1}{2}\Big(x_n + Tx_n\Big), n \geq 0, \tag{0.1}$$

then the iterative sequence converges to the unique fixed point. In general, if E is a normed linear space and T is a nonexpansive mapping, the following generalization of (0.1) which has proved successful in the approximation of a fixed point of T (when it exists) was given by Schaefer [431]:

$$x_0 \in K, \quad x_{n+1} = (1 - \lambda)x_n + \lambda Tx_n, \quad n \geq 0, \quad \lambda \in (0, 1). \tag{0.2}$$

However, the most general iterative formula for approximation of fixed points of nonexpansive mappings, which is called the *Mann iteration formula* (in the light of Mann [319]), is the following:

$$x_0 \in K, \quad x_{n+1} = (1 - \alpha_n)x_n + \alpha_n Tx_n, \quad n \geq 0, \tag{0.3}$$

where $\{\alpha_n\}$ is a sequence in the interval $(0, 1)$ satisfying the following conditions: (i) $\lim_{n \to \infty} \alpha_n = 0$ and (ii) $\sum_{n=1}^{\infty} \alpha_n = \infty$. The recursion formula (0.2) is consequently called the *Krasnoselskii-Mann (KM)* formula for finding fixed points of *nonexpansive (ne)* mappings. The following quotation indicates part of the interest in iterative approximation of fixed points of *nonexpansive mappings*.

- "Many well-known algorithms in signal processing and image reconstruction are iterative in nature .... A wide variety of iterative procedures used in signal processing and image reconstruction and elsewhere are special cases of the KM iteration procedure, for particular choices of the ne operator..." (Charles Byrne, [63]).

For the past 30 years or so, the study of the Krasnoselskii-Mann iterative procedures for the approximation of fixed points of nonexpansive mappings and fixed points of some of their generalizations, and approximation of zeros of accretive-type operators have been flourishing areas of research for many mathematicians. Numerous applications of analogues of (∗) and (∗∗) to nonlinear iterations involving various classes of nonlinear operators have since then been topics of intensive research. Today, substantial definitive results have been proved, some of the methods have reached their boundaries while others are still subjects of intensive research activity. However, it is apparent that the theory has now reached a level of maturity appropriate for an examination of its central themes.

The aim of this monograph is to present an in-depth and up-to-date coverage of the main ideas, concepts and most important results on iterative algorithms for approximation of fixed points of nonlinear nonexpansive mappings and some of their important generalizations; iterative approximation of zeros of accretive-type operators; iterative approximation of solutions of variational inequality problems involving these operators; iterative algorithms for solutions of Hammerstein integral equations; and iterative approximation of common fixed points (and common zeros) of families of these mappings. Furthermore, some important open questions related to these selected topics are included.

We assume familiarity with basic concepts of analysis and topology. The monograph is addressed to graduate students of mathematics, computer science, statistics, informatics, engineering, to mathematicians interested in learning about the subject, and to numerous specialists in the area.

I have great pleasure in thanking Professor Giovanni Vidossich of Institute for Advanced Studies, SISSA, Trieste, Italy for his constant encouragement, and Professor Billy Rhoades of the Department of Mathematics, Indiana University, Bloomington, Indiana, USA, who read a version of the first draft and whose comments spurred me on. Professors Vasile Berinde and Naseer Shahzad helped with putting my Latex files into the Spinger LNM format. I am very grateful to them for this. Very special thanks go to the staff of the Publications Department of The Abdus Salam ICTP, for their patience and ever-ready assistance in typesetting the original version of the monograph. Finally, I have great pleasure in expressing my sincere gratitude to my wife, Ify, and to our children for their encouragement and understanding.

The monograph was written, over a period of several years at The Abdus Salam International Centre for Theoretical Physics, Trieste. I am grateful to the various Directors: Professor Abdus Salam (Founding Director), Professor M.A. Virasoro (Second Director), Professor K.R. Sreenivasan (Present Director), and to UNESCO and IAEA for hospitality.

Trieste,   *Charles Ejike Chidume*
June 2008

Contents

Chapter 1
Some Geometric Properties of Banach Spaces

1.1 Introduction

In the first part of this monograph (Chapters 1 to 5), we explore selected geometric properties of Banach spaces that will play crucial roles in our study of iterative algorithms for nonlinear operators in various Banach spaces.

In this chapter, we introduce the classes of *uniformly convex* and *strictly convex* spaces, and in Chapter 2, we shall introduce the class of *smooth spaces*. All the results presented in these two chapters are well-known and standard and can be found in several books on geometry of Banach spaces, for example, in Diestel [206], or in Lindenstrauss and Tzafriri [312]. Consequently, we shall skip some details and long proofs.

It is well known that if E is a real normed space, the following identities hold

$$||x + y||^2 = ||x||^2 + 2\langle x, y \rangle + ||y||^2, \tag{1.1}$$

$$||\lambda x + (1 - \lambda)y||^2 = \lambda||x||^2 + (1 - \lambda)||y||^2 - \lambda(1 - \lambda)||x - y||^2, \tag{1.2}$$

for all $x, y \in E, \lambda \in (0, 1)$ if and only if E is a real inner product space.

These geometric identities which characterize inner product spaces make numerous problems posed in real Hilbert spaces more manageable than those posed in arbitrary real Banach spaces. Consequently, to extend some of the Hilbert space techniques to more general Banach spaces, analogues of these identities have to be developed in such Banach spaces.

In Chapter 3, we introduce the *duality map* which has become a most important tool in nonlinear functional analysis. We compute the duality map explicitly for some specific Banach spaces. In arbitrary normed spaces, the duality map will serve as the analogue of the inner product in Hilbert spaces.

C. Chidume, *Geometric Properties of Banach Spaces and Nonlinear Iterations*,
Lecture Notes in Mathematics 1965,
© Springer-Verlag London Limited 2009

In Chapters 4 and 5, we present the analogues of the identities (1.1) and
(1.2) in uniformly convex and uniformly smooth Banach spaces, respectively.
Most of the results presented in these chapters were developed in 1991 or
later.

At the end of Chapter 5, we characterize uniformly smooth spaces and
spaces with uniformly Gâteaux differentiable norm in terms of *uniform continuity of the normalized duality map on bounded sets*. These characterizations
will be used extensively in the monograph. We begin with some basic notions.

In 1936, A.J. Clarkson [191] published his famous paper on *uniform convexity* (defined below). This work signalled the beginning of extensive research
efforts on the geometry of Banach spaces and its applications in functional
analysis.

1.2 Uniformly Convex Spaces

Let X be an arbitrary normed space and for fixed $x_0 \in X$, let $S_r(x_0)$ denote
the *sphere* centred at x_0 with radius $r > 0$, that is,

$$S_r(x_0) := \{x \in X : \|x - x_0\| = r\}.$$

Definition 1.1. A normed space X is called *uniformly convex* if for any $\varepsilon \in$
$(0, 2]$ there exists a $\delta = \delta(\varepsilon) > 0$ such that if $x, y \in X$ with $\|x\| = 1, \|y\| = 1$
and $\|x - y\| \geq \varepsilon$, then $\left\| \frac{1}{2}(x + y) \right\| \leq 1 - \delta$.

Thus, a normed space is uniformly convex if for any two distinct points x and
y on the unit sphere centred at the origin the midpoint of the line segment
joining x and y is never on the sphere but is close to the sphere only if x and
y are sufficiently close to each other.

We note immediately that the following definition is also used: A normed
space X is *uniformly convex* if for any $\varepsilon \in (0, 2]$ there exists a $\delta = \delta(\varepsilon) >$
0 such that if $x, y \in X$ with $\|x\| \leq 1, \|y\| \leq 1$ and $\|x - y\| \geq \varepsilon$, then
$\|\frac{1}{2}(x + y)\| \leq 1 - \delta$. In the sequel we shall use either of the two definitions.

Theorem 1.2. L_p *spaces,* $1 < p < \infty$, *are uniformly convex.*

Proof. See e.g., Diestel [206].

Theorem 1.3. *Let* X *be a uniformly convex space. Then, for any* $d > 0$, $\varepsilon > 0$
and arbitrary vectors $x, y \in X$ *with* $\|x\| \leq d, \|y\| \leq d, \|x-y\| \geq \varepsilon$, *there exists*
a $\delta > 0$ *such that*

$$\left\| \frac{1}{2}(x + y) \right\| \leq \left[1 - \delta\left(\frac{\varepsilon}{d}\right)\right] d .$$

Proof. For arbitrary $x, y \in X$, let $z_1 = \frac{x}{d}, z_2 = \frac{y}{d}$, and set $\bar{\varepsilon} = \frac{\varepsilon}{d}$. Obviously
$\bar{\varepsilon} > 0$. Moreover, $\|z_1\| \leq 1, \|z_2\| \leq 1$ and $\|z_1 - z_2\| = \frac{1}{d}\|x - y\| \geq \frac{\varepsilon}{d} = \bar{\varepsilon}$.
Now, by uniform convexity, we have for some $\delta = \delta(\frac{\varepsilon}{d}) > 0$,

$$\left\|\frac{1}{2}\left(z_1 + z_2\right)\right\| \le 1 - \delta(\bar{\varepsilon}),$$

that is,

$$\left\|\frac{1}{2d}\left(x + y\right)\right\| \le 1 - \delta\left(\frac{\varepsilon}{d}\right),$$

which implies,

$$\left\|\frac{1}{2}\left(x + y\right)\right\| \le \left[1 - \delta\left(\frac{\varepsilon}{d}\right)\right]d.$$

The proof is complete. □

Proposition 1.4. *Let X be a uniformly convex space and let $\alpha \in (0,1)$ and $\varepsilon > 0$. Then for any $d > 0$, if $x, y \in X$ are such that $\|x\| \le d, \|y\| \le d$, $\|x - y\| \ge \varepsilon$, then there exists $\delta = \delta\left(\frac{\varepsilon}{d}\right) > 0$ such that*

$$\|\alpha x + (1 - \alpha)y\| \le \left[1 - 2\delta\left(\frac{\varepsilon}{d}\right)\min\{\alpha, 1 - \alpha\}\right]d.$$

Proof. See Exercises 1.1, Problem 3.

1.3 Strictly Convex Banach Spaces

Definition 1.5. A normed space E is called *strictly convex* if for all $x, y \in E, x \ne y, \|x\| = \|y\| = 1$, we have $\|\lambda x + (1 - \lambda)y\| < 1 \; \forall \; \lambda \in (0,1)$.

Theorem 1.6. *Every uniformly convex space is strictly convex.*

Proof. See Exercises 1.1, Problem 4.

Theorem 1.6 gives a large class of strictly convex spaces. However, we shall see later that some well known Banach spaces are *not* strictly convex.

We first give two examples of Banach spaces which are *strictly convex* but not *uniformly convex*.

Example 1.7. (Goebel and Kirk, [230]). Fix $\mu > 0$ and let $C[0,1]$ be endowed with the norm $\|.\|_\mu$ defined as follows,

$$\|x\|_\mu := \|x\|_0 + \mu\left(\int_0^1 x^2(t)dt\right)^{\frac{1}{2}},$$

where $\|.\|_0$ is the usual supremum norm. Then,

$$\|x\|_0 \le \|x\|_\mu \le (1 + \mu)\|x\|_0, \quad x \in C[0,1],$$

and the two norms are equivalent with $\|.\|_\mu$ near $\|.\|_0$ for small μ. However, $(C[0,1], \|.\|_0)$ is not strictly convex while for any $\mu > 0$, $(C[0,1], \|.\|_\mu)$ is. On the other hand, for any $\varepsilon \in (0,2]$ there exist functions $x, y \in C[0,1]$

with $||x||_\mu = ||y||_\mu = 1, ||x - y|| = \varepsilon$, and $||\frac{x+y}{2}||$ arbitrarily near 1. Thus, $(C[0,1], ||.||_\mu)$ is not uniformly convex.

Example 1.8. (Goebel and Kirk, [230]). Let $\mu > 0$ and let $c_0 = c_0(\mathbb{N})$ be given the norm $||.||_\mu$ defined for $x = \{x_n\} \in c_0$ by

$$||x||_\mu := ||x||_{c_0} + \mu\Big(\sum_{i=1}^\infty \Big(\frac{x_i}{i}\Big)^2\Big)^{\frac{1}{2}},$$

where $||.||_{c_0}$ is the usual l_∞ norm. As in Example 1.7, the spaces $(c_0, ||.||_\mu)$ for $\mu > 0$ are strictly convex but not uniformly convex, while c_0 with its usual norm is not strictly convex.

We now return to uniformly convex spaces.

Although Theorem 1.2 provides examples of large classes of spaces which are uniformly convex, some well known spaces are not uniformly convex. We, in fact, show that these spaces are not strictly convex.

Example 1 *The space ℓ_1 is not strictly convex.* To see this, take $\varepsilon = 1$ and choose $\bar{x} = (1,0,0,0,\dots), \bar{y} = (0,-1,0,0,\dots)$. Clearly $\bar{x}, \bar{y} \in \ell_1$ and $||\bar{x}||_{\ell_1} = 1 = ||\bar{y}||_{\ell_1}, ||\bar{x} - \bar{y}||_{\ell_1} = 2 > \varepsilon$. However, $||\frac{1}{2}(\bar{x} + \bar{y})|| = 1$, showing that ℓ_1 is not strictly convex.

Example 2 *The space ℓ_∞ is not strictly convex.*
Consider $\bar{u} = (1,1,0,0,0,\dots)$ and $\bar{v} = (-1,1,0,0,0,\dots)$. Both $\bar{u}, \bar{v} \in \ell_\infty$. Take $\varepsilon = 1$. Then $||\bar{u}||_\infty = 1 = ||\bar{v}||_\infty$ and $||\bar{u} - \bar{v}||_\infty = 2 > \varepsilon$. However, $||\frac{1}{2}(\bar{u} + \bar{v})||_\infty = 1$ and so ℓ_∞ is not strictly convex.

Example 3 Consider $C[a,b]$, the space of real–valued continuous functions on the compact interval $[a,b]$, with the "sup norm". Then $C[a,b]$ is not strictly convex. To see this, choose two functions f, g defined as follows:

$$f(t) := 1 \quad for\ all \quad t \in [a,b], \ g(t) := \frac{b-t}{b-a} \quad for\ each \quad t \in [a,b].$$

Take $\varepsilon = \frac{1}{2}$. Clearly $f, g \in C[a,b], ||f|| = ||g|| = 1$ and $||f - g|| = 1 > \varepsilon$. Also $||\frac{1}{2}(f + g)|| = 1$ and so $C[a,b]$ is not strictly convex.

Example 4 The spaces L_1, L_∞ and c_0 are not strictly convex. The verifications are left as easy exercises for the reader.

1.4 The Modulus of Convexity

In this section, we shall define a function called the *modulus of convexity* of a normed space, X (denoted by $\delta_X, \delta_X : (0,2] \to [0,1]$) and prove three important properties of the function that will be used in the sequel, namely:

1. *(Lemma 1.12): For every normed space, the function $\frac{\delta_X(\varepsilon)}{\varepsilon}$ is nondecreasing on $(0,2]$.*
2. *(Theorem 1.13): The modulus of convexity of a normed space X, δ_X, is a convex and continuous function.*
3. *(Corollary 1.15): In a uniformly convex space X, the modulus of convexity, δ_X, is a strictly increasing function.*

We begin with the notion of *convex functions*.

A real-valued function f is called *convex* if it satisfies the following inequality:

$$f(\lambda x + (1-\lambda)y) \le \lambda f(x) + (1-\lambda)f(y),$$

for every $\lambda \in [0,1]$ and every $x, y \in D(f)$, the domain of f, that *we demand to be a convex set*. Good references for the study of convex functions are Krasnosel'skii and Ruticklii [292] and Phelps [383]. From these references we obtain the next result that will be useful later.

Lemma 1.9. *Every convex function f with convex domain in $\mathbb{R}$ is continuous.*

To motivate the definition of the *modulus of convexity*, we begin with some properties of inner product spaces. In an inner product space H, we consider the parallelogram law: For $x, y \in H$,

$$\|x + y\|^2 + \|x - y\|^2 = 2(\|x\|^2 + \|y\|^2).$$

In the particular case $\|x\| = \|y\| = 1$ we get the expression

$$\left\|\frac{x+y}{2}\right\|^2 = 1 - \frac{1}{4}\|x - y\|^2.$$

From this equality we can determine the distance between the midpoint of the segment joining x and y from the unit sphere $S := \{x \in H : \|x\| = 1\}$ in H by

$$1 - \left\|\frac{x+y}{2}\right\| = 1 - \sqrt{1 - \frac{\|x - y\|^2}{4}}.$$

Evidently this distance always lies between 0 and 1. If $\varepsilon \le \|x - y\|$ then

$$1 - \left\|\frac{x+y}{2}\right\| \ge 1 - \sqrt{1 - \frac{\varepsilon^2}{4}}.$$

The idea behind these formulas is the *convexity of the unit ball* in an inner product space, i.e., if the distance between two points x and y in the unit sphere is larger than ε, then the midpoint of the segment joining x and y remains in the unit ball with $1 - \frac{\varepsilon^2}{4} \ge \|\frac{x+y}{2}\|^2$. Motivated by this, we extend this notion to spaces, not with an inner product, but with a norm and study "how much convex" the unit ball is.

Definition 1.10. Let X be a normed space with $\dim X \geq 2$. The *modulus of convexity* of X is the function $\delta_X : (0,2] \rightarrow [0,1]$ defined by

$$\delta_X(\varepsilon) := \inf\left\{ 1 - \left\| \frac{x+y}{2} \right\| : \|x\| = \|y\| = 1 \,;\, \varepsilon = \|x-y\| \right\}.$$

In the particular case of an inner product space H, we have

$$\delta_H(\varepsilon) = 1 - \sqrt{1 - \frac{\varepsilon^2}{4}}.$$

Several authors have given various definitions of $\delta_X(\varepsilon)$. In order to state the relation between some of these definitions, we present the following lemmas due to M.M. Day, [196]. The proofs of these two results use some geometric ideas that we do not develop here, they can be found in [196].

Lemma 1.11. *Let X be a normed space with $\dim X \geq 2$. Then*

$$\delta_X(\varepsilon) = \inf\left\{ 1 - \left\| \frac{x+y}{2} \right\| : \|x\| \leq 1 \,;\, \|y\| \leq 1 \,;\, \varepsilon \leq \|x-y\| \right\}$$

$$= \inf\left\{ 1 - \left\| \frac{x+y}{2} \right\| : \|x\| \leq 1 \,;\, \|y\| \leq 1 \,;\, \varepsilon = \|x-y\| \right\}.$$

Note that from this lemma, it is evident that $\delta_X(0) = 0$.

Lemma 1.12. *For every normed space X, the function $\delta_X(\varepsilon)/\varepsilon$ is non-decreasing on $(0,2]$.*

Proof. Fix $0 < \eta \leq 2$ with $\eta \leq \varepsilon$ and x, y in X such that $\|x\| = 1 = \|y\|$ and $\|x - y\| = \varepsilon$. Suffices to prove $\frac{\delta_X(\eta)}{\eta} \leq \frac{\delta_X(\epsilon)}{\epsilon}$. Consider

$$u = \frac{\eta}{\varepsilon}x + \left(1 - \frac{\eta}{\varepsilon}\right)\frac{x+y}{\|x+y\|} \quad \text{and} \quad v = \frac{\eta}{\varepsilon}y + \left(1 - \frac{\eta}{\varepsilon}\right)\frac{x+y}{\|x+y\|}.$$

Then $u - v = \frac{\eta}{\varepsilon}(x - y)$, $\|u - v\| = \eta$ and

$$\frac{u+v}{2} = \frac{x+y}{\|x+y\|}\left(1 - \frac{\eta}{\varepsilon} + \frac{\eta\|x+y\|}{2\varepsilon}\right),$$

which implies that

$$\left\| \frac{x+y}{\|x+y\|} - \frac{u+v}{2} \right\| = \frac{\eta}{\varepsilon} - \frac{\eta\|x+y\|}{2\varepsilon}$$

$$= 1 - \left(1 - \frac{\eta}{\varepsilon} + \frac{\eta\|x+y\|}{2\varepsilon}\right)$$

$$= 1 - \frac{\|u+v\|}{2}.$$

Note that

$$\left\| \frac{x+y}{\|x+y\|} - \frac{x+y}{2} \right\| = \|x+y\| \left(\frac{1}{\|x+y\|} - \frac{1}{2} \right) = 1 - \frac{\|x+y\|}{2}. \quad (i)$$

Now

$$\frac{\left\| \frac{x+y}{\|x+y\|} - \frac{u+v}{2} \right\|}{\|u-v\|} = \frac{1}{\eta} \left(\frac{\eta}{\varepsilon} - \frac{\eta\|x+y\|}{2\varepsilon} \right)$$

$$= \frac{1}{\varepsilon} \left(1 - \frac{\|x+y\|}{2} \right) = \frac{\left\| \frac{x+y}{\|x+y\|} - \frac{x+y}{2} \right\|}{\|x-y\|}, \quad (ii)$$

and then

$$\frac{\delta_X(\eta)}{\eta} \leq \frac{1 - \frac{\|u+v\|}{2}}{\|u-v\|} = \frac{\left\| \frac{x+y}{\|x+y\|} - \frac{u+v}{2} \right\|}{\|u-v\|}$$

$$= \frac{\left\| \frac{x+y}{\|x+y\|} - \frac{x+y}{2} \right\|}{\|x-y\|} = \frac{1 - \frac{\|x+y\|}{2}}{\|x-y\|}$$

$$= \frac{1 - \frac{\|x+y\|}{2}}{\varepsilon}.$$

By taking the infimum over all possible x and y with $\varepsilon = \|x - y\|$ and $\|x\| = \|y\| = 1$, we obtain that $\frac{\delta_X(\eta)}{\eta} \leq \frac{\delta_X(\varepsilon)}{\varepsilon}$. □

We now present the following known and interesting theorem. A proof can be found in Lindenstrauss and Tzafriri, [312].

Theorem 1.13. *The modulus of convexity of a normed space X, δ_X, is a convex and continuous function.*

1.5 Uniform Convexity, Strict Convexity and Reflexivity

The following result characterizes the uniformly convex spaces in terms of δ_X.

Theorem 1.14. *A normed space X is uniformly convex if and only if $\delta_X(\varepsilon) > 0$ for all $\varepsilon \in (0, 2]$.*

Proof. If X is uniformly convex, given $\varepsilon > 0$ there exists $\delta > 0$ such that $\delta \leq 1 - \left\| \frac{x+y}{2} \right\|$ for every x and y such that $\|x\| = \|y\| = 1$ and $\varepsilon \leq \|x - y\|$. Therefore $\delta_X(\varepsilon) > 0$. For the converse, assume $0 < \delta_X(\varepsilon)$ for every $\varepsilon \in (0, 2]$. Fix $\varepsilon \in (0, 2]$ and take x, y with $\|x\| = \|y\| = 1$ and $\varepsilon \leq \|x - y\|$, then

$$0 < \delta_X(\varepsilon) \leq 1 - \left\| \frac{x+y}{2} \right\|$$

and therefore $\left\|\frac{x+y}{2}\right\| \leq 1 - \delta$ with $\delta = \delta_X(\varepsilon)$ which does not depend on x or y.   □

Corollary 1.15 *In a uniformly convex space X, the modulus of convexity is a strictly increasing function.*

Proof. By Lemma 1.12, for $s < t \leq 2$ we have $t\delta_X(s) \leq s\delta_X(t) < t\delta_X(t)$ since $\delta_X(t) > 0$. Thus we get $\delta_X(s) < \delta_X(t)$.   □

Until now, we know that a normed space is uniformly convex if $\delta_X(\varepsilon) > 0$ and that in the particular case of an inner product space H, $\delta_H(\varepsilon) = 1 - \sqrt{1 - \frac{\varepsilon^2}{4}}$. The following theorem (Day-Nörlander's theorem) asserts that the inner product spaces are the "most convex" of all uniformly convex spaces.

Theorem 1.16. *If X is an arbitrary uniformly convex space, then*

$$\delta_X(\varepsilon) \leq 1 - \sqrt{1 - \frac{\varepsilon^2}{4}}.$$

The proof can be found in Nörlander [360]. We skip it since it uses arguments of coordinates systems and curves in Banach spaces that we do not explore in our approach.

An interesting result about uniform convexity is the Milman–Pettis' theorem. The proof that we give here is due to J. Ringrose and is cited in Diestel [206].

Theorem 1.17. *(Milman-Pettis' theorem). If X is a uniformly convex Banach space, then X is reflexive.*

Proof. Suppose that X is a non-reflexive, uniformly convex Banach space. Then for some $\varepsilon > 0$ there exists x^{**} in X^{**} with $\|x^{**}\| = 1$ and such that the distance between x^{**} and $B(X)$, the closed unit ball of X is 2ε.

Let δ be chosen such that if x and y are in X with $\|x\| \leq 1$, $\|y\| \leq 1$ and $2 - \delta \leq \|x + y\|$ then $\|x - y\| \leq \varepsilon$. Take x^* in X^*, with $\|x^*\| = 1$ such that $\langle x^{**}, x^* \rangle = 1$.

Let V be the weak-star neighborhood of x^{**} given by

$$V = \left\{ u^{**} \in X^{**} \, |\langle x^*, u^{**} \rangle - 1| < \frac{\delta}{2} \right\}.$$

If x and y are in the closed unit ball of X belonging to V (under identification by the canonical injection) then $|\langle x^*, x + y \rangle| > 2 - \delta$ so $2 - \delta \leq \|x + y\|$. Hence $\|x - y\| \leq \varepsilon$. Fixing x we conclude that $V \cap B(X) \subset x + \varepsilon B(X^{**})$. By Goldstein's theorem (Goldstein's theorem says that "$J(B(X))$ is weak-star dense in $B(X^{**})$" where J is the canonical injection of X into X^{**}), we know that $V \cap B(X)$ is weak-star dense in $V \cap B(X^{**})$ which, since $x + \varepsilon B(X^{**})$ is weak-star closed, yields x^{**} belongs to $x + \varepsilon B(X^{**})$. But this means that the distance between x^{**} and $B(X)$ is less than or equal to ε, contradicting our choice of x^{**}.   □

SUMMARY

For the ease of reference, we summarize the key results obtained in this chapter. Here δ_X denotes the modulus of convexity of a normed space X.

S1

(a) *Every uniformly convex space is strictly convex.*
(b) *Every uniformly convex space is reflexive.*
(c) *X is uniformly convex if and only if $\delta_X(\varepsilon) > 0$ for all $\varepsilon \in (0, 2]$.*

S2

(a) $\frac{\delta_X(\varepsilon)}{\varepsilon}$ *is a nondecreasing function on $(0, 2]$.*
(b) $\delta_X : (0, 2] \to [0, 1]$ *is a convex and continuous function.*
(c) $\delta_X : (0, 2] \to [0, 1]$ *is a strictly increasing function.*

EXERCISES 1.1

1. Prove that L_p (or l_p) spaces, $1 < p < \infty$, are uniformly convex.
   (Hint: see e.g., Diestel [206]).
2. Justify the statement made in Example 4 that none of the spaces L_1, L_∞ and c_0 (with their usual norms) is strictly convex.
3. Prove Proposition 1.4.

   (Hint: Without loss of generality, assume $\alpha \in (0, \frac{1}{2}]$. Observe that

   $$\|\alpha x + (1 - \alpha)y\| = \|\alpha(x + y) + (1 - 2\alpha)y\| \le 2\alpha \|\tfrac{1}{2}(x + y)\| + (1 - 2\alpha)\|y\|.$$

   Apply Proposition 1.3).
4. Prove Theorem 1.6.

   (Hint: Apply Proposition 1.4 with $d = 1$, $\epsilon := \|\lambda x + (1 - \lambda)y\|$).
5. Prove that $\delta_H(\epsilon) = 1 - \sqrt{1 - \frac{\epsilon^2}{4}}$, where the symbols have the usual meanings.
6. Justify the following claims made in example 1.7. (a) $(C[0, 1], \|.\|_0)$ is not strictly convex. (b) For any $\mu > 0, (C[0, 1], \|.\|_\mu)$ is strictly convex. (c) $(C[0, 1], \|.\|_\mu)$ is not uniformly convex.

1.6 Historical Remarks

The strictly convex Banach spaces were introduced in 1936 by Clarkson, [191], who also studied the concept of uniform convexity. The uniform convexity of L_p spaces, $1 < p < \infty$, was established by Clarkson [191].

Chapter 2
Smooth Spaces

2.1 Introduction

In this chapter, we introduce the class of *smooth spaces*. We remark immediately that there is a *duality relationship* between uniform smoothness and uniform convexity. In the sequel, we shall examine this relationship. We begin with the following definition.

Definition 2.1. A normed space X is called smooth if for every $x \in X, ||x|| = 1$, there exists a *unique x^** in X^* such that $||x^*|| = 1$ and $\langle x, x^* \rangle = ||x||$.

2.2 The Modulus of Smoothness

In this section, we shall define a function called the *modulus of smoothness* of a normed space X (denoted by $\rho_X : [0, \infty) \rightarrow [0, \infty)$) and prove three important properties of the function that will be used in the sequel, namely:

1. *(Proposition 2.3) For every normed space X, the modulus of smoothness, ρ_X, is a convex and continuous function.*
2. *(Theorem 2.5) A normed space X is uniformly smooth if and only if*
$$\lim_{t \to 0^+} \frac{\rho_X(t)}{t} = 0.$$
3. *(Corollary 2.8) For every normed space, $\frac{\rho_X(t)}{t}$ is a nondecreasing function on $[0, \infty)$ and $\rho_X(t) \leq t \ \forall \ t \geq 0$.*

There exists a complete dual notion to uniformly smooth space which plays a central role in the structure of Banach spaces (see, Diestel [206]). In this section, we present this duality. Recall that a Banach space is called *smooth* if for every x in X with $||x|| = 1$, there exists a *unique x^** in X^* such that $||x^*|| = \langle x, x^* \rangle = 1$. Assume now that X is not smooth and take x in X and u^*, v^* in X^* such that $||x|| = ||u^*|| = ||v^*|| = \langle x, u^* \rangle = \langle x, v^* \rangle = 1$ and

C. Chidume, *Geometric Properties of Banach Spaces and Nonlinear Iterations,*
Lecture Notes in Mathematics 1965,
© Springer-Verlag London Limited 2009

$u^* \neq v^*$. Let y in X be such that $\|y\| = 1$, $\langle y, u^* \rangle > 0$ and $\langle y, v^* \rangle < 0$. Then for every $t > 0$ we have

$$1 + t\langle y, u^* \rangle = \langle x + ty, u^* \rangle \leq \|x + ty\|,$$

$$1 - t\langle y, v^* \rangle = \langle x - ty, v^* \rangle \leq \|x - ty\|$$

which imply $2 < 2 + t(\langle y, u^* \rangle - \langle y, v^* \rangle) \leq \|x + ty\| + \|x - ty\|$ or, equivalently

$$0 < t\left(\frac{\langle y, u^* \rangle - \langle y, v^* \rangle}{2}\right) \leq \frac{\|x + ty\| + \|x - ty\|}{2} - 1.$$

With this motivation we introduce the following definition.

Definition 2.2. Let X be a normed space with $\dim X \geq 2$. The *modulus of smoothness* of X is the function $\rho_X : [0, \infty) \to [0, \infty)$ defined by

$$\rho_X(\tau) := \sup\left\{\frac{\|x + y\| + \|x - y\|}{2} - 1 : \|x\| = 1; \|y\| = \tau\right\}$$

$$= \sup\left\{\frac{\|x + \tau y\| + \|x - \tau y\|}{2} - 1 : \|x\| = 1 = \|y\|\right\}.$$

Note that evidently $\rho_X(0) = 0$. The following results explore some properties of the modulus of smoothness. We begin with the following important proposition.

Proposition 2.3. *For every normed space X, the modulus of smoothness, ρ_X, is a convex and continuous function.*

Proof. Fix x and y with $\|x\| = \|y\| = 1$ and consider

$$f_{x,y}(t) := \frac{\|x + ty\| + \|x - ty\|}{2} - 1.$$

Then for λ in $[0, 1]$

$$f_{x,y}(\lambda t + (1 - \lambda)s)$$

$$= \frac{\|x + (\lambda t + (1 - \lambda)s)y\| + \|x - (\lambda t + (1 - \lambda)s)y\|}{2} - 1$$

$$\leq \frac{\lambda\|x + ty\| + (1 - \lambda)\|x + sy\| + \lambda\|x - ty\| + (1 - \lambda)\|x - sy\|}{2} - 1$$

$$= \lambda f_{x,y}(t) + (1 - \lambda)f_{x,y}(s).$$

Therefore $f_{x,y}$ is a convex function, for every choice of x and y.

Now for arbitrary $\varepsilon > 0$, there exist x, y with $\|x\| = \|y\| = 1$ such that

$$\rho_X(\lambda t + (1-\lambda)s) - \varepsilon \le f_{x,y}(\lambda t + (1-\lambda)s)$$
$$\le \lambda f_{x,y}(t) + (1-\lambda)f_{x,y}(s)$$
$$\le \lambda \rho_X(t) + (1-\lambda)\rho_X(s).$$

Therefore ρ_X is a convex function since ε is arbitrary. The continuity follows from lemma 1.9. $\qquad\square$

Definition 2.4. A normed space X is said to be *uniformly smooth* whenever given $\varepsilon > 0$ there exists $\delta > 0$ such that for all $x, y \in X$ with $\|x\| = 1$ and $\|y\| \le \delta$, then

$$\|x + y\| + \|x - y\| < 2 + \varepsilon\|y\|.$$

As the modulus of convexity characterizes the uniformly convex spaces, the modulus of smoothness can be used to characterize the uniformly smooth spaces. This is the manner of the following theorem.

Theorem 2.5. *A normed space X is uniformly smooth if and only if*

$$\lim_{t \to 0^+} \frac{\rho_X(t)}{t} = 0.$$

Proof. If X uniformly smooth and $\varepsilon > 0$ then, there exists $\delta > 0$ such that

$$\frac{\|x + y\| + \|x - y\|}{2} - 1 < \frac{\varepsilon}{2}\|y\|,$$

for every $x, y \in X$ such that $\|x\| = 1$, $\|y\| = \delta$. This implies $\rho_X(t) < \frac{\varepsilon}{2}t$ for every $t < \delta$.

Conversely, let $\varepsilon > 0$, and suppose there exists $\delta > 0$ such that $\rho_X(t) < \frac{1}{2}\varepsilon t$, for every $t < \delta$. Let $\|x\| = 1$, $\|y\| = \delta$. Then with $t = \|y\|$ we have $\|x + y\| + \|x - y\| < 2 + \varepsilon\|y\|$ and the space is uniformly smooth. $\qquad\square$

Proposition 2.6. *Every uniformly smooth normed space X is smooth.*

Proof. Suppose that X is not smooth, then there exist x_0 in X and x_1^*, x_2^* in X^* such that $x_1^* \ne x_2^*$, $\|x_1^*\| = \|x_2^*\| = 1$ and $\langle x_0, x_1^* \rangle = \|x_0\| = \langle x_0, x_2^* \rangle$. Evidently we can assume $\|x_0\| = 1$. Let y_0 in X be such that $\|y_0\| = 1$ and $\langle y_0, x_1^* - x_2^* \rangle > 0$. For every $t > 0$ we have

$$0 < t\langle y_0, x_1^* - x_2^* \rangle$$
$$= t(\langle y_0, x_1^* \rangle - \langle y_0, x_2^* \rangle)$$
$$= \frac{\langle x_0 + ty_0, x_1^* \rangle + \langle x_0 - ty_0, x_2^* \rangle}{2} - 1$$
$$\le \frac{\|x_0 + ty_0\| + \|x_0 - ty_0\|}{2} - 1,$$

therefore, $0 < \langle y_0, x_1^* - x_2^* \rangle \le \frac{\rho_X(t)}{t}$ for any $t > 0$ and then X is not uniformly smooth. $\qquad\square$

2.3 Duality Between Spaces

As usual in mathematics, the study of one concept is in close relation with another which reflects its characteristics. This kind of "duality" is present in our case. Now we state one of the fundamental links between the *Lindenstrauss duality formulas*.

Proposition 2.7. *Let X be a Banach space. For every $\tau > 0$, x in X, $\|x\| = 1$ and x^* in X^* with $\|x^*\| = 1$ we have*

$$\rho_{X^*}(\tau) = \sup\left\{\frac{\tau\varepsilon}{2} - \delta_X(\varepsilon) : 0 < \varepsilon \leq 2\right\},$$

$$\rho_X(\tau) = \sup\left\{\frac{\tau\varepsilon}{2} - \delta_{X^*}(\varepsilon) : 0 < \varepsilon \leq 2\right\}.$$

Proof. Let $\tau > 0$, $0 < \varepsilon \leq 2$, x, y in X with $\|x\| = \|y\| = 1$. By Hahn-Banach Theorem there exist x_0^*, y_0^* in X^* with $\|x_0^*\| = \|y_0^*\| = 1$ such that

$$\langle x + y, x_0^* \rangle = \|x + y\| \quad \text{and} \quad \langle x - y, y_0^* \rangle = \|x - y\|.$$

Then

$$\|x + y\| + \tau\|x - y\| - 2 = \langle x + y, x_0^* \rangle + \tau\langle x - y, y_0^* \rangle - 2$$

$$= \langle x, x_0^* + \tau y_0^* \rangle + \langle y, x_0^* - \tau y_0^* \rangle - 2$$

$$\leq \|x_0^* + \tau y_0^*\| + \|x_0^* - \tau y_0^*\| - 2$$

$$\leq \sup\{\|x^* + \tau y^*\| + \|x^* - \tau y^*\| - 2 :$$

$$\|x^*\| = \|y^*\| = 1\} = 2\rho_{X^*}(\tau).$$

If $\varepsilon \leq \|x - y\|$ then, from the last inequality,

$$\frac{\tau\varepsilon}{2} - \rho_{X^*}(\tau) \leq 1 - \frac{\|x + y\|}{2}.$$

Therefore $\frac{\tau\varepsilon}{2} - \rho_{X^*}(\tau) \leq \delta_X(\varepsilon)$ and since $0 < \varepsilon \leq 2$ is arbitrary, we get

$$\sup\left\{\frac{\tau\varepsilon}{2} - \delta_X(\varepsilon) : 0 < \varepsilon \leq 2\right\} \leq \rho_{X^*}(\tau).$$

Now let x^*, y^* in X^* with $\|x^*\| = \|y^*\| = 1$ and let $\delta > 0$. For a given $\tau > 0$ there exist x_0, y_0 in X with $x_0 \neq y_0$ and $\|x_0\| = \|y_0\| = 1$ such that

$$\|x^* + \tau y^*\| \leq \langle x_0, x^* + \tau y^* \rangle + \delta \quad \text{and} \quad \|x^* - \tau y^*\| \leq \langle y_0, x^* - \tau y^* \rangle + \delta$$

(from the definition of $\|.\|$ in X^*). From these inequalities, we obtain

$$
\begin{aligned}
&\|x^* + \tau y^*\| + \|x^* - \tau y^*\| - 2 \\
&\leq \langle x_0, x^* + \tau y^* \rangle + \langle y_0, x^* - \tau y^* \rangle - 2 + 2\delta \\
&= \langle x_0 + y_0, x^* \rangle + \tau \langle x_0 - y_0, y^* \rangle - 2 + 2\delta \\
&\leq \|x_0 + y_0\| - 2 + \tau |\langle x_0 - y_0, y^* \rangle| + 2\delta.
\end{aligned}
$$

Hence, if we define $\varepsilon_0 := |\langle x_0 - y_0, y^* \rangle|$, then $0 < \varepsilon_0 \leq \|x_0 - y_0\| \leq 2$ and

$$
\frac{\|x^* + \tau y^*\| + \|x^* - \tau y^*\|}{2} - 1 \leq \frac{\tau \varepsilon_0}{2} + \delta - \delta_X(\varepsilon_0)
$$

$$
\leq \delta + \sup \left\{ \frac{\tau \varepsilon}{2} - \delta_X(\varepsilon) : 0 < \varepsilon \leq 2 \right\}.
$$

As $\delta > 0$ is arbitrary, we have

$$
\rho_{X^*}(\tau) \leq \sup \left\{ \frac{\tau \varepsilon}{2} - \delta_X(\varepsilon) : 0 < \varepsilon \leq 2 \right\},
$$

and the first equation is proved.

In order to prove the second equation, let $\tau > 0$ and let x^*, y^* in X^* with $\|x^*\| = \|y^*\| = 1$. For any $\eta > 0$, from the definition of $\|.\|$ in X^* there exist x_0, y_0 in X with $\|x_0\| = \|y_0\| = 1$ such that

$$
\|x^* + y^*\| - \eta \leq \langle x_0, x^* + y^* \rangle, \quad \|x^* - y^*\| - \eta \leq \langle y_0, x^* - y^* \rangle.
$$

Then

$$
\begin{aligned}
&\|x^* + y^*\| + \tau \|x^* - y^*\| - 2 \\
&\leq \langle x_0, x^* + y^* \rangle + \tau \langle y_0, x^* - y^* \rangle - 2 + \eta(1 + \tau) \\
&= \langle x_0 + \tau y_0, x^* \rangle + \langle x_0 - \tau y_0, y^* \rangle - 2 + \eta(1 + \tau).
\end{aligned}
$$

Since in a Banach space, $\|x\| = \sup\{|\langle x, x^* \rangle| : \|x^*\| = 1\}$ we have

$$
\begin{aligned}
&\|x^* + y^*\| + \tau \|x^* - y^*\| - 2 \\
&\leq \|x_0 + \tau y_0\| + \|x_0 - \tau y_0\| - 2 + \eta(1 + \tau) \\
&\leq \sup\{\|x + \tau y\| + \|x - \tau y\| - 2 : \|x\| = \|y\| = 1\} \\
&\quad + \eta(1 + \tau) \\
&= 2\rho_X(\tau) + \eta(1 + \tau).
\end{aligned}
$$

If $0 < \varepsilon \leq \|x^* - y^*\| \leq 2$ we have

$$
\frac{\tau \varepsilon}{2} - \rho_X(\tau) - \eta(1 + \tau) \leq 1 - \left\| \frac{x^* + y^*}{2} \right\|
$$

which implies that

$$\frac{\tau\varepsilon}{2} - \rho_X(\tau) - \eta(1+\tau) \leq \delta_{X^*}(\varepsilon).$$

Now, since η is arbitrary we conclude that

$$\frac{\tau\varepsilon}{2} - \rho_X(\tau) \leq \delta_{X^*}(\varepsilon)$$

for every ε in $(0,2]$ and therefore

$$\sup\left\{\frac{\tau\varepsilon}{2} - \delta_{X^*}(\varepsilon) : 0 < \varepsilon \leq 2\right\} \leq \rho_X(\tau).$$

Now let x, y be in X with $\|x\| = \|y\| = 1$ and let $\tau > 0$. By Hahn-Banach Theorem there exist x_0^*, y_0^* in X^* with $\|x_0^*\| = \|y_0^*\| = 1$ and such that

$$\langle x + \tau y, x_0^* \rangle = \|x + \tau y\|, \quad \langle x - \tau y, y_0^* \rangle = \|x - \tau y\|.$$

Then,

$$\begin{aligned}
\|x + \tau y\| + \|x - \tau y\| - 2 &= \langle x + \tau y, x_0^* \rangle + \langle x - \tau y, y_0^* \rangle - 2 \\
&= \langle x, x_0^* + y_0^* \rangle + \tau\langle y, x_0^* - y_0^* \rangle - 2 \\
&\leq \|x_0^* + y_0^*\| + \tau|\langle y, x_0^* - y_0^* \rangle| - 2.
\end{aligned}$$

Hence, if we define $\varepsilon_0 = |\langle y, x_0^* - y_0^* \rangle|$, then $0 < \varepsilon_0 \leq \|x_0 - y_0\| \leq 2$ and

$$\begin{aligned}
\frac{\|x + \tau y\| + \|x - \tau y\|}{2} - 1 &\leq \frac{\|x_0^* + y_0^*\| + \tau|\langle y, x_0^* - y_0^* \rangle|}{2} - 1 \\
&= \frac{\tau\varepsilon_0}{2} - \left(1 - \frac{\|x_0^* + y_0^*\|}{2}\right) \\
&\leq \frac{\tau\varepsilon_0}{2} - \delta_{X^*}(\varepsilon_0) \\
&\leq \sup\left\{\frac{\tau\varepsilon}{2} - \delta_{X^*}(\varepsilon) : 0 < \varepsilon \leq 2\right\}.
\end{aligned}$$

Therefore

$$\rho_X(\tau) \leq \sup\left\{\frac{\tau\varepsilon}{2} - \delta_{X^*}(\varepsilon) : 0 < \varepsilon \leq 2\right\}$$

and the second equality holds.   □

From the second formula of Proposition 2.7 we get the following corollaries.

Corollary 2.8 *For every Banach space X, the function $\frac{\rho_X(t)}{t}$ is non-decreasing and $\rho_X(t) \leq t$.*

Corollary 2.9 *For every Banach space X and for every Hilbert space H we have*

$$\rho_H(\tau) = \sup\left\{\frac{\tau\varepsilon}{2} - 1 + \sqrt{1 - \frac{\varepsilon^2}{4}} : 0 < \varepsilon \leq 2\right\} = \sqrt{1 + \tau^2} - 1 \leq \rho_X(\tau).$$

The following result gives the duality between uniformly convex and uniformly smooth spaces.

Theorem 2.10. *Let X be a Banach space.*
(a) X *is uniformly smooth if and only if X^* is uniformly convex.*
(b) X *is uniformly convex if and only if X^* is uniformly smooth.*

Proof. (a) $\to$. If X^* is not uniformly convex, there exists ε_0 in $(0, 2]$ with $\delta_{X^*}(\varepsilon_0) = 0$, and by the second formula in Proposition 2.7, we obtain for every $\tau > 0$,

$$0 < \frac{\varepsilon_0}{2} \leq \frac{\rho_X(\tau)}{\tau}$$

which means that X is not uniformly smooth.
(a) $\leftarrow$. Assume that X is not uniformly smooth. Then

$$\lim_{t \to 0^+} \frac{\rho_X(t)}{t} \neq 0,$$

that means, there exists $\varepsilon > 0$ such that for every $\delta > 0$ we can find t_δ with $0 < t_\delta < \delta$ and $t_\delta \varepsilon \leq \rho_X(t_\delta)$. Then there exists a sequence $(\tau_n)_n$ such that $0 < \tau_n < 1$, $\tau_n \to 0$ and $\rho_X(\tau_n) > \frac{\varepsilon}{2}\tau_n$. By the second formula in Proposition 2.7, for every n there exists ε_n in $(0, 2]$ such that

$$\frac{\varepsilon}{2}\tau_n \leq \frac{\tau_n \varepsilon_n}{2} - \delta_{X^*}(\varepsilon_n)$$

which implies

$$0 < \delta_{X^*}(\varepsilon_n) \leq \frac{\tau_n}{2}(\varepsilon_n - \varepsilon),$$

in particular $\varepsilon < \varepsilon_n$ and $\delta_{X^*}(\varepsilon_n) \to 0$. Given the fact that δ_{X^*} is a nondecreasing function we have $\delta_{X^*}(\varepsilon) \leq \delta_{X^*}(\varepsilon_n) \to 0$. Therefore X^* is not uniformly convex. Note that, interchanging the roles of X and X^* in this proof, and using the first formula in Proposition 2.7, we get the proof of part (b). $\qquad\square$

The last part of the proof of Theorem 2.10 proves the following result.

Corollary 2.11 *Every uniformly smooth space is reflexive.*

SUMMARY

For the ease of reference, we summarize the key results obtained in this chapter. Here ρ_X denotes the modulus of smoothness of a Banach space X.

S1

(a) *Every uniformly smooth space is smooth.*
(b) *Every uniformly smooth space is reflexive.*
(c) *X is uniformly smooth if and only if X^* is uniformly convex.*

S2

(a) *X is uniformly smooth if and only if $\lim\limits_{t \to 0^+} \dfrac{\rho_X(t)}{t} = 0$.*
(b) *$\rho_X : [0.\infty) \to [0, \infty)$ is a convex and continuous function.*
(c) *$\dfrac{\rho_X(t)}{t}$ is a nondecreasing function on $[0, \infty)$.*
(d) *$\rho_X(t) \leq t$ for all $t \geq 0$.*

EXERCISES 2.1

1. Prove Corollaries 2.8 and 2.9.
2. Verify that L_p (or l_p) spaces, $1 < p < \infty$, are uniformly smooth (and are therefore smooth).
3. Prove Corollary 2.11.

   (Hint: A Banach space X is reflexive whenever the dual space X^* is).
4. Establish the following inequalities. In L_p (or l_p) spaces, $1 < p < \infty$,

$$\rho_{L_p}(\tau) = \begin{cases} (1 + \tau^p)^{\frac{1}{p}} - 1 < \frac{1}{p}\tau^p, & 1 < p < 2, \\ \frac{p-1}{2}\tau^2 + o(\tau^2) < \frac{p-1}{2}\tau^2, & p \geq 2. \end{cases}$$

(Hint: See Lindenstrauss and Tzafriri [312]).

2.4 Historical Remarks

Results in this chapter can be found in Beauzamy [26] or Diestel [206]. The Lindenstrauss duality formulas are proved in Lindenstrauss and Tzafriri [312].

Chapter 3
Duality Maps in Banach Spaces

3.1 Motivation

In trying to develop analogue of the identity (1.1) in Banach spaces more general than Hilbert spaces, one has to find a suitable replacement for inner product, $\langle .,. \rangle$. In this chapter, we present the notion of *duality mappings* which will provide us with a pairing between elements of a normed space E and elements of its dual space E^*, which we shall also denote by $\langle .,. \rangle$ and will serve as a suitable analogue of the inner product in Hilbert spaces.

In a given Hilbert space H, for any x^* in H^*, by Riesz representation theorem, there exists a unique x in H such that $\|x\| = \|x^*\|$ and $\langle y, x^* \rangle = \langle y, x \rangle$ for every y in H where the first bracket is the evaluation of x^* in y. In the particular case $y = x$, we have $\langle x, x^* \rangle = \langle x, x \rangle = \|x\|\|x^*\|$. For a general Banach space X, by Hahn-Banach theorem, for a given x in X, there exists at least one x^* in X^* such that $\langle x, x^* \rangle = \|x\|\|x^*\|$. The notion of *duality maps* (defined below) was motivated by these facts.

First we introduce a useful concept, the *gauge function*.

Definition 3.1. A continuous and strictly increasing function $\phi : \mathbb{R}^+ \to \mathbb{R}^+$ such that $\phi(0) = 0$ and $\lim\limits_{t\to\infty} \phi(t) = \infty$ is called a *gauge function*.

Lemma 3.2. *Let ϕ be a gauge function and*

$$\psi(t) = \int_0^t \phi(s)ds,$$

then ψ is a convex function on $\mathbb{R}^+$.

Proof. For $h > 0$ and $t > 0$ we have

$$\frac{\psi(t+h) - \psi(t)}{h} = \frac{1}{h} \int_t^{t+h} \phi(s) \, ds$$

$$\geq \frac{\phi(t)}{h} \int_t^{t+h} ds = \phi(t)$$

C. Chidume, *Geometric Properties of Banach Spaces and Nonlinear Iterations*,
Lecture Notes in Mathematics 1965,
© Springer-Verlag London Limited 2009

and

$$\frac{\psi(t) - \psi(t-h)}{h} = \frac{1}{h} \int_{t-h}^{t} \phi(s) \, ds \leq \phi(t).$$

Let $0 \leq t_1 < t_2$, $\lambda \in [0,1]$ and $t = \lambda t_1 + (1-\lambda)t_2$; then

$$\lambda = \frac{t_2 - t}{t_2 - t_1} \quad and \quad 1 - \lambda = \frac{t - t_1}{t_2 - t_1}$$

and the above two inequalities yield

$$\psi(t_2) - \psi(t) \geq (t_2 - t)\phi(t),$$

$$\psi(t) - \psi(t_1) \leq (t - t_1)\phi(t).$$

Multiplying first inequality by $(1-\lambda)$ and the second by $-\lambda$, we get

$$(1-\lambda)\psi(t_2) - (1-\lambda)\psi(t) \geq (1-\lambda)(t_2 - t)\phi(t)$$
$$= \lambda(1-\lambda)(t_2 - t_1)\phi(t)$$

and

$$-\lambda\psi(t) + \lambda\psi(t_1) \geq -\lambda(t - t_1)\phi(t) = -\lambda(1-\lambda)(t_2 - t_1)\phi(t).$$

Summing the above two inequalities we get

$$\lambda\psi(t_1) + (1-\lambda)\psi(t_2) - \psi(t) \geq 0,$$

that is,

$$\psi(t) \leq \lambda\psi(t_1) + (1-\lambda)\psi(t_2)$$
$$\Rightarrow \quad \psi(\lambda t_1 + (1-\lambda)t_2) \leq \lambda\psi(t_1) + (1-\lambda)\psi(t_2).$$

$\square$

Definition 3.3. Given a gauge function ϕ, the mapping $J_\phi : X \rightarrow 2^{X^*}$ defined by

$$J_\phi x := \{\, u^* \in X^* : \langle x, u^* \rangle = \|x\| \|u^*\|; \|u^*\| = \phi(\|x\|) \,\}$$

is called the *duality map* with gauge function ϕ where X is any normed space.

In the particular case $\phi(t) = t$, the duality map $J = J_\phi$ is called the *normalized duality map*.

Lemma 3.4. *In a normed linear space X, for every gauge function ϕ, $J_\phi x$ is not empty for any x in X.*

Proof. The case $x = 0$ is trivial by taking $u^* = 0$ in X^*. For $x \neq 0$ in X, then $x\phi(\|x\|) \neq 0$. Consider, by Hahn-Banach theorem, x^* in X^* such that

$||x^*|| = 1$ and $\langle x\phi(||x||), x^* \rangle = ||x||\phi(||x||)$. Note that $||\phi(||x||)x^*|| = \phi(||x||)$ and

$$\langle x, \phi(||x||)x^* \rangle = \langle x\phi(||x||), x^* \rangle = ||x||\phi(||x||) = ||x|| \, ||\phi(||x||)x^*||,$$

therefore $u^* := \phi(||x||)x^*$ is in $J_\phi x$. □

As a consequence of this result, from now on, we will work on normed linear spaces to ensure that for each $x \in X$, $J_\phi(x)$ is not empty.

An interesting result that indicates the importance of the duality map is stated in the following theorem which can be found in Reich [411].

Theorem 3.5. *Let X be a Banach space. For $0 < \varepsilon \le 2$, let*

$$\gamma(\varepsilon) = inf\{1 - \langle y, j(x) \rangle : ||x|| = ||y|| = 1, ||x - y|| \ge \varepsilon, j(x) \in J(x)\},$$

then X is uniformly convex if and only if $\gamma(\varepsilon)$ is positive.

Proposition 3.6. *In a real Hilbert space H, the normalized duality map is the identity map.*

Proof. Since H is a Hilbert space, we identify $H = H^*$ as usual. Let $x \in H$, $x \ne 0$. Since $\langle x, x \rangle = ||x||.||x||$, trivially x is in Jx. If y is in Jx, then $\langle x, x \rangle = ||x||.||x||$, and $||y|| = ||x||$. Then,

$$||x - y||^2 = \langle x - y, x - y \rangle = \langle x, x \rangle - \langle y, x \rangle - \langle x, y \rangle + \langle y, y \rangle = 0.$$

Therefore, $y = x$ and $J(y)$ is the singleton $\{x\}$. This is enough to conclude that the normalized duality map can be considered as the identity map in the case of real Hilbert spaces. □

Proposition 3.7. *If $Jx = \{x^*\}$, a singleton for every x in a Banach space X, and if J is **linear** in the sense that*

$$\lambda x^* + y^* = (\lambda x + y)^*$$

for every $x, y \in X$ and scalar λ, then X is a Hilbert space.

In a general Banach space, we have the following proposition.

Proposition 3.8. *In a Banach space X, let J_ϕ be a duality map of gauge function ϕ. Then for every x in X, $x \ne 0$ and every λ in $\mathbb{R}$,*

$$J_\phi(\lambda x) = sign(\lambda)\frac{\phi(|\lambda| ||x||)}{\phi(||x||)} J_\phi(x).$$

Corollary 3.9 *Let X be a real Banach space and J be the normalized duality map on X. Then, $J(\lambda x) = \lambda J(x) \, \forall \lambda \in \mathbb{R} \, \forall \, x \in X$.*

Proposition 3.10. *Let J be a duality mapping on a real Banach space X associated with a gauge function ϕ. Then,*

(i) for every $x \in X$, the set Jx is convex and weak* closed in X^*.
(ii) J is monotone in the sense that $\langle x - y, x^* - y^* \rangle \geq 0 \; \forall x, y \in X$ and $x^* \in Jx, \; y^* \in Jy$.

In particular, if J is single-valued and is denoted by j, then $\langle x - y, j(x) - j(y) \rangle \geq 0 \; \forall x, y \in X$.

We now prove the following important result, *a theorem of Kato*, which will be central in many applications.

Proposition 3.11. *(Kato, [275]) Let X be a real Banach space and let J be the normalized duality mapping on X; then for every $x, y \in X$ the following statements are equivalent:*

(i) $\|x\| \leq \|x + \lambda y\|, \quad \forall \lambda > 0$.
(ii) there exists $u^* \in Jx$ such that $\langle y, u^* \rangle \geq 0$.

Proof. $(i) \Rightarrow (ii)$ For $\lambda > 0$, consider $x_\lambda^* \in J(x + \lambda y), x^* \neq 0$, and define

$$y_\lambda^* := \frac{x_\lambda^*}{\|x_\lambda^*\|}.$$

Clearly $\|y_\lambda^*\| = 1$ so that $y_\lambda^* \nrightarrow 0$ as $\lambda \to 0^+$. Now

$$y_\lambda^* := \frac{x_\lambda^*}{\|x_\lambda^*\|} \in \frac{1}{\|x_\lambda^*\|} J(x + \lambda y).$$

So we have, since $\|y_\lambda^*\| = 1$,

$$\begin{aligned}
\|x\| \leq \|x + \lambda y\| &= \langle x + \lambda y, x_\lambda^* \rangle \frac{1}{\|x_\lambda^*\|} \\
&= \langle x + \lambda y, y_\lambda^* \rangle = \langle x, y_\lambda^* \rangle + \lambda \langle y, y_\lambda^* \rangle \\
&\leq \|x\| + \lambda \langle y, y_\lambda^* \rangle.
\end{aligned} \tag{3.1}$$

We know from *Banach-Alaoglu theorem* that the unit ball is *weak* − compact* in X^*. Therefore the net $\{y_\lambda^*\}_{\lambda \in \mathbb{R}^+}$ has a weak* limit point y^* as $\lambda \to 0^+$ which by (3.1) and the fact that the net is in the unit ball of X^* satisfies (as $\lambda \to 0^+$),

$$\|y^*\| \leq 1 \, , \; \langle x, y^* \rangle \geq \|x\| \quad and \quad \langle y, y^* \rangle \geq 0.$$

But then

$$\|x\| \leq \langle x, y^* \rangle \leq \|x\| \|y^*\| \leq \|x\|$$

so we get that

$$\langle x, y^* \rangle = \|x\| \quad and \quad \|y^*\| = 1.$$

Let $\quad u^* = y^*||x||$, then $\quad u^* \in Jx$ and $\quad \langle y, u^* \rangle \geq 0$.

$(ii) \Rightarrow (i)$ Suppose that for $x, y \in X, x \neq 0$, there is $u^* \in Jx$ such that $\langle y, u^* \rangle \geq 0$. Then for $\lambda > 0$, we may write

$$||x||^2 = \langle x, u^* \rangle \leq \langle x, u^* \rangle + \lambda \langle y, u^* \rangle$$
$$= \langle x + \lambda y, u^* \rangle \leq ||u^*||||x + \lambda y||$$
$$= ||x + \lambda y||||x||,$$
$$\Rightarrow \qquad ||x|| \leq ||x + \lambda y||.$$

This inequality also holds for $x = 0$. The proof is complete. $\qquad\square$

3.2 Duality Maps of Some Concrete Spaces

Proposition 3.12. *Let* $a = (a_1, a_2) \in \mathbb{R}^2$. *Then,* $J(a) = A_a$, *where*

$$A_a := \{\phi_a \in (\mathbb{R}^2)^* : \phi_a(x) := \langle a, x \rangle \forall x \in \mathbb{R}^2\}. \tag{3.2}$$

Proof. By the definition of the normalized duality map $J : E \to 2^{E^*}$,

$$J(a) := \{\phi \in (\mathbb{R}^2)^* : \phi(a) = ||a||.||\phi||, \ ||a|| = ||\phi||\}.$$

Let $\phi_a \in J(a)$. Then,

$$\phi_a(a) = ||a||.||\phi_a||; ||a|| = ||\phi_a||. \tag{3.3}$$

Since $\phi_a \in (\mathbb{R}^2)^*$, by Riesz representation theorem, there exists a unique $y \in \mathbb{R}^2$ such that

$$\phi_a(x) = \langle y, x \rangle \ \forall x \in \mathbb{R}^2; \ ||y|| = ||\phi_a||.$$

In particular,

$$\phi_a(a) = \langle y, a \rangle; ||y|| = ||\phi_a||. \tag{3.4}$$

From (3.3) and (3.4), we obtain the following system of equations:

$$\phi_a(a) = \langle y, a \rangle = ||y||.||\phi_a||; \ ||y|| = ||\phi_a||,$$

so that

$$\langle y, a \rangle = ||a||^2; \ ||y|| = ||a||. \tag{3.5}$$

Solving equations (3.5) for y we obtain that $y = a$. This implies that

$$J(a) \subset \{\phi_a : \phi_a(x) = \langle a, x \rangle \forall x \in \mathbb{R}^2\} = A_a.$$

But if $\phi_a \in A_a$, $\phi_a(a) = \langle a, a \rangle = ||a||.||a||$ and so $\phi_a \in J(a)$. Hence, $A_a \subset J(a)$, so that

$$J(a) = A_a \subset (\mathbb{R}^2)^*.$$

The proof is complete.   □

Proposition 3.13. *Let E be a real Hilbert space. Let $J : H \to 2^{H^*}$ be the normalized duality map. Then, for each $a \in H$, $J(a) = A_a$, where*

$$A_a = \{\phi_a \in H^* : \phi_a(x) = \langle a, x \rangle \ \forall \ x \in H\}. \tag{3.6}$$

Proof. By the definition of the normalized duality map $J : H \to 2^{H^*}$,

$$J(a) := \{\phi_a \in H^* : \phi_a(a) = ||a||.||\phi_a||, \ ||a|| = ||\phi_a||\}.$$

Let $\phi_a \in J(a)$. Then,

$$\phi_a(a) = ||a||.||\phi_a||; ||a|| = ||\phi_a||. \tag{3.7}$$

Since $\phi_a \in H^*$, by Riesz representation theorem, there exists a unique $y \in H$ such that

$$\phi_a(x) = \langle y, x \rangle, \ \forall x \in H; \ ||y|| = ||\phi_a||. \tag{3.8}$$

In particular,

$$\phi_a(a) = \langle y, a \rangle; \ ||y|| = ||\phi_a||. \tag{3.9}$$

From (3.7) and (3.9), we obtain the following system of equations:

$$\phi_a(a) = \langle y, a \rangle = ||a||.||\phi_a|| \ by \ (3.7); \ ||y|| = ||\phi_a|| = ||a||,$$

so that

$$\langle y, a \rangle = ||a||^2; \tag{3.10}$$

$$||y|| = ||a||. \tag{3.11}$$

Solving equations (3.10) and (3.11) for y we obtain that

$$y = a.$$

This implies that

$$J(a) \subset \{\phi_a : \phi_a(x) = \langle a, x \rangle \forall x \in H\} = A_a. \tag{3.12}$$

But if $\phi_a \in A_a$, $\phi_a(a) = \langle a, a \rangle = ||a||.||a||$ and so $\phi_a \in J(a)$. Hence, $A_a \subset J(a)$, so that

$$J(a) = \{\phi_a \in H^* : \phi_a(x) := \langle a, x \rangle, \forall x \in H\}.$$

The proof is complete. □

Proposition 3.14. *The normalized duality mapping on $L^p([0,1])$ is given by*

$$A_g = \Big\{ \phi_g \in (L^p)^* = L^q : \phi_g(f)$$

$$= \int_0^1 g(t)f(t)dt,$$

$$g(t) := \frac{|f|^{p-1} \operatorname{sign} f(t)}{||f||_p^{p-2}}, f \in L^p([0,1]) \Big\}. \tag{3.13}$$

Proof. From the definition of the normalized duality map $J : L_p \to 2^{L_p^*}$,

$$J(f) := \{\phi \in (L_p)^* : \phi(f) = ||f|| \cdot ||\phi||, \ ||f|| = ||\phi||\}. \tag{3.14}$$

Let $\phi \in (L_p)^*$. By Riesz representation theorem, there exists a unique $g \in L_q, \frac{1}{p} + \frac{1}{q} = 1, p, q > 1$ such that

$$\phi(f) = \langle g, f \rangle = \int_0^1 g(t)f(t)dt; \ ||\phi|| = ||g||.$$

Set $\phi \equiv \phi_g$. Then, this equation becomes

$$\phi_g(f) = \langle g, f \rangle = \int_0^1 g(t)f(t)dt; \ ||\phi_g|| = ||g||. \tag{3.15}$$

Then, from (3.14) and (3.15), we get

$$\phi_g(f) = \int_0^1 g(t)f(t)dt = ||\phi_g|| \cdot ||f||; \ ||f|| = ||\phi_g|| = ||g||. \tag{3.16}$$

Hence, we obtain the following set of equations;

$$\int_0^1 f(t)g(t)dt = ||f||^2; \ ||g||_{L_q} = ||f||_{L_p}. \tag{3.17}$$

We now solve equations (3.17) for g.

To fix ideas, let us first solve equations (3.17) for the case in which $p = 2 = q$. From equations (3.17), we obtain the following equations:

$$\int_0^1 f(t)g(t)dt = \int_0^1 f(t)f(t)dt. \tag{3.18}$$

$$\int_0^1 g(t)g(t)dt = \int_0^1 f(t)f(t)dt.$$ (3.19)

From equations (3.18) and (3.19), we obtain the following equations,

$$\int_0^1 f(t)(g(t) - f(t))dt = 0.$$ (3.20)

$$\int_0^1 (g(t) - f(t))g(t)dt = 0.$$ (3.21)

From (3.20) and (3.21), we get

$$\int_0^1 (g(t) - f(t))(g(t) - f(t))dt = 0,$$

so that

$$\int_0^1 |g(t) - f(t)|^2 dt = 0.$$

This implies $g(t) = f(t)$ a.e. Consequently, $J(f) \subset \{\phi_f : \phi_f(h) = \int_0^1 f(t)h(t)dt\}$. On the other hand, $\phi_f \in A_g$. For,

$$\phi_f(f) = \int_0^1 f(t)f(t) = \int_0^1 |f(t)|^2 dt = ||f||.||f||.$$

So,

$$\left\{\phi_f : \phi_f(h) = \int_0^1 f(t)h(t)dt\right\} \subset J(f).$$

Hence,

$$J(f) = \left\{\phi_f : \phi_f(h) = \int_0^1 f(t)h(t)dt\right\} = \{\phi_f\},$$

a singleton.

Claim. The general solution of (3.17) is given by (Exercises 3.1, Problem 6(b)),

$$g(t) = |f(t)|^{(p-1)} sign \frac{f(t)}{||f||_p^{(p-2)}},$$ (3.22)

where $sign\, f(t) = \frac{f(t)}{|f(t)|}, f \neq 0$, and $sign f(t) = 0, f = 0$.

Corollary 3.15 *Let H be a real Hilbert space. Then, for each $a \in H$,*

$$J(a) = \phi_a \in H^*,$$

where $\phi_a(x) := \langle a, x \rangle \; \forall \; x \in H$.

EXERCISES 3.1

1. Prove that the following are convex functions.

   (a) $f(x) = ||x||$; (b) $f(x) = \frac{1}{p}||x||^p \ \forall \ p > 1$.

2. Prove that for $x \neq 0, J_p(x) = ||x||^{p-2}J(x), p > 1$ where J is the single-valued normalized duality map.

3. Let X be a real Banach space and $T : D(T) \subseteq X \to X$ be a map such that $\forall x, y \in D(T)$, the following inequality is satisfied:

$$||x - y|| \leq ||(1 + r)(x - y) - rt(Tx - Ty)||,$$

   for $r > 0$ and some $t > 1$. By applying Proposition 3.11 or otherwise, prove that there exists $j(x - y) \in J(x - y)$ such that

$$\langle (I - T)x - (I - T)y, j(x - y) \rangle \geq k||x - y||^2$$

   for all $x, y \in D(T)$, where $k := (\frac{t-1}{t})$.

4. Let E be a real normed linear space and $A : E \to E$ be a mapping such that

$$\langle Ax - Ay, j(x - y) \rangle \geq 0,$$

   for all $x, y \in E$, where j is the single-valued normalized duality map on E. Assume $(A + I) : E \to E$ is surjective, where I denotes the identity map on E. Prove:

   (a) $(I + A)^{-1}$ exists.
   (b) $||(I + A)^{-1}x - (I + A)^{-1}y|| \leq ||x - y|| \ \forall \ x, y \in E$.

   (Hint: Start by expanding $\langle (I + A)x - (I + A)y, j(x - y) \rangle$.)

5. Let E be a real normed linear space, $j : E \to E^*$ be the normalized duality map. Assume that j is norm-to weak* uniformly continuous on bounded subsets of E. Let $\{x_n\}$ and $\{u_n\}$ be sequences in E such that

   (i) $\{x_n\}$ is bounded;
   (ii) $u_m \to x^*$ in norm (i.e., $||u_m - x^*|| \to 0$ as $m \to \infty$);
   (iii) $|\langle x^* - u_m, j(x_n - u_m) \rangle| \to 0$ as $n, m \to \infty$.

   Prove that given any $\epsilon > 0$, there exists an integer $N > 0$ such that

   (1) $|\langle u - u_m, j(x_n - u_m) \rangle - \langle u - x^*, j(x_n - x^*) \rangle| < \epsilon \ \forall \ n, m > N$.

   (Hint: Subtract and add: $\langle u - x^*, j(x_n - u_m) \rangle$ in the LHS of (1), use the given conditions (i) to (iii) and the condition that j is norm-to-weak* uniformly continuous on bounded subsets of E).

6. (a) Solve equations (3.5) to obtain $y = a$. Let $a = (-3, 2)$. Compute $J(a)$.
   (b) Verify that g defined by (3.22) satisfies equations (3.17).

Hint: Note that $|g(t)|^q = \frac{|f(t)|^{q(p-1)}}{||f||_p^{q(p-2)}}$ so that

$$\left(\int_0^1 |g(t)|^q dt\right)^{\frac{1}{q}} = \frac{1}{||f||_p^{p-2}}\left(\int_0^1 |f(t)|^{q(p-1)} dt\right)^{\frac{1}{q}}, \quad \frac{1}{p} + \frac{1}{q} = 1, i.e.,$$

$$||g||_q = \frac{1}{||f||_p^{p-2}}\left(\int_0^1 |f(t)|^p dt\right)^{\frac{1}{q}}, \text{ or}$$

$$||g||_q = \frac{1}{||f||_p^{p-2}} ||f||_p^{\frac{p}{q}}, \frac{p}{q} = p - 1.$$

So, $||g||_q = \frac{||f||_p^{p-1}}{||f||_p^{p-2}} = ||f||_p.$

Also, for $f \neq 0, \phi \in J(f) \Rightarrow$

$$\int_0^1 f(t)g(t)dt = \int_0^1 f(t)\frac{|f(t)|^{p-1}}{||f||_p^{p-2}}\frac{f(t)}{|f(t)|}dt = \int_0^1 |f(t)|^2 \frac{|f(t)|^{p-1}}{|f(t)|}\frac{1}{||f||^{p-2}}dt$$

$$= \left(\int_0^1 |f(t)|^p dt\right)\frac{1}{||f||_p^{p-2}} = \frac{||f||_p^p}{||f||_p^{p-2}} = ||f||_p^2 = ||f||_p||f||_p$$

$$= ||f||_p||g||_q = ||f||_p||\varphi||.$$

Hence, $J(f) \subseteq \left\{\varphi_g \text{ where } \varphi_g(t) = \langle g, f\rangle\right.$
$= \int_0^1 f(t)g(t)dt$ and $g(t) := \frac{|f(t)|^{p-1}}{||f||_p^{p-2}} sign f(t)\right\} := A_g$, i.e., $J(f) \subset A_g.$
On the other hand, if $\varphi \in A_g$, then $\varphi = \varphi_g$ and $\varphi_g(h) = \int_0^1 h(t)g(t)dt$ with
$||g||_q = ||\varphi||$. If we take $h = f$, then, since $f(t)sign f(t) = |f(t)|$,

$$\varphi_g(f) = \int_0^1 f(t).\frac{|f(t)|^{p-1}}{||f||_p^{p-2}} sign f(t)dt = \int_0^1 \frac{|f(t)||f(t)|^{p-1}}{||f||_p^{p-2}}dt$$

$$= \frac{1}{||f||_p^{p-2}}.\int_0^1 |f(t)|^p dt = \frac{||f||_p^p}{||f||_p^{p-2}} = ||f||_p^2.$$

i.e., $\varphi_g(f) = ||f||_p.||f||_p$ and $||f||_p = ||g||_q.$
It follows that $||\varphi|| = ||f||_p$. Therefore, $A_g \subset J(f).$

Conclusion: $J(f) = \left\{\varphi_g \in L_q, \frac{1}{p} + \frac{1}{q} = 1, \text{ where } g(t) = \frac{|f(t)|^{p-1}}{||f||_p^{p-2}} sign f(t)\right\}.$
7. Justify Corollary 3.15.

3.3 Historical Remarks

The concept of duality map was introduced in 1962 by Beurling and Livingston [30] and was further developed by Asplund [17], Browder [41, 47], Browder and De Figueiredo [48], De Figueiredo [197]. General properties of the duality map can be found in De Figueiredo [197].

Chapter 4
Inequalities in Uniformly Convex Spaces

4.1 Introduction

Among all Banach spaces, Hilbert spaces are generally regarded as the ones with the simplest geometric structures. The reason for this observation is that certain geometric properties which characterize Hilbert spaces (e.g., the existence of *inner product; the parallelogram law* or equivalently the *polarization identity*; and the fact that the *proximity map* or *nearest point mapping* in Hilbert spaces is Lipschitz *with constant 1*) make certain problems posed in Hilbert spaces *comparatively* straightforward and relatively easy to solve. In several applications, however, many problems fall naturally in Banach spaces more general than Hilbert spaces. Therefore, to extend the techniques of solutions of problems in Hilbert spaces to more general Banach spaces, one needs to establish identities or inequalities in general Banach spaces analogous to the ones in Hilbert spaces. As shown by recent works, several authors have conducted worthwhile research in this direction (e.g., Al'ber ([3], [4], [9], Beauzamy [26], Bynum [61, 62], Clarkson [191], Lindenstrauss ([309], [310]), Hanner [247], Kay [276], Lim [306, 303], Lindenstrauss and Tzafriri [311], Prus and Smarzewski [387], Reich [408], Tribunov [491], Xu [509], Xu [523], Xu and Roach [525], and a host of other authors). In this chapter (and also in Chapter 5), we shall describe some of the results obtained primarily within the last thirty years or so. Applications of these results to iterative solutions of nonlinear equations in Banach spaces will be given in subsequent chapters. For now, we shall examine inequalities obtained in various Banach spaces as analogues of the following Hilbert space identities:

For each $x, y \in H$ (a real Hilbert space), we have the *polarization identity*

$$\|x + y\|^2 = \|x\|^2 + 2\langle x, y \rangle + \|y\|^2, \tag{4.1}$$

and for $\lambda \in [0, 1]$, we also have the identity

$$\|\lambda x + (1 - \lambda)y\|^2 = \lambda\|x\|^2 + (1 - \lambda)\|y\|^2 - \lambda(1 - \lambda)\|x - y\|^2. \tag{4.2}$$

C. Chidume, *Geometric Properties of Banach Spaces and Nonlinear Iterations*,
Lecture Notes in Mathematics 1965,
© Springer-Verlag London Limited 2009

4.2 Basic Notions of Convex Analysis

Notation Throughout this section, X will denote a real Banach space and D will denote a convex nonempty subset of X, unless otherwise stated. We shall assume $p > 1$.

Definition 4.1. Let $f : X \to (-\infty, \infty)$ be a map. Then $D(f) := \{x \in X : f(x) < +\infty\}$ is called the **effective domain of f**. The function f is called **proper** if $D(f) \neq \emptyset$.

In this section D will denote $D(f)$, the domain of a convex function, f.

Definition 4.2. A function $f : X \to (-\infty, \infty)$ is said to be **lower-semi continuous** (l.s.c.) at $x_0 \in X$, if $\{x_n\}$ is a sequence in X such that $x_n \to x_0$ and $f(x_n) \to y$, then $f(x_0) \leq y$.

Definition 4.3. Let $f : X \to (-\infty, \infty)$ be a proper functional. Recall that f is said to be **convex** on D if

$$f[\lambda x + (1 - \lambda)y] \leq \lambda f(x) + (1 - \lambda)f(y) \qquad \forall\, x, y \in D, \qquad 0 \leq \lambda \leq 1.$$

Definition 4.4. Let $\mathbb{R}^+ = [0, \infty)$. A convex function f on D is said to be **uniformly convex on** D if there exists a function $\mu : \mathbb{R}^+ \longrightarrow \mathbb{R}^+$ with $\mu(t) = 0 \Leftrightarrow t = 0$ such that

$$f[\lambda x + (1 - \lambda)y] \leq \lambda f(x) + (1 - \lambda)f(y) - \lambda(1 - \lambda)\mu(\|x - y\|) \quad \forall \lambda \in [0, 1],$$

$\forall\, x, y \in D$. If the above relation holds for all $x, y \in D$ when $\lambda = \frac{1}{2}$, then f is said to be **uniformly convex at centre** on D.

A convex function f is uniformly convex on D if and only if f is uniformly convex at centre on D (see e.g., Zalinescu ([537], p.352)).

Definition 4.5. The *sub-differential* of a function f is a map

$$\partial f : X \to 2^{X^*}$$

defined by

$$\partial f(x) = \{x^* \in X^* : f(y) \geq f(x) + \langle y - x, x^* \rangle \,\forall\, y \in X\}.$$

Definition 4.6. For each $p > 1$, let $\phi(t) = t^{p-1}$ be a gauge function. Following Definition 3.3, we define the **generalized duality map** $J_p : X \longrightarrow 2^{X^*}$ by

$$J_{\phi(t)} : = J_p(x) = \{x^* \in X^* : \langle x, x^* \rangle = \|x\|.\|x^*\|,$$
$$\|x^*\| = \phi(\|x\|) = \|x\|^{p-1}\}.$$

We observe that for $p = 2$, we write $J_p = J_2 = J$ which is the *normalized duality map* on X defined in Chapter 3.

Proposition 4.7. *For every $x \neq 0$ in a Banach space X,*

$$\partial \|x\| = \{u^* \in X^* : \langle x, u^* \rangle = \|x\| = \|u^*\|; \|u^*\| = 1\}.$$

Proof. Note that

$$\partial \|x\| := \{u^* \in X^* : \langle y - x, u^* \rangle \leq \|y\| - \|x\| \ \forall \ y \in X\}.$$

Therefore,

$$\{u^* \in X^* : \langle x, u^* \rangle = \|x\| = \|u^*\|; \|u^*\| = 1\} \subset \partial \|x\|$$

since $\langle y, u^* \rangle \leq \|y\|$.
Conversely, if u^* is in $\partial \|x\|$ then

$$\langle y, u^* \rangle = \langle (y + x) - x, u^* \rangle \leq \|y + x\| - \|x\| \leq \|y\|,$$

$$\langle y, u^* \rangle \leq \|y + x\| - \|x\| \leq \|y\|.$$

From this $\|u^*\| \leq 1$ and with $y = 0$ in the definition of $\partial \|x\|$, we have $\|x\| \leq \langle x, u^* \rangle$. Therefore $\|u^*\| = 1$, $\|x\| = \langle x, u^* \rangle$ and the result holds. $\quad \square$

Now we state the relation between J_ϕ and ∂.

Theorem 4.8. $J_\phi x = \partial \psi(\|x\|)$ *for each x in a Banach space X, where* $\psi(\|x\|) = \int_0^{\|x\|} \phi(s) ds$.

Proof. Since ϕ is strictly increasing and continuous, ψ is differentiable and $\psi'(t) = \phi(t)$. By the convexity of ψ, if $s \neq t$ then, $(s - t)\psi'(t) \leq \psi(s) - \psi(t)$. Now let u^* in J_ϕ and let y in X. If $\|y\| > \|x\|$,

$$\|u^*\| = \phi(\|x\|) = \psi'(\|x\|) \leq \frac{\psi(\|y\|) - \psi(\|x\|)}{\|y\| - \|x\|},$$

which implies that

$$\langle y - x, u^* \rangle = \langle y, u^* \rangle - \langle x, u^* \rangle \leq \|u^*\|\|y\| - \|u^*\|\|x\| \leq \psi(\|y\|) - \psi(\|x\|).$$

In a similar way, if $\|y\| < \|x\|$, $\langle y - x, u^* \rangle \leq \psi(\|y\|) - \psi(\|x\|)$. If $\|y\| = \|x\|$, $\langle y - x, u^* \rangle \leq \|u^*\|(\|y\| - \|x\|) = 0 = \psi(\|y\|) - \psi(\|x\|)$. Therefore, for every y in X, $\langle y - x, u^* \rangle \leq \psi(\|y\|) - \psi(\|x\|)$. Hence

$$u^* \in \partial \phi(\|x\|) = \{x^* \in X^* : \langle y - x, x^* \rangle \leq \psi(\|y\|) - \psi(\|x\|); \forall y \in X\}.$$

We have proved $J_\psi x \subset \partial \phi(\|x\|)$.

Conversely, consider u^* in $\partial\psi(\|x\|)$, with $x \neq 0$.

$$
\begin{aligned}
\|u^*\|\|x\| &= \sup\{\langle y, u^*\rangle\|x\| : \|y\| = 1\} \\
&= \sup\{\langle y, u^*\rangle : \|y\| = \|x\|\} \\
&\leq \sup\{\langle x, u^*\rangle + \psi(\|y\|) - \psi(\|x\|) : \|y\| = \|x\|\} \\
&\leq \|u^*\|\|x\|,
\end{aligned}
$$

which implies $\langle x, u^*\rangle = \|u^*\|\|x\|$. Now we want to prove that $\|u^*\| = \phi(\|x\|) = \psi'(\|x\|)$. If $t > \|x\| > 0$, then

$$
\begin{aligned}
\|u^*\|(t - \|x\|) &= t\|u^*\| - \|x\|\|u^*\| = \langle x, u^*\rangle\left(\frac{t}{\|x\|} - 1\right) \\
&= \left\langle \frac{xt}{\|x\|} - x, u^*\right\rangle \leq \psi(t) - \psi(\|x\|),
\end{aligned}
$$

which implies

$$
\|u^*\| \leq \frac{\psi(t) - \psi(\|x\|)}{t - \|x\|}.
$$

In a similar way, with $0 < t < \|x\|$ we have

$$
\frac{\psi(t) - \psi(\|x\|)}{t - \|x\|} \leq \|u^*\|.
$$

By letting $t \to \|x\|$ we have, if $x \neq 0$, $\|u^*\| = \psi'(\|x\|) = \phi(\|x\|)$.

In the case $x = 0$, $\partial\psi(0) = \{u^* \in X^* : \langle y, u^*\rangle \leq \psi(\|y\|); \forall y \in X\}$ and we must prove that $\partial\psi(0) = \{0\}$. Now,

$$
\langle y, u^*\rangle \leq \psi(\|y\|) = \int_0^{\|y\|} \phi(t)dt \leq \phi(\|y\|)\|y\|,
$$

which implies $\|u^*\| \leq \phi(\|y\|)$ for all y in X. Therefore $u^* = 0$ and the result holds. $\square$

Proposition 4.9. *For $p \geq 1$, J_p is the sub-differential of the functional $\frac{1}{p}\|.\|^p$.*

Proof. Following the definition of J_p, we note that $J_p = J_{\phi(s)}$ where $\phi(s) = s^{p-1}$. By Theorem 4.8, setting $\phi(s) = s^{p-1}, p \geq 1$, we have

$$
J_p(x) = J_{\phi(s)}(x) = \partial \int_0^{\|x\|} \phi(s)ds = \partial \int_0^{\|x\|} s^{p-1}ds = \partial\left(\frac{1}{p}\|x\|^p\right),
$$

completing the proof. $\square$

Using the notation of sub-differentials, we first establish the following inequality which will be used in the sequel and which is valid in *arbitrary real normed spaces*.

Theorem 4.10. *Let E be a real normed space, and $J_p : E \to 2^{E^*}, 1 < p < \infty$, be the generalized duality map. Then, for any $x, y \in E$, the following inequality hold*

$$||x + y||^p \leq ||x||^p + p\langle y, j_p(x + y)\rangle \qquad (4.3)$$

for all $j_p(x + y) \in J_p(x + y)$. In particular, if $p = 2$, then

$$||x + y||^2 \leq ||x||^2 + 2\langle y, j(x + y)\rangle \qquad (4.4)$$

for all $j(x + y) \in J(x + y)$.

Proof. By Proposition 4.9, J_p is the sub-differential of $\frac{1}{p}||.||^p$, if $p > 1$. Hence, by the sub-differential inequality, for all $x, y \in E$ and $j_p(x + y) \in J_p(x + y)$, we obtain that

$$\frac{1}{p}||x||^p - \frac{1}{p}||x + y||^p \geq \langle x - (x + y), j_p(x + y)\rangle$$

so that

$$||x + y||^p \leq ||x||^p + p\langle y, j_p(x + y)\rangle,$$

as required. The case for $p = 2$ follows trivially. $\qquad \square$

We note that inequality (4.4) actually holds in any normed linear space simply from the definition of the normalized duality map and triangle inequality. For,

$$||x + y||^2 = \langle x + y, j(x + y)\rangle$$
$$\leq ||x||.||x + y|| + \langle y, j(x + y)\rangle$$
$$\leq \frac{1}{2}(||x||^2 + ||x + y||^2) + \langle y, j(x + y)\rangle,$$

so that $||x + y||^2 \leq ||x||^2 + 2\langle y, j(x + y)\rangle$, as required. $\qquad \square$

The following lemma will also be useful.

Lemma 4.11. *Let $f(x) = (1 - x)^\alpha$, $\alpha \geq 1$. Then if $x \in (0, 1)$ we have*

$$(1 - x)^\alpha \geq 1 - \alpha x.$$

Proof. Given $f(x) = (1 - x)^\alpha$, there exists $\zeta \in (0 , x)$ such that

$$f(x) = f(0) + xf'(0) + \frac{1}{2}x^2 f''(\zeta).$$

This implies $f(x) \geq f(0) + xf'(0)$ which gives $(1 - x)^\alpha \geq 1 - \alpha x$, as required. $\qquad \square$

4.3 p-uniformly Convex Spaces

Throughout this section, $\delta_X : (0, 2] \to (0, 1]$ denotes the modulus of convexity of a normed space X.

Definition 4.12. Let $p > 1$ be a real number. Then a normed space X is said to be **p-uniformly convex** if there is a constant $c > 0$ such that

$$\delta_X(\varepsilon) \geq c\varepsilon^p.$$

Example 4.13. If $E = L_p$ (or l_p), $1 < p < \infty$, then

(a) $\delta_E(\varepsilon) \geq \frac{1}{2^{p+1}}\varepsilon^2$, if $1 < p < 2$,
(b) $\delta_E(\varepsilon) \geq \varepsilon^p$, if $2 \leq p < \infty$.

To develop further inequalities, we shall make use of the following lemma whose very long proof can be found in Zalinescu [537].

Lemma 4.14. *Let X be a real Banach space. Then, $\delta_X(\epsilon) \geq c.\epsilon^p$ if and only if there exists a constant $c > 0$ such that*

$$\frac{1}{2}\left(\|x + y\|^p + \|x - y\|^p\right) \geq \|x\|^p + c\|y\|^p \qquad \forall\, x, y \in X. \qquad (4.5)$$

Using this lemma, we now prove the following proposition.

Proposition 4.15. *Let X be a real Banach space. Then for some constant $c > 0$,*

$$\frac{1}{2}\left(\|x + y\|^p + \|x - y\|^p\right) \geq \|x\|^p + c\|y\|^p \qquad \forall\, x, y \in X$$

if and only if $\|.\|^p$ is uniformly convex at centre on X.

Proof. Suppose for some constant $c > 0$,

$$\frac{1}{2}\left(\|x + y\|^p + \|x - y\|^p\right) \geq \|x\|^p + c\|y\|^p \qquad \forall\, x, y \in X.$$

Let $x = u + v$ and $y = u - v$, then the above inequality becomes

$$\|u + v\|^p + c\|u - v\|^p \leq 2^{p-1}(\|u\|^p + \|v\|^p).$$

So,

$$\left\|\frac{u + v}{2}\right\|^p \leq \frac{1}{2}(\|u\|^p + \|v\|^p) - c2^{-p}\|u - v\|^p$$
$$= \frac{1}{2}(\|u\|^p + \|v\|^p) - \frac{1}{4}\mu(\|u - v\|)$$

where $\mu : \mathbb{R}^+ \to \mathbb{R}^+$ is defined by $\mu(t) = t^p \, c \, 2^{-p+2}$. Hence $\|.\|^p$ is uniformly convex at centre.

Conversely suppose $\|.\|^p$ is uniformly convex at centre. So,

$$\left\| \frac{x+y}{2} \right\|^p \leq \frac{1}{2}(\|x\|^p + \|y\|^p) - \frac{1}{4} \cdot \mu(\|x - y\|) \ \forall \ x, y \in X$$

$$= \frac{1}{2}(\|x\|^p + \|y\|^p) - c2^{-p}\|x - y\|^p.$$

Set $x = u + v$ and $y = u - v$ for arbitrary $u, v \in X$ to obtain the desired inequality. $\qquad\square$

Theorem 4.16. *Let $p > 1$ be a fixed real number. Then the functional $\|.\|^p$ is uniformly convex on the Banach space X if and only if X is $p-$uniformly convex i.e., if and only if there exists a constant $c > 0$ such that*

$$\delta_X(\varepsilon) \geq c.\varepsilon^p \quad for \quad 0 < \varepsilon \leq 2.$$

Proof. Suppose the functional $\|.\|^p$ is uniformly convex on X. We show that X is p-uniformly convex. Since $\|.\|^p$ is uniformly convex on X we have, by definition 4.4,

$$\|\lambda x + (1 - \lambda)y\|^p \leq \lambda \|x\|^p + (1 - \lambda)\|y\|^p - \lambda(1 - \lambda)\, \overline{\mu}(\|x - y\|)$$

for some $\overline{\mu} : \mathbb{R}^+ \to \mathbb{R}^+$ satisfying

$$\overline{\mu}(t) = 0 \ \Leftrightarrow \ t = 0.$$

Now we define, for each $t > 0$, $\mu : \mathbb{R}^+ \to \mathbb{R}^+$ by

$$\mu(t) := \inf \left\{ \frac{\lambda \|x\|^p + (1 - \lambda)\|y\|^p - \|\lambda x + (1 - \lambda)y\|^p}{W_p(\lambda)} : \right.$$

$$\left. 0 < \lambda < 1, x, y \in X, \|x - y\| = t \right\},$$

where $W_p(\lambda) := \lambda^p(1 - \lambda) + \lambda(1 - \lambda)^p$. Clearly, $\mu(0) = 0$.

We claim that $\mu(ct) = c^p \mu(t) \quad \forall \, c, \, t > 0$.
In fact,

$$\mu(ct) = \inf \left\{ \frac{\lambda \|u\|^p + (1 - \lambda)\|v\|^p - \|\lambda u + (1 - \lambda)v\|^p}{W_p(\lambda)} \right.$$

$$\left. : u, v \in X, \ \|u - v\| = ct \right\}.$$

Put $\frac{u}{c} = x$ and $\frac{v}{c} = y$. Then

$$\mu(ct) = c^p \inf \left\{ \frac{\lambda \|x\|^p + (1-\lambda)\|y\|^p - \|\lambda x + (1-\lambda)y\|^p}{W_p(\lambda)} : \right.$$

$$\left. 0 < \lambda < 1, \ x, y \in X, \ \|x - y\| = t \right\}.$$

So, $\mu(ct) = c^p.\mu(t)$. In particular, we can write,

$$\mu(t) = \mu(t.1) = t^p \mu(1) \text{ for } t > 0.$$

From the definition of $\mu(t)$, we get for $t = \|x - y\|$,

$$W_p(\lambda).\mu(t) \leq \lambda\|x\|^p + (1-\lambda)\|y\|^p - \|\lambda x + (1-\lambda)y\|^p$$

so that

$$\|\lambda x + (1-\lambda)y\|^p \leq \lambda\|x\|^p + (1-\lambda)\|y\|^p - W_p(\lambda)\mu(t)$$

$$= \lambda\|x\|^p + (1-\lambda)\|y\|^p - W_p(\lambda) \, c \, \|x - y\|^p$$

where $c = \mu(1) > 0$. Therefore, for $0 \leq \lambda \leq 1$ and for all $x, y \in X$, we have

$$\|\lambda x + (1-\lambda)y\|^p \leq \lambda\|x\|^p + (1-\lambda)\|y\|^p - W_p(\lambda)c\|x - y\|^p. \qquad (4.6)$$

In particular, for $\lambda = \frac{1}{2}$ we get,

$$\left\|\frac{x+y}{2}\right\|^p \leq \frac{1}{2}(\|x\|^p + \|y\|^p) - 2^{-p}c\|x - y\|^p.$$

Let $\|x\| = 1$, $\|y\| = 1$, $\|x - y\| \geq \varepsilon$; $\quad 0 < \varepsilon \leq 2$. Then

$$\left\|\frac{x+y}{2}\right\| \leq (1 - 2^{-p}c\varepsilon^p)^{1/p}.$$

so that

$$1 - \left\|\frac{x+y}{2}\right\| \geq 1 - (1 - 2^{-p}c\varepsilon^p)^{\frac{1}{p}}.$$

Hence, using Lemma 4.11 we get,

$$\delta_X(\varepsilon) \geq 1 - (1 - p^{-1}2^{-p}c\varepsilon^p) \geq c_1\varepsilon^p,$$

where $c_1 := p^{-1}2^{-p}c > 0$. Therefore X is p-uniformly convex as desired.

Conversely, suppose X is p-uniformly convex. We show $\|.\|^p$ is uniformly convex on the Banach space X. By Lemma 4.14, there exists a constant $c > 0$ such that inequality (4.5) is satisfied, and by Proposition 4.15, $\|.\|^p$ is uniformly convex at centre on X. Hence by the remark of Zalinescu [537], p. 352, referred to at the end of Definition 4.4, since $\|.\|^p$ is convex, it is uniformly convex on X, as required. □

Corollary 4.17 *Let $p > 1$ be a given real number. Then the following are equivalent in a Banach space X:*

(i) X *is p-uniformly convex.*

(ii) *There is a constant $c_1 > 0$ such that for every $x, y \in X$, $f_x \in J_p(x)$, the following inequality holds:*

$$\|x + y\|^p \geq \|x\|^p + p\langle y , f_x \rangle + c_1 \|y\|^p.$$

(iii) *There is a constant $c_2 > 0$ such that for every $x, y \in X$, $f_x \in J_p(x)$, $f_y \in J_p(y)$, the following inequality holds:*

$$\langle x - y , f_x - f_y \rangle \geq c_2 \|x - y\|^p.$$

Proof. $(i) \Rightarrow (ii)$: Since X is p-uniformly convex, by Theorem 4.16, $\|.\|^p$ is uniformly convex on X i.e. for all $\lambda \in [0, 1]$,

$$\|\lambda x + (1 - \lambda)y\|^p \leq \lambda\|x\|^p + (1 - \lambda)\|y\|^p - \lambda(1 - \lambda)\mu(\|x - y\|), \quad (4.7)$$

where $\mu(t) = \mu(1)t^p$ for all $x, y \in X$. Moreover, since J_p is the sub-differential of the functional $\frac{1}{p}.\|.\|^p$, we have for $f_x \in J_p(x)$ that

$$p\langle y - x , f_x \rangle \leq \|y\|^p - \|x\|^p \qquad \forall \, y \in X.$$

Replace y by $x + \lambda y$, $0 < \lambda < 1$, then,

$$
\begin{aligned}
p\langle y , f_x \rangle &\leq \frac{\|x + \lambda y\|^p - \|x\|^p}{\lambda} \\
&= \frac{\|(1 - \lambda)x + \lambda(x + y)\|^p - \|x\|^p}{\lambda} \\
&\leq \frac{\lambda\|x + y\|^p + (1 - \lambda)\|x\|^p - \lambda(1 - \lambda)\mu(\|y\|) - \|x\|^p}{\lambda} \\
&= \|x + y\|^p - \|x\|^p - (1 - \lambda)\mu(1)\|y\|^p.
\end{aligned}
$$

Letting $\lambda \to 0$, we obtain,

$$p\langle y , f_x \rangle \leq \|x + y\|^p - \|x\|^p - c_1\|y\|^p,$$

where $c_1 := \mu(1)$. Therefore,

$$\|x + y\|^p \geq \|x\|^p + p\langle y , f_x \rangle + c_1\|y\|^p.$$

Hence, (ii) is proved.

$(ii) \Rightarrow (iii)$: For $x, y \in X$, $f_x \in J_p(x)$ and $f_y \in J_p(y)$ we have,

$$\|x + y\|^p \geq \|x\|^p + p\langle y , f_x \rangle + c_1\|y\|^p, \tag{4.8}$$

$$\|x + y\|^p \geq \|y\|^p + p\langle x , f_y \rangle + c_1\|x\|^p. \tag{4.9}$$

Replace y by $y - x$ in (4.8) and replace x by $x - y$ in (4.9), to get,

$$\|y\|^p \geq \|x\|^p + p\langle y - x \,, \, f_x \rangle + c_1 \|y - x\|^p, \tag{4.10}$$

$$\|x\|^p \geq \|y\|^p + p\langle x - y \,, \, f_y \rangle + c_1 \|x - y\|^p. \tag{4.11}$$

Adding (4.10) and (4.11) we get,

$$p\langle y - x \,, \, f_x - f_y \rangle + 2c_1 \|x - y\|^p \leq 0.$$

This gives,

$$\langle x - y \,, \, f_x - f_y \rangle \geq \frac{2c_1}{p} \|x - y\|^p,$$

and so,

$$\langle x - y \,, \, f_x - f_y \rangle \geq c_2 \|x - y\|^p,$$

where $c_2 := \frac{2c_1}{p}$. Hence, (iii) follows.

$(iii) \Rightarrow (i)$: Given $\langle x - y \,, \, f_x - f_y \rangle \geq c_2 \|x - y\|^p$, we first show that there exists a constant $c_1^* > 0$ such that

$$\|x + y\|^p \geq \|x\|^p + p\langle y \,, \, f_x \rangle + c_1^* \|y\|^p \quad \forall \; x, y \in X.$$

So, for $x, y \in X$ and $0 \leq \lambda \leq 1$, let $g : [0, 1] \longrightarrow \mathbb{R}^+$ be defined by

$$g(t) = \|x + ty\|^p.$$

Therefore, using the fact that J_p is the sub-differential of the functional $\frac{1}{p}\|.\|^p$, we get,

$$\begin{aligned}
g(t + h) - g(t) &= \|x + (t + h)y\|^p - \|x + ty\|^p \\
&\geq hp\langle y \,, \, f_{x+ty} \rangle, \quad f_{x+ty} \in J_p(x + ty).
\end{aligned}$$

So,

$$g'_+(t) = \lim_{h \to 0^+} \frac{g(t + h) - g(t)}{h} \geq p\langle y \,, \, f_{x+ty} \rangle.$$

Now,

$$\begin{aligned}
\|x + y\|^p - \|x\|^p &= g(1) - g(0) \\
&= \int_0^1 g'_+(t)dt \\
&\geq p \int_0^1 \langle y \,, \, f_{x+ty} \rangle dt
\end{aligned}$$

$$= p\langle y\,,\,f_x\rangle + p\int_0^1 \frac{1}{t}\langle x + ty - x, f_{x+ty} - f_x\rangle dt$$

$$\geq p\langle y\,,\,f_x\rangle + pc_2\int_0^1 \|x + ty - x\|^p \frac{dt}{t},\ by\ (iii)$$

$$= p\langle y\,,\,f_x\rangle + c_2\|y\|^p.$$

Hence,

$$\|x + y\|^p - \|x\|^p \geq p\langle y\,,\,f_x\rangle + c_2\|y\|^p \qquad \forall\, x, y \in X. \qquad (4.12)$$

Replace
x by $\lambda x + (1 - \lambda)y$ and y by $(1 - \lambda)(x - y)$ in (4.12) to get,

$$\|x\|^p - \|\lambda x + (1 - \lambda)y\|^p \geq p(1 - \lambda)\langle x - y\,,\,f_{\lambda x+(1-\lambda)y}\rangle$$
$$+ c_2.(1 - \lambda)^p.\|x - y\|^p. \qquad (4.13)$$

Next, we replace x by $\lambda x + (1 - \lambda)y$ and y by $\lambda(y - x)$ in (4.12) to get,

$$\|y\|^p - \|\lambda x + (1 - \lambda)y\|^p \geq p\lambda\langle y - x\,,\,f_{\lambda x+(1-\lambda)y}\rangle + c_2\lambda^p\|y - x\|^p. \quad (4.14)$$

Performing $\lambda(4.13) + (1 - \lambda)(4.14)$, we get

$$\lambda\|x\|^p + (1 - \lambda)\|y\|^p \geq \|\lambda x + (1 - \lambda)y\|^p + c_2 W_p(\lambda)\|x - y\|^p$$
$$\geq \|\lambda x + (1 - \lambda)y\|^p + \lambda(1 - \lambda)2^{-p}c_2\|x - y\|^p.$$

Therefore, setting $\mu(t) = 2^{-p}t^p c_2$, we obtain that

$$\lambda\|x\|^p + (1 - \lambda)\|y\|^p \geq \|\lambda x + (1 - \lambda)y\|^p + \lambda(1 - \lambda).\mu(\|x - y\|)$$

which shows that $\|.\|^p$ is uniformly convex on X and hence X is p-uniformly convex. $\qquad \square$

4.4 Uniformly Convex Spaces

For uniformly convex Banach spaces, the following theorem has been proved.

Theorem 4.18. *(Xu,[509]) Let $p > 1$ and $r > 0$ be two fixed real numbers. Then a Banach space X is uniformly convex if and only if there exists a continuous, strictly increasing and convex function*

$$g : \mathbb{R}^+ \to \mathbb{R}^+,\ g(0) = 0$$

such that for all $x, y \in B_r$ and $0 \leq \lambda \leq 1$,

$$\|\lambda x + (1 - \lambda)y\|^p \leq \lambda \|x\|^p + (1 - \lambda)\|y\|^p - W_p(\lambda)g(\|x - y\|), \quad (4.15)$$

where $W_p(\lambda) := \lambda^p(1 - \lambda) + \lambda(1 - \lambda)^p$ and $B_r := \{x \in X : \|x\| \leq r\}$.

Corollary 4.19 *Let $p > 1$ and $r > 0$ be two fixed real numbers and X be a Banach space. Then the following are equivalent.*
(i) X is uniformly convex.
(ii) There is a continuous strictly increasing convex function

$$g : \mathbb{R}^+ \to \mathbb{R}^+, g(0) = 0$$

such that

$$\|x + y\|^p \geq \|x\|^p + p\langle y, f_x \rangle + g(\|y\|)$$

for every $x, y \in B_r$ and $f_x \in J_p(x)$.
(iii) There is a continuous, strictly increasing convex function

$$g : \mathbb{R}^+ \to \mathbb{R}^+, \; g(0) = 0$$

such that

$$\langle x - y, f_x - f_y \rangle \geq g(\|x - y\|)$$

for every $x, y \in B_r$ and $f_x \in J_p(x), \; f_y \in J_p(y)$.

Proof. $(i) \Rightarrow (ii)$: Assume X is uniformly convex. Then by Theorem 4.18, there exists a continuous, strictly increasing and convex function

$$g : \mathbb{R}^+ \to \mathbb{R}^+, \; g(0) = 0$$

such that for all $x, y \in B_r$, we have

$$\|\lambda x + (1 - \lambda)y\|^p \leq \lambda \|x\|^p + (1 - \lambda)\|y\|^p - W_p(\lambda).g(\|x - y\|). \quad (4.16)$$

Moreover, since J_p is the sub-differential of the functional $\frac{1}{p}.\|.\|^p$, we have for $f_x \in J_p(x)$, the sub-differential inequality

$$\|y\|^p \geq \|x\|^p + p\langle y - x, f_x \rangle \quad \forall y \in X.$$

Replace y by $x + \lambda y$ for $0 < \lambda < 1$, then

$$p\langle y, f_x \rangle \leq \frac{\|x + \lambda y\|^p - \|x\|^p}{\lambda}$$

$$= \frac{\|\lambda(x + y) + (1 - \lambda)x\|^p - \|x\|^p}{\lambda}$$

$$\leq \|x + y\|^p - \|x\|^p - \lambda^{-1}W_p(\lambda)g(\|y\|). \quad (4.17)$$

Letting $\lambda \to 0$ we get,

$$\|x + y\|^p \geq \|x\|^p + p\langle y, f_x \rangle + g(\|y\|).$$

Hence (ii) follows.

$(ii) \Rightarrow (iii)$: Given, there exists a continuous strictly increasing convex function

$$g : \mathbb{R}^+ \longrightarrow \mathbb{R}^+, \ g(0) = 0$$

such that

$$\|x + y\|^p \geq \|x\|^p + p\langle y, f_x \rangle + g(\|y\|), \tag{4.18}$$

$$\|x + y\|^p \geq \|y\|^p + p\langle x, f_y \rangle + g(\|x\|), \tag{4.19}$$

$\forall x, y \in B_r$, $f_x \in J_p(x)$, $f_y \in J_p(y)$.

Replace y by $\frac{y-x}{2}$, x by $\frac{x}{2}$ in inequality (4.18) and replace x by $\frac{x-y}{2}$, y by $\frac{y}{2}$ in inequality (4.19) to get

$$\left\|\frac{y}{2}\right\|^p \geq \left\|\frac{x}{2}\right\|^p + \frac{p}{2}\langle y - x, f_{\frac{x}{2}} \rangle + g\left(\frac{1}{2}\|x - y\|\right) \tag{4.20}$$

and

$$\left\|\frac{x}{2}\right\|^p \geq \left\|\frac{y}{2}\right\|^p + \frac{p}{2}\langle x - y, f_{\frac{y}{2}} \rangle + g\left(\frac{1}{2}\|x - y\|\right). \tag{4.21}$$

Replacing $\frac{x}{2}$ by x and $\frac{y}{2}$ by y and combining (4.20) and (4.21) we get

$$\langle x - y, f_x - f_y \rangle \geq \frac{2}{p}g(\|x - y\|).$$

Let $g_* = \frac{2}{p}g$, then g_* is continuous, strictly increasing convex and $g_*(0) = 0$. Moreover $\langle x - y, f_x - f_y \rangle \geq g_*(\|x - y\|)$. Hence g_* is the required function which satisfies (iii).

$(iii) \Rightarrow (i)$: Given that there exists $g : \mathbb{R}^+ \to \mathbb{R}^+$ a continuous, strictly increasing convex function with $g(0) = 0$ such that

$$\langle x - y, f_x - f_y \rangle \geq g(\|x - y\|)$$

$\forall x, y \in B_r$, $f_x \in J_p(x)$, $f_y \in J_p(y)$, we first show that $(iii) \Rightarrow (ii)$.

For $x, y \in B_r$ and $0 \leq t \leq 1$, define $G : [0, 1] \longrightarrow \mathbb{R}^+$ by $G(t) = \|x + ty\|^p$. Then, since J_p is the sub-differential of the functional $\frac{1}{p}\|.\|^p$ and for $f_{x+ty} \in J_p(x + ty)$, we have the sub-differential inequality

$$G(t + h) - G(t) \geq ph\langle y, f_{x+ty} \rangle.$$

So,

$$G'_+(t) := \lim_{t \to 0_+} \frac{G(t+h) - G(t)}{h} \geq p\langle y , \, f_{x+ty}\rangle. \tag{4.22}$$

Now, by using $g(0) = 0$, the convexity of g and (iii) we have,

$$\|x + y\|^p - \|x\|^p = G(1) - G(0)$$

$$= \int_0^1 G'_+(t)dt$$

$$\geq p \int_0^1 \langle y, \, f_{x+ty}\rangle dt$$

$$= p\langle y, \, f_x\rangle + p \int_0^1 \langle y, \, f_{x+ty} - f_x\rangle dt$$

$$= p\langle y, \, f_x\rangle + p \int_0^1 \langle x + ty - x , \, f_{x+ty} - f_x\rangle \frac{dt}{t}$$

$$\geq p\langle y, \, f_x\rangle + pg(\|y\|) \int_0^1 dt$$

$$= p\langle y, \, f_x\rangle + pg(\|y\|).$$

Let $g^* = pg$. Then we have,

$$\|x + y\|^p - \|x\|^p \geq p\langle y , \, f_x\rangle + g^*(\|y\|). \tag{4.23}$$

Hence (ii) is satisfied for the function g^*. Replace x by $\frac{\lambda x + (1-\lambda)y}{2}$ and y by $\frac{(1-\lambda)(x-y)}{2}$ in (4.23) to obtain

$$\|x\|^p - \|\lambda x + (1 - \lambda)y\|^p \geq p(1 - \lambda)2^{p-1}\langle x - y, \, f_{\frac{\lambda x + (1-\lambda)y}{2}} \rangle$$

$$+ 2^p g^*(\delta\|x - y\|) \tag{4.24}$$

and replace x by $\frac{\lambda x + (1-\lambda)y}{2}$ and y by $\frac{\lambda(y-x)}{2}$ in (4.23) to get

$$|y\|^p - \|\lambda x + (1 - \lambda)y\|^p \geq p\lambda 2^{p-1}\langle y - x, f_{\frac{\lambda x + (1-\lambda)y}{2}} \rangle$$

$$+ 2^p g^*(\delta\|x - y\|), \tag{4.25}$$

where $\delta = \frac{1}{2}\min\{1 - \lambda, \lambda\}$. Multiplying (4.24) by λ, (4.25) by $(1 - \lambda)$ and adding the two expressions we get, since $p > 1, W_p(\lambda) < 1$, that

$$\|\lambda x + (1 - \lambda)y\|^p \leq \lambda\|x\|^p + (1 - \lambda)\|y\|^p - W_p(\lambda).g^{**}(\|x - y\|),$$

where $g^{**}(t) := g^*(\delta t)$. So by Theorem 4.18, X is uniformly convex. $\qquad\square$

SUMMARY

For ease of reference, we now summarize the key inequalities obtained in this chapter.

Let $p \in (1, \infty), \lambda \in [0,1]$ and $W_p(\lambda) := \lambda^p(1-\lambda) + \lambda(1-\lambda)^p$. Then, we have the following results.

S1. (Theorem 4.10) Let E be *a real normed space*, and $J_p : E \to 2^{E^*}, 1 < p < \infty$, be the generalized duality map. Then, for any $x, y \in E$, the following inequality holds:

$$||x + y||^p \le ||x||^p + p\langle y, j_p(x+y)\rangle \tag{4.26}$$

for all $j_p(x+y) \in J_p(x+y)$. In particular, if $p = 2$, i.e., if j denotes the normalized duality map, then

$$||x + y||^2 \le ||x||^2 + 2\langle y, j(x+y)\rangle \tag{4.27}$$

holds for all $j(x+y) \in J(x+y)$.

S2. (Xu, [509]). Let E be a $p-$*uniformly convex space*. Then, there exist constants $d_p > 0, c_p > 0$ such that for every $x, y \in E, f_x \in J_p(x), f_y \in J_p(y)$, the following inequalities hold:

$$||x + y||^p \ge ||x||^p + p\langle y, f_x\rangle + d_p||y||^p, \tag{4.28}$$

$$||\lambda x + (1-\lambda)y||^p \le \lambda||x||^p + (1-\lambda)||y||^p - c_p W_p(\lambda)||x - y||^p, \tag{4.29}$$

for all $\lambda \in [0,1]$, and

$$\langle x - y, f_x - f_y\rangle \ge c_2||x - y||^p. \tag{4.30}$$

S3. (Xu, [509]) Let E be *a real uniformly convex Banach space*. For arbitrary $r > 0$, let $B_r(0) := \{x \in E : ||x|| \le r\}$. Then, there exists a continuous strictly increasing convex function

$$g : [0, \infty) \to [0, \infty), g(0) = 0$$

such that for every $x, y \in B_r(0), f_x \in J_p(x), f_y \in J_p(y)$, the following inequalities hold:

$$||x + y||^p \ge ||x||^p + p\langle y, f_x\rangle + g(||y||), \tag{4.31}$$

$$||\lambda x + (1-\lambda)y||^p \le \lambda||x||^p + (1-\lambda)||y||^p - W_p(\lambda)g(||x - y||), \tag{4.32}$$

and,

$$\langle x - y, f_x - f_y\rangle \ge g(||x - y||). \tag{4.33}$$

EXERCISES 4.1

1. Prove that Hilbert spaces H and the Banach spaces l_p, L_p, and $W^{m,p}(1 < p < \infty)$ are all uniformly convex, and establish the following estimates:

$$\delta_H(\varepsilon) = 1 - \sqrt{1 - \frac{1}{4}\varepsilon^2}.$$

$$\delta_{l^p}(\varepsilon) = \delta_{L^p}(\varepsilon) = \delta_{W_m^p}(\varepsilon) = \begin{cases} \frac{p-1}{8}\varepsilon^2 + o(\varepsilon^2) > \frac{p-1}{8}\varepsilon^2, & 1 < p < 2 \\ 1 - \left[1 - (\frac{\varepsilon}{2})^p\right]^{\frac{1}{p}} > \frac{1}{p}\left(\frac{\varepsilon}{2}\right)^p, & p \geq 2. \end{cases}$$

(Hint: See Lindenstrauss and Tzafriri, [312]).

4.5 Historical Remarks

The second inequality of Theorem 4.10 was used by Downing [214]. Theorem 4.16, Corollary 4.17 and Theorem 4.18 are due to Xu [509]. Lim [306] also derived inequalities similar to those of Xu for L_p spaces. If $q \geq 2$, a Banach space is called *uniformly convex with modulus of convexity of power type q* (see e.g., Prus and Smarzewski [387], Lindenstrauss and Tzafriri [312]) if there exists a constant $c > 0$ such that $\delta_E(\epsilon) \geq c\epsilon^q, (0 < \epsilon \leq 2)$. Clearly, this coincides basically with the definition of $q-$uniformly convex Banach spaces. Prus and Smarzewski [387], using the notion of modulus of convexity of power type also obtained inequality (4.29) and deduced that L_p has modulus of convexity of power type 2 if $1 < p \leq 2$, and has modulus of convexity of power type p, if $p > 2$. The characterization of duality mapping as the subdifferential of the convex function $f(x) = \frac{1}{p}||x||^p$ is due to Asplund [17].

Chapter 5
Inequalities in Uniformly Smooth Spaces

5.1 Definitions and Basic Theorems

In this chapter, we obtain analogues of the identities (1.1) and (1.2) in smooth spaces. We begin with the following definitions.

Definition 5.1. For $q > 1$, a Banach space X is said to be **q-uniformly smooth** if there exists a constant $c > 0$ such that

$$\rho_X(t) \leq c\, t^q,\ t > 0.$$

Definition 5.2. Let $f : X \to \mathbb{R}$ be a convex function. Then, the *conjugate of* f, $f^* : X^* \to \mathbb{R}$, is defined by

$$f^*(x^*) = sup\{\langle x,\, x^* \rangle - f(x)\ :\ x \in X\}.$$

Proposition 5.3. *If* $f(x) = \frac{1}{p}\|x\|^p, p > 1$ *then*

$$f^*(x^*) = \frac{1}{q}\|x^*\|^q, \quad \frac{1}{p} + \frac{1}{q} = 1.$$

Proof. Since J_p is the sub-differential of the functional $\frac{1}{p}\|.\|^p$, we have

$$\langle x,\, x^* \rangle = \|x\|^p, \quad \|x^*\| = \|x\|^{p-1}.$$

So,

$$\langle x,\, x^* \rangle - \frac{1}{p}\|x\|^p = \|x\|^p - \frac{1}{p}\|x\|^p = \left(1 - \frac{1}{p}\right)\|x\|^p = \frac{1}{q}\|x\|^p.$$

Now $\|x^*\| = \|x\|^{p-1}$, so $\|x^*\|^q = \|x\|^{q(p-1)} = \|x\|^p$. Therefore $\langle x,\, x^* \rangle - \frac{1}{p}\|x\|^p = \frac{1}{q}\|x^*\|^q$ so that

$$\sup_{x \in X}\{\langle x,\, x^* \rangle - \frac{1}{p}\|x\|^p\} = \frac{1}{q}\|x^*\|^q.$$

Hence $f^*(x^*) = \frac{1}{q}\|x^*\|^q$, when $f(x) = \frac{1}{p}\|x\|^p$. $\qquad\square$

C. Chidume, *Geometric Properties of Banach Spaces and Nonlinear Iterations*, 45
Lecture Notes in Mathematics 1965,
© Springer-Verlag London Limited 2009

Proposition 5.4. *For any constant* $c > 0$, $(cf)^*(x^*) = cf^*(c^{-1}x^*)$.

Proof.

$$
\begin{aligned}
(cf)^*(x^*) &:= \sup\{\langle x, x^* \rangle - (cf)(x) : x \in X\} \\
&= \sup\{\langle x, x^* \rangle - cf(x) : x \in X\} \\
&= c\sup\{c^{-1}\langle x, x^* \rangle - f(x) : x \in X\} \\
&= c\sup\{\langle x, c^{-1}x^* \rangle - f(x) : x \in X\} \\
&= cf^*(c^{-1}x^*),
\end{aligned}
$$

completing the proof. □

We shall make use of the following lemma due to Zalinescu [537].

Lemma 5.5. *(Zalinescu, [537], Theorem 2.2) Let X be a reflexive Banach space, $f : X \to \mathbb{R}$ be a convex functional and $g : \mathbb{R}^+ \to \mathbb{R}^+$ be a proper l.s.c. convex function whose domain is not a singleton. Then, the following are equivalent:*

(i) $f(y) \geq f(x) + \langle y - x, x^* \rangle + g(\|x - y\|) \ \forall\, x, y \in X, \ x^* \in \partial f(x)$,
(ii) $f^*(y^*) \leq f^*(x^*) + \langle x, y^* - x^* \rangle + g^*(\|x^* - y^*\|) \forall x \in X, \ \forall\ x^*, y^* \in X^*$.

5.2 q−uniformly Smooth Spaces

Proposition 5.6. *Let X be a real Banach space.*

(a) X is p-uniformly convex if and only if X^* is q-uniformly smooth,
(b) X is q-uniformly smooth if and only if X^* is p-uniformly convex, $\frac{1}{p} + \frac{1}{q} = 1$.

Proof. This is an immediate consequence of Theorem 2.10.

Theorem 5.7. *Let $q > 1$ be a fixed real number and X be a smooth Banach space. Then X is q-uniformly smooth if and only if there exists a constant $c_q > 0$ such that*

$$\|\lambda x + (1 - \lambda)y\|^q \geq \lambda\|x\|^q + (1 - \lambda)\|y\|^q - W_q(\lambda)c_q\|x - y\|^q \qquad (5.1)$$

for all $x, y \in X; 0 \leq \lambda \leq 1$.

Proof. Suppose X is q-uniformly smooth. Then by Proposition 5.6, X^* is p-uniformly convex. Then by Corollary 4.17 there exists a constant $c_1 > 0$ such that for $x^*, y^* \in X^*, f_{x^*} \in J_p(x^*)$ we have

$$\|x^* + y^*\|^p \geq \|x^*\|^p + p\langle y^*, f_{x^*} \rangle + c_1\|y^*\|^p. \qquad (5.2)$$

Note that the conjugate of $\frac{1}{p}\|x^*\|^p$ is $\frac{1}{q}\|x\|^q$. So from inequality (5.2), using Lemma 5.5, we get

$$\|x + y\|^q \leq \|x\|^q + q\langle y,\ J_q(x)\rangle + d\|y\|^q \tag{5.3}$$

for every $x, y \in X$ and some constant d given in this case by $d = c_1^{1-q}$. Replace x by $\lambda x + (1 - \lambda)y$ and y by $(1 - \lambda)(x - y)$ in (5.3) to get

$$\|x\|^q \leq \|\lambda x + (1 - \lambda)y\|^q + q(1 - \lambda)\langle x - y,\ J_q(\lambda x + (1 - \lambda)y)\rangle \\ + d(1 - \lambda)^q\|x - y\|^q. \tag{5.4}$$

Again replace x by $\lambda x + (1 - \lambda)y$ and y by $\lambda(y - x)$ in (5.3) to get

$$\|y\|^q \leq \|\lambda x + (1 - \lambda)y\|^q + q\lambda\langle y - x,\ J_q(\lambda x + (1 - \lambda)y)\rangle \\ + d\lambda^q\|y - x\|^q. \tag{5.5}$$

Multiplying (5.4) by λ; (5.5) by $(1-\lambda)$ and adding the two expressions we get

$$\|\lambda x + (1 - \lambda)y\|^q \geq \lambda\|x\|^q + (1 - \lambda)\|y\|^q - dW_q(\lambda)\|y - x\|^q$$

which is the desired inequality.

Conversely, suppose (5.1) holds. Since $J_q(x)$ is the sub-differential of the functional $\frac{1}{q}\|x\|^q$, for $f_x \in J_q(x)$ we have

$$\|y\|^q \geq \|x\|^q + q\langle y - x,\ f_x\rangle \qquad \forall\ y \in X.$$

Replace y by $x + \lambda y$, for $0 < \lambda < 1$. Then for $c > 0$,

$$\lambda q\langle y,\ f_x\rangle \leq \|x + \lambda y\|^q - \|x\|^q \\ = \|\lambda(x + y) + (1 - \lambda)x\|^q - \|x\|^q \\ \geq \lambda\|x + y\|^q - \lambda\|x\|^q - W_q(\lambda)c_q\|y\|^q,$$

so that

$$q\langle y,\ f_x\rangle \geq \|x + y\|^q - \|x\|^q - \lambda^{-1}W_q(\lambda)c_q\|y\|^q.$$

Letting $\lambda \to 0^+$, we get

$$\|x + y\|^q \leq \|x\|^q + q\langle y,\ f_x\rangle + c_q\|y\|^q.$$

Again by using Lemma 5.5 we get

$$\|x^* + y^*\|^p \geq \|x^*\|^p + p\langle y^*,\ f_{x^*}\rangle + d\|y^*\|^p$$

for $x^*, y^* \in X^*$, $f_{x^*} \in J_p(x^*)$ and some constant $d > 0$. Therefore by Corollary 4.17, X^* is p-uniformly convex and hence, X is q-uniformly smooth. This proves the theorem. $\square$

Corollary 5.8 *Let $q > 1$ be a fixed real number and X be a smooth Banach space. Then the following statements are equivalent:*

(i) X is q-uniformly smooth.

(ii) There is a constant $c > 0$ such that for all $x, y \in X$

$$\|x + y\|^q \le \|x\|^q + q\langle y, \, J_q(x)\rangle + c\|y\|^q.$$

(iii) There is a constant $c_1 > 0$ such that

$$\langle x - y, \, J_q(x) - J_q(y)\rangle \le c_1\|x - y\|^q \quad \forall \ x, y \in X.$$

Proof. $(i) \Leftrightarrow (ii)$: This follows implicitly from Theorem 5.7.
$(ii) \Rightarrow (iii)$: From (ii), for $f_x \in J_q(x)$ and $f_y \in J_q(y)$ we have

$$\|y\|^q \le \|x\|^q + q\langle y - x, \, J_q(x)\rangle + c\|x - y\|^q$$

and (replacing y by $(x - y)$ and x by y),

$$\|x\|^q \le \|y\|^q + q\langle x - y, \, J_q(y)\rangle + c\|x - y\|^q.$$

Adding these two inequalities we get

$$\langle x - y, \, J_q(x) - J_q(y)\rangle \le c_1\|x - y\|^q$$

where $c_1 = 2cq^{-1}$. This proves (iii).

$(iii) \Rightarrow (i)$: Since $(i) \Leftrightarrow (ii)$ so it is sufficient to show that $(iii) \Rightarrow (ii)$.
Define $g : [0, 1] \longrightarrow \mathbb{R}^+$ by $g(t) = \|x + ty\|^q \ \forall\, x, y \in X$. Let $h < 0$. Then since $J_q(x)$ is the sub-differential of the functional $\frac{1}{q}\|x\|^q$ we get

$$g(t + h) - g(t) = \|x + (t + h)y\|^q - \|x + ty\|^q$$
$$\ge hq\langle y, \, J_q(x + ty)\rangle.$$

So,

$$g'_-(t) := \lim_{h \to 0_-} \frac{g(t + h) - g(t)}{h} \le q\langle y, \, J_q(x + ty)\rangle.$$

Now

$$\|x + y\|^q - \|x\|^q = g(1) - g(0)$$
$$= \int_0^1 g'_-(t)dt$$
$$\le q \int_0^1 \langle y, \, J_q(x + ty)\rangle dt$$
$$\le q\langle y, \, J_q(x)\rangle$$

$$+q \int_0^1 \langle x + ty - x, \ J_q(x+ty) - J_q(x) \rangle \frac{dt}{t}$$

$$\leq q \langle y, J_q(x) \rangle + q c_1 \int_0^1 \|ty\|^q \frac{dt}{t}$$

$$= q \langle y, J_q(x) \rangle + q c_1 \|y\|^q \int_0^1 t^{q-1} dt$$

$$= q \langle y, J_q(x) \rangle + c_1 \|y\|^q.$$

This completes the proof of the corollary. $\qquad\qquad\qquad\qquad\square$

5.3 Uniformly Smooth Spaces

Theorem 5.9. *Let $q > 1$ and $r > 0$ be two fixed real numbers. Then a Banach space X is uniformly smooth if and only if there exists a continuous, strictly increasing and convex function*

$$g^* : \mathbb{R}^+ \rightarrow \mathbb{R}^+, \ g^*(0) = 0$$

such that

$$\|\lambda x + (1 - \lambda) y\|^q \geq \lambda \|x\|^q + (1 - \lambda) \|y\|^q - W_q(\lambda) g^*(\|x - y\|) \quad (5.6)$$

for all $x, y \in B_r, 0 \leq \lambda \leq 1$, where $W_q(\lambda) := \lambda^q (1 - \lambda) + \lambda(1 - \lambda)^q$.

Proof. Suppose X is uniformly smooth. Then X^* is uniformly convex. So by Theorem 4.18 there exists a continuous strictly increasing and convex function $g : \mathbb{R}^+ \longrightarrow \mathbb{R}^+$, $g(0) = 0$ such that

$$\|\lambda x^* + (1 - \lambda) y^*\|^p \leq \lambda \|x^*\|^p + (1 - \lambda) \|y^*\|^p - W_p(\lambda) g^*(\|x^* - y^*\|), \ (5.7)$$

for all $x^*, y^* \in B_r^*$ and $0 \leq \lambda \leq 1$. By using Lemma 5.5 in the above expression we get inequality (5.6). Conversely, suppose inequality (5.6) holds. Then we have for $x, y \in B_r$ and $t \in (0, 1)$ that

$$\frac{\|x + ty\|^q - \|x\|^q}{t} = \frac{\|(1 - t)x + t(x + y)\|^q - \|x\|^q}{t}$$

$$\geq \frac{t\|x + y\|^q + (1 - t)\|x\|^q - W_q(t) g(\|y\|) - \|x\|^2}{t}$$

$$= \|x + y\|^q - \|x\|^q - t^{-1} W_q(t) g(\|y\|).$$

Using this,

$$q \langle y, J_q(x) \rangle = \lim_{t \to 0^+} \frac{\|x + ty\|^q - \|x\|^q}{t}$$

$$\geq \|x + y\|^q - \|x\|^q - g(\|y\|),$$

so that we have

$$\|x + y\|^q \leq \|x\|^q + q\langle y,\ J_q(x)\rangle + g(\|y\|). \qquad (5.8)$$

Using Lemma 5.5 in above inequality we get

$$\|x^* + y^*\|^p \geq \|x^*\|^p + p\langle y^*,\ J_p(x^*)\rangle + g^*(\|y^*\|). \qquad (5.9)$$

This shows (by Corollary 4.19) that X^* is uniformly convex and hence (by Theorem 2.10) X is uniformly smooth as desired. This completes the proof. $\square$

Corollary 5.10 *Let $q > 1$ and $r > 0$ be two fixed real numbers and X be a smooth Banach space. Then the following are equivalent:*

(i) X is uniformly smooth.

(ii) There is a continuous strictly increasing and convex function

$$g : \mathbb{R}^+ \longrightarrow \mathbb{R}^+, \ \ g(0) = 0$$

such that for every $x, y \in B_r$ we get

$$\|x + y\|^q \leq \|x\|^q + q\langle y,\ J_q(x)\rangle + g(\|y\|).$$

(iii) There is a continuous, strictly increasing and convex function

$$g : \mathbb{R}^+ \longrightarrow \mathbb{R}^+, \ \ g(0) = 0$$

such that for all $x, y \in B_r$ we have

$$\langle x - y,\ J_q(x) - J_q(y)\rangle \leq g(\|x - y\|).$$

Proof. (see, Xu, [509]).

We state the following important inequality which holds in uniformly smooth spaces.

Theorem 5.11. *(Reich, [411]) Let X be a uniformly smooth Banach space and let $\beta(t)$ be defined for $t > 0$ by*

$$\beta(t) := \sup\left\{ \frac{(\|x + ty\|^2 - \|x\|^2)}{t} - 2\mathrm{Re}\langle y, J(x)\rangle :\ \|x\| \leq 1, \|y\| \leq 1 \right\}.$$

Then, $\lim_{t \to 0^+} \beta(t) = 0$ *and*

$$\|x + y\|^2 \leq \|x\|^2 + 2\mathrm{Re}\langle y, J(x)\rangle + \max\{\|x\|, 1\}\|y\|\beta(\|y\|)$$

for all $x, y \in X$.

5.4 Characterization of Some Real Banach Spaces by the Duality Map

In this section, we present characterizations of *uniformly smooth Banach spaces* and *Banach spaces with uniformly Gâteaux differentiable norms* in terms of the normalized duality maps.

5.4.1 Duality Maps on Uniformly Smooth spaces

In this subsection we give a characterization of uniformly smooth Banach spaces in terms of the normalized duality map.

Theorem 5.12. *Let X be a real uniformly smooth Banach space. Then the normalized duality map $J : X \to X^*$ is norm-to-norm uniformly continuous on the unit ball of X.*

Proof. X is uniformly smooth implies X^* is uniformly convex, i.e., given $\epsilon > 0$, there exists $\delta > 0$ such that if $x^*, y^* \in X^*$, $||x^*|| = ||y^*|| = 1, ||x^* + y^*|| > 2 - \delta$, then we have $||x^* - y^*|| < \epsilon$. Now, let $x, y \in X, ||x|| = ||y|| = 1$ and suppose $||x - y|| < \delta$. Then,

$$||Jx + Jy|| \geq \langle y, Jx + Jy \rangle$$
$$= \langle x, Jx \rangle + \langle y, Jy \rangle - \langle x - y, Jx \rangle$$
$$> 2 - ||x - y|| > 2 - \delta.$$

Hence, $||Jx - Jy|| < \epsilon$, i.e., J is norm-to-norm uniformly continuous on the unit ball of X. $\square$

5.4.2 Duality Maps on Spaces with Uniformly Gâteaux Differentiable Norms

Proposition 5.13. *(see e.g., Cudia [195]) If a Banach space E has a uniformly Gâteaux differentiable norm, then $j : E \to E^*$ is uniformly continuous on bounded subsets of E from the strong topology of E to the weak* topology of E^*.*

Proof. If the result were not true, then there exist sequences $\{x_n\}$ and $\{z_n\}$, a point $y_0 \in E$, and a positive ε such that for all $n \in \mathbb{N}$,

$$||x_n|| = ||z_n|| = ||y_0|| = 1, \ z_n - x_n \to 0, \ and \ \langle y_0, j(z_n) - j(x_n) \rangle \geq \varepsilon.$$

Let

$$a_n := \frac{||x_n + ty_0|| - ||x_n|| - t\langle y_0, j(x_n)\rangle}{t},$$

$$b_n := \frac{||z_n + ty_0|| - ||z_n|| - t\langle y_0, j(z_n)\rangle}{t}.$$

If $t > 0$ is sufficiently small, both a_n and b_n are less than $\frac{1}{2}\varepsilon$, i.e., $a_n + b_n < \varepsilon$. On the other hand, we have

$$a_n \geq \frac{\left\langle x_n + ty_0, j(z_n)\right\rangle - \left\langle x_n + ty_0, j(x_n)\right\rangle}{t}$$

$$= \left\langle y_0, j(z_n) - j(x_n)\right\rangle + \frac{\left\langle x_n, j(z_n) - j(x_n)\right\rangle}{t}$$

and

$$b_n \geq \frac{\left\langle z_n - ty_0, j(x_n)\right\rangle - \left\langle z_n - ty_0, j(z_n)\right\rangle}{t}$$

$$= \left\langle y_0, j(z_n) - j(x_n)\right\rangle - \frac{\left\langle z_n, j(z_n) - j(x_n)\right\rangle}{t}.$$

Hence,

$$a_n + b_n \geq 2\langle y_0, j(z_n) - j(x_n)\rangle - \frac{\langle x_n - z_n, j(z_n) - j(x_n)\rangle}{t}$$

$$\geq 2\varepsilon - 2\frac{||x_n - z_n||}{t}.$$

We arrive at a contradiction by choosing $t = \frac{2}{\varepsilon}||x_n - z_n||$ for sufficiently large n. □

We now summarize the key inequalities obtained in this chapter. We shall include also an inequality of Xu and Roach [525] (**S4** below). The proof of the inequality is rather long and is omitted. The interested reader may consult Xu and Roach, [525].

SUMMARY

Let $q \in (1, \infty), \lambda \in [0, 1], W_q(\lambda) := \lambda^q(1 - \lambda) + \lambda(1 - \lambda)^q$. We have the following results.

S1. (Xu, [509]) Let E be a real q-uniformly smooth Banach space. Then, there exist constants $c_q > 0, d_q > 0, d > 0$ such that for all $x, y \in E$,

$$||x + y||^q \leq ||x||^q + q\langle y, j_q(x)\rangle + d_q||y||^q, \tag{5.10}$$

$$||\lambda x + (1 - \lambda)y||^q \geq \lambda||x||^q + (1 - \lambda)||y||^q - c_q W_q(\lambda)||x - y||^q, \quad (5.11)$$

and,

$$\langle x - y, j_q(x) - j_q(y) \rangle \leq d||x - y||^q. \tag{5.12}$$

S2. (Xu, [509]) Let E be a real uniformly smooth Banach space. For arbitrary $r > 0$, let $B_r(0) := \{x \in E : ||x|| \leq r\}$. Then, there exists a continuous strictly increasing and convex function

$$g : [0, \infty) \rightarrow [0, \infty), g(0) = 0$$

such that for every $x, y \in B_r(0)$, the following inequalities hold:

$$||x + y||^q \leq ||x||^q + q\langle y, j_q(x) \rangle + g(||y||), \tag{5.13}$$

$$||\lambda x + (1 - \lambda)y||^q \geq \lambda||x||^q + (1 - \lambda)||y||^q - W_q(\lambda)g(||x - y||), \tag{5.14}$$

and,

$$\langle x - y, j_q(x) - j_q(y) \rangle \leq g(||x - y||). \tag{5.15}$$

S3. (Reich, [408]) Let E be a real uniformly smooth Banach space. Then, there exists a nondecreasing continuous function

$$\beta : [0, \infty) \rightarrow [0, \infty),$$

satisfying the following conditions:

(i) $\beta(ct) \leq c\beta(t) \ \forall \ c \geq 1$;
(ii) $\lim_{t \rightarrow 0^+} \beta(t) = 0$ and,

$$||x + y||^2 \leq ||x||^2 + 2\mathrm{Re}\langle y, j(x) \rangle$$
$$+ \max\{||x||, 1\}||y||\beta(||y||)\forall x, y \in E. \tag{5.16}$$

S4. (Xu and Roach, [525]) Let E be a real uniformly smooth Banach space. Then, there exist constants D and C such that for all $x, y \in E, j(x) \in J(x)$, the following inequality holds:

$$||x + y||^2 \leq ||x||^2 + 2\langle y, j(x) \rangle$$
$$+ D\max\{||x|| + ||y||, \frac{1}{2}C\}\rho_E(||y||), \tag{5.17}$$

where ρ_E denotes the modulus of smoothness of E.

S5. Let X be a real uniformly smooth Banach space with dual X^*. Then, the normalized duality map $J : X \rightarrow X^*$ is norm-to-norm uniformly continuous on bounded subsets of X.

S6. Let X be a real normed Banach space space with a uniformly Gâteaux differentiable norm. Then, the duality mapping $J : X \rightarrow 2^{X^*}$ is norm-to-weak* uniformly continuous on bounded subsets of X.

EXERCISES 5.1

1. Give the details of the proof of Proposition 5.6.
2. Prove that Hilbert spaces H and the Banach spaces l_p, L_p, and $W^{m,p}$ ($1 < p < \infty$) are all uniformly smooth, and that the following estimates hold:

$$\rho_H(\tau) = (1 + \tau^2)^{\frac{1}{2}} - 1.$$

$$\rho_{l^p}(\tau) = \rho_{L^p}(\tau) = \rho_{W_m^p}(\tau) = \begin{cases} (1 + \tau^p)^{\frac{1}{p}} - 1 < \frac{1}{p}\tau^p, & 1 < p < 2 \\ \frac{p-1}{2}\tau^2 + o(\tau^2) < \frac{p-1}{2}\tau^2, & p \geq 2. \end{cases}$$

(Hint: (Lindenstrauss and Tzafriri, [312])).
3. Assume $1 < p < \infty$ and t_p is the unique solution of the equation

$$(p - 1)t^{p-1} + (p - 1)t^{p-2} - 1 = 0, \ 0 < t < 1.$$

Let $c_p = (1 + t_p^{p-1})(1 + t_p)^{-(p-1)}$. Then we have the following results:

(i) If $2 < p < \infty$, then we have for all x, y in L_p and $0 \leq \lambda \leq 1$, the following inequalities:

$$\|\lambda x + (1 - \lambda)y\|^p \leq \lambda\|x\|^p + (1 - \lambda)\|y\|^p - W_p(\lambda)c_p\|x - y\|^p.$$

$$\|\lambda x + (1 - \lambda)y\|^2 \geq \lambda\|x\|^2 + (1 - \lambda)\|y\|^2 - \lambda(1 - \lambda)(p - 1)\|x - y\|^2.$$

$$\|x + y\|^p \geq \|x\|^p + p\langle y, j_p(x)\rangle + c_p\|y\|^p.$$

$$\|x + y\|^2 \leq \|x\|^2 + 2\langle y, j(x)\rangle + (p - 1)\|y\|^2.$$

$$\langle x - y, J_p(x) - J_p(y)\rangle \geq 2p^{-1}c_p\|x - y\|^p.$$

$$\langle x - y, J(x) - J(y)\rangle \leq (p - 1)\|x - y\|^2.$$

(ii) If $1 < q \leq 2$, then we have for all x, y in L_q and $0 \leq \lambda \leq 1$, the following inequalities:

$$\|\lambda x + (1 - \lambda)y\|^2 \leq \lambda\|x\|^2 + (1 - \lambda)\|y\|^2 - \lambda(1 - \lambda)(q - 1)\|x - y\|^2.$$

$$\|\lambda x + (1 - \lambda)y\|^q \geq \lambda\|x\|^q + (1 - \lambda)\|y\|^q - W_q(\lambda)c_q\|x - y\|^q.$$

$$||x + y||^2 \geq ||x||^2 + 2\langle y, j(x) \rangle + c_q ||y||^2.$$

$$||x + y||^q \leq ||x||^q + q\langle y, j_q(x) \rangle + c_q ||y||^q.$$

$$\langle x - y, J(x) - J(y) \rangle \geq (q - 1)||x - y||^2.$$

$$\langle x - y, J_q(x) - J_q(y) \rangle \leq 2q^{-1} c_q ||x - y||^q.$$

(Hint: (Xu, [509], p. 1132, Corollary 2)).

Chapter 6
Iterative Method for Fixed Points of Nonexpansive Mappings

6.1 Introduction

We begin this chapter with the following well known definition and theorem.

Definition 6.1. Let (M, ρ) be a metric space. A mapping $T : M \to M$ is called a *contraction* if there exists $k \in [0, 1)$ such that $\rho(Tx, Ty) \leq k\rho(x, y)$ for all $x, y \in M$. If $k = 1$, then T is called *nonexpansive*.

Theorem 6.2. *(Banach Contraction Mapping Principle). Let (M, ρ) be a complete metric space and $T : M \to M$ be a contraction. Then T has a unique fixed point, i.e. there exists a unique $x^* \in M$ such that $Tx^* = x^*$. Moreover, for arbitrary $x_0 \in M$, the sequence $\{x_n\}$ defined iteratively by $x_{n+1} = Tx_n$, $n \geq 0$, converges to the unique fixed point of T.*

Apart from being an obvious generalization of the contraction mappings, nonexpansive maps are important, as has been observed by Bruck [59], mainly for the following two reasons:

- Nonexpansive maps are intimately connected with the monotonicity methods developed since the early 1960's and constitute one of the first classes of nonlinear mappings for which fixed point theorems were obtained by using the fine geometric properties of the underlying Banach spaces instead of compactness properties.
- Nonexpansive mappings appear in applications as the transition operators for initial value problems of differential inclusions of the form $0 \in \frac{du}{dt} + T(t)u$, where the operators $\{T(t)\}$ are, in general, set-valued and are *accretive* or *dissipative* and *minimally continuous*.

The following fixed point theorem has been proved for nonexpansive maps on uniformly convex spaces.

Theorem 6.3. *(Kirk, [283]) Let E be a reflexive Banach space and let K be a nonempty closed bounded and convex subset of E with normal structure.*

C. Chidume, *Geometric Properties of Banach Spaces and Nonlinear Iterations*,
Lecture Notes in Mathematics 1965,

Let $T : K \to K$ be a nonexpansive mapping of K into itself. Then T has a fixed point.

Unlike in the case of the Banach contraction mapping principle, trivial examples show that the sequence of successive approximations $x_{n+1} = Tx_n$, $x_0 \in K$, $n \geq 0$, (where K is a nonempty closed convex and bounded subset of a real Banach space E), for a nonexpansive mapping $T : K \to K$ even with a unique fixed point, may fail to converge to the fixed point. It suffices, for example, to take for T, a rotation of the unit ball in the plane around the origin of coordinates. More precisely, we have the following example.

Example 6.4. Let $B := \{x \in \mathbb{R}^2 : ||x|| \leq 1\}$ and let T denote an anticlockwise rotation of $\frac{\pi}{4}$ about the origin of coordinates. Then T is nonexpansive with the origin as the only fixed point. Moreover, the sequence $\{x_n\}$ defined by $x_{n+1} = Tx_n, x_0 = (1,0) \in B, n \geq 0$, does not converge to zero.

Krasnoselskii [291], however, showed that in this example, one can obtain a convergent sequence of successive approximations if instead of T one takes the auxiliary nonexpansive mapping $\frac{1}{2}(I + T)$, where I denotes the identity transformation of the plane, i.e., if the sequence of successive approximations is defined by $x_0 \in K$,

$$x_{n+1} = \frac{1}{2}(x_n + Tx_n), \quad n = 0, 1, ... \tag{6.1}$$

instead of by the usual so-called *Picard iterates*, $x_{n+1} = Tx_n$, $x_0 \in K$, $n \geq 0$. It is easy to see that the mappings T and $\frac{1}{2}(I + T)$ have the same set of fixed points, so that the limit of a convergent sequence defined by (6.1) is necessarily a fixed point of T.

More generally, if X is a normed linear space and K is a convex subset of X, a generalization of equation (6.1) which has proved successful in the approximation of fixed points of nonexpansive mappings $T : K \to K$ (when they exist), is the following scheme: $x_0 \in K$,

$$x_{n+1} = (1 - \lambda)x_n + \lambda Tx_n, \quad n = 0, 1, 2, \dots ; \lambda \in (0, 1) , \tag{6.2}$$

λ constant (see, e.g., Schaefer [431]). However, the most general *Mann-type* iterative scheme now studied is the following: $x_0 \in K$,

$$x_{n+1} = (1 - C_n)x_n + C_n Tx_n, \quad n = 0, 1, 2, ... \tag{6.3}$$

where $\{C_n\}_{n=1}^\infty \subset (0, 1)$ is a real sequence satisfying appropriate conditions (see, e.g., Chidume [87], Edelstein and O'Brian [218], Ishikawa [259]). Under the following additional assumptions (i) $\lim C_n = 0$; and (ii) $\sum\limits_{n=0}^\infty C_n = \infty$, the sequence $\{x_n\}$ generated by (6.3) is generally referred to as the *Mann sequence* in the light of Mann [319]. The recursion formula (6.2) is consequently called the *Krasnoselskii-Mann (KM)* formula for finding fixed points of ne (*nonexpansive*) mappings. The following quotation further shows the importance of iterative methods for approximating fixed points of nonexpansive mappings.

- "Many well-known algorithms in signal processing and image reconstruction are iterative in nature .... A wide variety of iterative procedures used in signal processing and image reconstruction and elsewhere are special cases of the KM iteration procedure, for particular choices of the ne operator...." (Charles Byrne , [63]).

Definition 6.5. Let K be a subset of a normed linear space E. Let $T : K \to E$ be a map such that $F(T) := \{x \in K : Tx = x\} \neq \emptyset$. $F(T)$ is called the *fixed point set* of T. The map T is called *quasi-nonexpansive* if $\|Tx - Tx^*\| \leq \|x - x^*\|$ holds for all $x \in K$ and $x^* \in F(T)$.

It is clear that every nonexpansive map with a nonempty fixed point set, $F(T)$, is quasi-nonexpansive. In section 6.6, we will give an example of a quasi-nonexpansive map which is not nonexpansive.

6.2 Asymptotic Regularity

Let $T : K \to K$ be a nonexpansive self-mapping on a convex subset K of a normed linear space X. Let $S_\lambda := \lambda I + (1 - \lambda)T, \lambda \in (0,1)$, where I denotes the identity map of K. Then for fixed $x_0 \in K, \{S_\lambda^n(x_0)\}$ is defined by $S_\lambda^n(x_0) = \lambda x_n + (1 - \lambda)Tx_n$, where $x_n = S_\lambda^{n-1}(x_0)$. In [291], Krasnoselskii proved that if X is uniformly convex and K is compact then, for any $x_0 \in K$, the sequence $\{S_{\frac{1}{2}}^n(x_0)\}_{n=1}^\infty$, of iterates of x_0 under $S_{\frac{1}{2}} = \frac{1}{2}(I + T)$ converges to a fixed point of T. Schaefer [431] observed that the same holds for any $S_\lambda = \lambda I + (1 - \lambda)T$ with $0 < \lambda < 1$, and Edelstein [217] proved that strict convexity of X suffices. The important and natural question of whether or not strict convexity can be removed remained open for many years. In 1967, this question was resolved in the affirmative in the following theorem.

Theorem 6.6. *(Ishikawa, [259]) Let K be a subset of a Banach space X and let T be a nonexpansive mapping from K into X. For $x_0 \in K$, define the sequence $\{x_n\}_{n=1}^\infty$ by (6.3), where the real sequence $\{C_n\}_{n=0}^\infty$ satisfies: (a) $\sum_{n=0}^\infty C_n$ diverges, (b) $0 \leq C_n \leq b < 1$ for all positive integers n; and (c) $x_n \in K$ for all positive integers n. If $\{x_n\}_{n=1}^\infty$ is bounded, then $x_n - Tx_n \to 0$ as $n \to \infty$.*

One consequence of Theorem 6.6 is that if K is convex and compact, the sequence $\{x_n\}$ defined by (6.3) converges strongly to a fixed point of T (see Theorem 6.17 below). Another consequence of Theorem 6.6 is that for K convex and T mapping K into a bounded subset of X, the iterates of the map $S_\lambda = (1 - \lambda)I + \lambda T$, $\lambda \in (0,1)$ are *asymptotically regular* at x, i.e., $\|S_\lambda^{n+1}x - S_\lambda^n x\| \to 0$ as $n \to \infty$. The concept of asymptotic regularity was introduced by Browder and Petryshyn [51] and, as a metric notion, a mapping $T : M \to M$ is said to be asymptotically regular on M if it is so at each x_0 in M. The relevance of asymptotic regularity to the existence of a fixed point for T can clearly be seen from the following theorem.

Theorem 6.7. *Suppose M is a metric space and $T : M \to M$ is continuous and asymptotically regular at x_0 in M. Then any cluster point of $\{T^n(x_0)\}_{n=1}^\infty$ is a fixed point of T.*

It follows that for *continuous* T, asymptotic regularity of S_λ at any x_0 in K implies $S_\lambda(p) = p$ for any cluster point p of $\{S_\lambda^n(x_0)\}_{n=1}^\infty$. Asymptotic regularity is not only useful in proving that fixed points exist but also in showing that in certain cases, the sequence of iterates at a point converges to a fixed point.

Proposition 6.8. *Let G be a linear mapping of a normed linear space E into itself and suppose G is power bounded (i.e., for some $k \geq 0$, $\|G^n\| \leq k$, ($n = 1, 2, ...$), and asymptotically regular. If, for some $x_0 \in E$, $\overline{co}\{G^n(x_0)\}$ contains a fixed point x^* of G, then $\{G^n(x_0)\}$ converges strongly to x^*.*

Proof. Let $\epsilon > 0$ be given and suppose that y is a point of $\overline{co}\{G^n(x_0)\}$ with $\|x^* - y\| < \frac{\epsilon}{2(k+1)}$. Setting $y = \sum_{j=1}^m \lambda_j G^j(x_0)$ we obtain, using the linearity of G,

$$G^n(x_0 - x^*) = G^n(x_0 - y) + G^n(y - x^*)$$

$$= G^n\left(x_0 - \sum_{j=1}^m \lambda_j G^j(x_0)\right) + G^n(y - x^*)$$

$$= \sum_{j=1}^m \lambda_j[G^n(x_0) - G^{n+j}(x_0)] + G^n(y - x^*),$$

since $\sum_{j=1}^m \lambda_j = 1$. Hence, $\left\|G^n(x_0 - x^*)\right\| \leq \left\|\sum_{j=1}^m \lambda_j[G^n(x_0) - G^{n+j}(x_0)]\right\| + \frac{k\epsilon}{2(k+1)}$ since $\|G^n(y - x^*)\| \leq \|G^n\|.\|y - x^*\| \leq \frac{k\epsilon}{2(k+1)}$. Now, by asymptotic regularity, there exists an integer $N_0 > 0$ such that for all $n \geq N_0$, $\|G^n(x_0) - G^{n+j}(x_0)\| \leq \frac{\epsilon}{2}, (j = 1, 2, ..., m)$. Hence $\|G^n(x_0 - x^*)\| < \sum_{j=1}^m \lambda_j\left(\frac{\epsilon}{2}\right) + \frac{\epsilon}{2} = \epsilon \ \forall \ n \geq N_0$. This implies $G^n(x_0 - x^*) = G^n(x_0) - x^* \to 0$ as $n \to \infty$, proving Proposition 6.8. □

Remark 6.9. In connection with Theorem 6.7, we note that if E is a normed linear space and K is a subset of E which is only assumed to be weakly compact, then, in general, the sequence $\{S_\lambda^n(x_0)\}$ will not have any strong cluster point as is shown in the following example.

Example 6.10. There is a closed bounded and convex set K in the Hilbert space l_2, a nonexpansive self-map T of K and a point $x_0 \in K$ such that $\{S_{\frac{1}{2}}^n(x_0)\}$ does not converge in the norm topology.

For details, see Genel and Lindenstrauss, [228].

Definition 6.11. A Banach space X is called an *Opial space* (see, e.g. Opial, [366]) if for all sequences $\{x_n\}_{n=0}^\infty$ in X such that $\{x_n\}_{n=0}^\infty$ converges weakly to some x in X, the inequality

$$\liminf_{n \to \infty} \|x_n - y\| > \liminf_{n \to \infty} \|x_n - x\|$$

holds for all $y \neq x$.

Every Hilbert space is an Opial space (see, e.g., Edelstein and O'Brian, [218], Opial, [367]). In fact, for any normed linear space X, the existence of a weakly sequentially continuous duality map implies X is an Opial space but the converse implication does not hold (see, e.g. Edelstein and O'Brian, [218]). In particular, $\ell_p(1 < p < \infty)$ spaces are Opial spaces but L_p $(1 < p < \infty, p \neq 2)$ spaces are not. Suppose now K is a weakly compact convex subset of a real Opial space X and T is a nonexpansive mapping of K into itself. While example 6.10 shows that we cannot, in general, get strong convergence of the sequence defined by (6.3) to a fixed point of T, Theorem 6.20 (below) allows us to conclude that the sequence converges *weakly* to a fixed point of T *if E is an Opial space.*

6.3 Uniform Asymptotic Regularity

Definition 6.12. Let K be a subset of a real normed linear space X. A mapping $U : K \to X$ is called *uniformly asymptotically regular* if for any $\epsilon > 0$, there exists an integer $N > 0$ such that for any $x_0 \in K$ and for all $n \geq N$, $\|U^{n+1}x_0 - U^n x_0\| < \epsilon$.

Definition 6.13. Given a set A and $x_0 \in A$, call a sequence $\{x_n\}_{n=0}^\infty$ *admissible* if there is a non-increasing sequence $\{C_n\}_{n=0}^\infty$ in $(0,1)$ such that (6.3) holds.

We now prove the following theorems.

Theorem 6.14. *Let K be a subset of a real normed linear space X and let f be a nonexpansive mapping from K into X. Suppose for $x_0 \in K$ there exists an admissible sequence $\{x_n\}_{n=0}^\infty \subseteq K$ which is bounded. Then $\lim_{n \to \infty} \|x_{n+1} - x_n\| = 0$. Moreover, if K is a bounded subset of X, then the above limit is uniform.*

Theorem 6.15. *With K, f and X as in Theorem 6.14 and $x_0 \in K$, suppose there exists an admissible sequence $\{x_n\}_{n=0}^{\infty} \subseteq K$ which is bounded and which is such that the non-increasing sequence $\{C_n\}_{n=0}^{\infty}$ also satisfies $0 < a \leq C_n < 1$ for all $n \geq 1$. Then $\lim_{n \to \infty} \|x_n - f(x_n)\| = 0$.*

The above theorems are easy consequences of the following more technical theorem.

Theorem 6.16. *Let K be a subset of a real normed linear space X and let f be a nonexpansive mapping from K into X. Suppose there exists a set $A \subseteq K$ such that for each $x_0 \in A$ there is an admissible sequence $\{x_n\}_{n=0}^{\infty} \subseteq A$, and suppose further that there exists some $\delta > 0$ such that for each positive integer N, and some admissible sequence $\{x_n\}_{n=0}^{\infty} \subseteq A$,*

$$\sup_{k \geq N} \|x_{k+1} - x_k\| > \delta. \tag{6.4}$$

Then, A is unbounded.

We prove Theorems 6.14 and 6.15 from Theorem 6.16.

Proof of Theorem 6.14. Both parts follow immediately from Theorem 6.16; the first by setting $\{x_n\}_{n=0}^{\infty} = A$ in the theorem and the second by setting $K = A$. $\qquad\square$

Proof of Theorem 6.15. Since f is nonexpansive we obtain,

$$\|x_{n+1} - f(x_{n+1})\| = \|(1 - C_n)(x_n - f(x_n)) + f(x_n) - f(x_{n+1})\|$$

$$\leq (1 - C_n)\|x_n - f(x_n)\|$$

$$+ \|x_n - ((1 - C_n)x_n + C_n f(x_n))\|$$

$$= \|x_n - f(x_n)\|.$$

Thus the sequence $\{\|x_n - f(x_n)\|\}_{n=0}^{\infty}$ is non-increasing and bounded below, so $\lim_{n \to \infty} \|x_n - f(x_n)\|$ exists. But from $x_{n+1} = (1 - C_n)x_n + C_n f(x_n)$,

$$\lim_{n \to \infty} \|x_n - f(x_n)\| = \lim_{n \to \infty} \frac{1}{C_n} \|x_{n+1} - x_n\| \leq \frac{1}{a} \lim_{n \to \infty} \|x_{n+1} - x_n\| = 0,$$

(by Theorem 6.16 since the admissible sequence $\{x_n\}_{n=0}^{\infty}$ is bounded), and this establishes Theorem 6.15. We next give the following proof.

Proof of Theorem 6.16. Assume by way of contradiction that A is bounded and let $\|x_n\| \leq \rho$ for each n. Let M be a fixed positive integer such that $(M - 1)\delta > 2\rho + 1$. Choose N, with $N > \max\{M, [(2\rho - \delta)M/(1 - C_1)^M C_1]\}$ (where here $[\cdot]$ denotes the greatest integer function) such that for some $\delta > 0$

and $x_0 \in A$, the corresponding admissible sequence $\{x_n\}_{n=0}^{\infty}$ in A satisfies $\|x_{N+1} - x_N\| > \delta$. Using the nonexpansiveness of f, we easily obtain the following:

$$\|x_{n+1} - x_n\| = C_n\|(1 - C_{n-1})(x_{n-1} - f(x_{n-1})) + f(x_{n-1}) - f(x_n)\|$$
$$\leq C_n(1 - C_{n-1})\|x_{n-1} - f(x_{n-1})\|$$
$$+\|x_{n-1} - [(1 - C_{n-1})x_{n-1} + C_{n-1}f(x_{n-1})]\|$$
$$= \frac{C_n}{C_{n-1}}\|x_n - x_{n-1}\| \leq \|x_n - x_{n-1}\| ,$$

the last equality following from (6.3) with n replaced by $(n-1)$ and T replaced by f while the last inequality follows since $\{C_n\}_{n=0}^{\infty}$ is a non-increasing sequence. Hence it follows that $\|x_{i+1} - x_i\| > \delta$ for all $i \leq N$, and furthermore we obtain the following:

$$\delta < \|x_{N+1} - x_N\| \leq \|x_N - x_{N-1}\| \leq \cdots \leq \|x_2 - x_1\| \leq 2\rho; \qquad (6.5)$$

$$\|f(x_{i+1}) - f(x_i)\| \leq \|x_{i+1} - x_i\| \text{ for all } i = 0, 1, \ldots, N; \qquad (6.6)$$

and $x_{i+1} = (1 - C_i)x_i + C_i f(x_i)$ so that

$$f(x_i) = \frac{x_{i+1}}{C_i} - \left(\frac{1 - C_i}{C_i}\right) x_i, \quad i = 1, 2, \ldots, N ; \qquad (6.7)$$

which implies,

$$\left\|\frac{1}{C_i}\{x_{i+1} - (1 - C_i)x_i\} - \frac{1}{C_{i-1}}\{x_i - (1 - C_{i-1})x_{i-1}\}\right\|$$

$$= \|f(x_i) - f(x_{i-1})\| \leq \|x_i - x_{i-1}\|,$$

and this reduces to

$$\left\|\frac{1}{C_i}[x_{i+1} - x_i] - \left(\frac{1 - C_{i-1}}{C_{i-1}}\right)[x_i - x_{i-1}]\right\| \leq \|x_i - x_{i-1}\| \qquad (6.8)$$

for all $i = 1, 2, \ldots, N$. Now set $I = [(2\rho - \delta)/(1 - C_1)^M C_1]$ and consider the collection of I intervals $[s_k, s_{k+1}]$ where

$$s_k = \begin{cases} \delta + k(1 - C_1)^M C_1, & k = 0, 1, \ldots, I - 1, \\ 2\rho, & k = I. \end{cases}$$

We claim that some one of these intervals must contain at least M of the numbers $\{\|x_i - x_{i+1}\|\}_{i=0}^{N-1} \subseteq [\delta, 2\rho]$. If this is not the case, then $N < MI = M\left[\frac{2\rho - \delta}{(1 - C_1)^M C_1}\right]$ contradicting our choice of N. Thus for some r, and some $s = s_k \in [\delta, 2\rho]$,

$$\|x_{r+i+1} - x_{r+i}\| \in [s, s + (1 - C_1)^M C_1] \tag{6.9}$$

for $i = 0, 1, \ldots, (M - 1)$. Define $\Delta x_i = x_i - x_{i-1}$, $i = 1, 2, \ldots, N$. Replacing i in (6.8) by $r + M - j - 1$, $(j = 0, 1, \ldots, M - 1)$ we see that (6.8) and (6.9) imply

$$\left\| \frac{1}{C_{r+M-j-1}} \Delta x_{r+M-j} - \left(\frac{1 - C_{r+M-j-2}}{C_{r+M-j-2}} \right) \Delta x_{r+M-j-1} \right\|$$
$$\leq s + (1 - C_1)^M C_1. \tag{6.10}$$

Choose $f^* \in X^*$ (the dual space of X) with $\|f^*\| = 1$ and $f^*(\Delta x_{r+M}) = \|\Delta x_{r+M}\|$. Then using (6.10) we obtain,

$$\left| \frac{1}{C_{r+M-j-1}} f^*(\Delta x_{r+M-j}) - \left(\frac{1 - C_{r+M-j-2}}{C_{r+M-j-2}} \right) f^*(\Delta x_{r+M-j-1}) \right|$$
$$\leq \|f^*\| \cdot \left\| \frac{1}{C_{r+M-j-1}} \Delta x_{r+M-j} - \left(\frac{1 - C_{r+M-j-2}}{C_{r+M-j-2}} \right) \Delta x_{r+M-j-1} \right\|$$
$$\leq s + (1 - C_1)^M C_1,$$

which yields

$$f^*(\Delta x_{r+M-j-1}) \geq \left(\frac{C_{r+M-j-2}}{C_{r+M-j-1}} \right) \left(\frac{1}{1 - C_{r+M-j-2}} \right) f^*(\Delta x_{r+M-j})$$
$$- \left(\frac{C_{r+M-j-2}}{1 - C_{r+M-j-2}} \right) (s + (1 - C_1)^M C_1). \tag{6.11}$$

Observe that since $\{C_i\}_{i=0}^\infty$ is non-increasing, for all $i \geq 1$, $(1 - C_i)^{-1} \leq (1 - C_1)^{-1}$ and $C_i(1 - C_i)^{-1} \leq C_1(1 - C_1)^{-1}$. Now for $j = 0$, using $f^*(\Delta x_{r+M}) = \|\Delta x_{r+M}\| \in [s, s + (1 - C_1)^M C_1]$ we obtain from (6.11),

$$f^*(\Delta x_{r+M-1}) \geq \left(\frac{1}{1 - C_{r+M-2}} \right) s - \left(\frac{C_{r+M-2}}{1 - C_{r+M-2}} \right) \{s + (1 - C_1)^M C_1\}$$
$$\geq s - C_1^2 (1 - C_1)^{M-1}. \tag{6.12}$$

We will show that (6.12) implies

$$f^*(\Delta x_{r+M-j-1}) \geq s - (1 - C_1)^{M-1} C_1^2 \sum_{t=0}^{j} \left(\frac{1}{1 - C_1} \right)^t, \tag{6.13}$$

for $j = 1, 2, \ldots, M - 1$. We establish this by induction. For $j = 0$, (6.13) reduces to (6.12). Suppose now (6.13) holds for $j \leq k$, for some $k \in \{1, 2, 3, \ldots, M - 2\}$. Then from (6.11) and the inductive hypothesis we obtain,

$$f^*(\Delta x_{r+M-(k+1)-1}) = f^*(\Delta x_{r+M-k-2})$$

$$\geq \left(\frac{C_{r+m-k-3}}{C_{r+M-k-2}}\right)\left(\frac{1}{1-C_{r+M-k-3}}\right) f^*(\Delta x_{r+M-k-1})$$

$$- \left(\frac{C_{r+M-k-3}}{1-C_{r+M-k-3}}\right)\{s + (1-C_1)^M C_1\}$$

$$\geq \left(\frac{1}{1-C_{r+M-k-3}}\right)\left[s - (1-C_1)^{M-1}C_1^2 \sum_{t=0}^{k}\left(\frac{1}{1-C_1}\right)^t\right]$$

$$- \left(\frac{C_{r+M-k-3}}{1-C_{r+M-k-3}}\right)\{s + (1-C_1)^M C_1\}$$

$$\geq s - \left(\frac{1}{1-C_1}\right)(1-C_1)^{M-1}C_1^2 \sum_{t=0}^{k}\left(\frac{1}{1-C_1}\right)^t$$

$$- \left(\frac{C_1}{1-C_1}\right)(1-C_1)^M C_1$$

$$= s - (1-C_1)^{M-1}C_1^2 \sum_{t=0}^{k+1}\left(\frac{1}{1-C_1}\right)^t ,$$

which completes the induction. Recalling that f^* is linear and summing (6.13), by telescoping, from $j = 0$ to $(M-2)$ yields:

$$f^*(x_{r+M-1} - x_r) = f^*(x_{r+M-1}) - f^*(x_r)$$

$$\geq (M-1)s - (1-C_1)^{M-1}C_1^2\left[1 + \left(1 + \frac{1}{1-C_1}\right)\right.$$

$$\left. +\cdots+ \left(1 + \frac{1}{1-C_1} +\cdots+ \left(\frac{1}{1-C_1}\right)^{M-2}\right)\right].$$

Set $\lambda = 1 - C_1$ so that,

$$f^*(x_{r+M-1} - x_r) = (M-1)s - \lambda^{M-1}(1-\lambda)^2\left[1 + \left(\frac{\lambda+1}{\lambda}\right)\right.$$

$$\left. +\cdots+ \left(\frac{\lambda^{M-2} +\cdots+ \lambda + 1}{\lambda^{M-2}}\right)\right]$$

$$= (M-1)s - \lambda(1-\lambda)\left[\lambda^{M-1}\left\{\left(\frac{1-\lambda}{\lambda}\right) + \left(\frac{1-\lambda^2}{\lambda^2}\right)\right.\right.$$

$$\left.\left. +\cdots+ \left(\frac{1-\lambda^{M-1}}{\lambda^{M-1}}\right)\right\}\right] \geq (M-1)s - 1,$$

the last inequality following since

$$\lambda(1-\lambda)\left[\lambda^{M-1}\left\{\left(\frac{1-\lambda}{\lambda}\right) + \left(\frac{1-\lambda^2}{\lambda^2}\right) +\cdots+ \left(\frac{1-\lambda^{M-1}}{\lambda^{M-1}}\right)\right\}\right]$$

$$< \lambda(1-\lambda)(\lambda^{M-2} +\cdots+ \lambda + 1) \leq 1 .$$

But $s \geq \delta$ implies $(M-1)s \geq (M-1)\delta > 2\rho+1$, so that $f^*(x_{r+M-1}-x_r) > \rho$. Also,

$$f^*(x_{r+M-1} - x_r) \leq |f^*(x_{r+M-1} - x_r)| \leq \|f^*\| \cdot \|x_{r+M-1} - x_r\|$$

$$= \|x_{r+M-1} - x_r\| \; .$$

Hence, $\|x_{r+M-1} - x_r\| > 2\rho$, contradicting the assumption that $\|x_n\| \leq \rho$ for each n, and completing the proof of Theorem 6.16. $\square$

6.4 Strong Convergence

Using the technique of Theorem 6.16 we are able to prove the following theorem.

Theorem 6.17. *With K, X and f as in Theorem 6.16, suppose for some $x_0 \in K$, the corresponding admissible sequence $\{x_n\}_{n=0}^{\infty} \subseteq K$ has a cluster point $q \in K$. Then $f(q) = q$ and $x_n \to q$. In particular, if the range of f is contained in a compact subset of K then $\{x_n\}_{n=0}^{\infty}$ converges strongly to a fixed point of f.*

Proof. In Edelstein [216], it was shown that q is also a cluster point of $\{f^n(q)\}$ and that $\|f^{n+1}(q) - f^n(q)\| = \|f(q) - q\|$ for all n. Thus if $f^i(q) := x_i$ and $\Delta x_i = x_i - x_{i-1}$, $i = 1, 2, \ldots$ then $\|\Delta x_{i+1}\| = \|\Delta x_i\|$ for all i. As in the proof of Theorem 6.16, we have, $\|\Delta x_{i+1}\| = \|x_{i+1} - x_i\| \leq \frac{C_i}{C_{i-1}}\|x_i - x_{i-1}\| \leq \|x_i - x_{i-1}\| = \|\Delta x_i\|$ and this implies $C_i = C_{i-1}$ for all i since $\|\Delta x_{i-1}\| = \|\Delta x_i\|$. Hence, from $x_{i+1} - x_i = (1 - C_i)x_i + C_i f(x_i) - (1 - C_{i-1})x_{i-1} - C_{i-1}f(x_{i-1})$ we obtain,

$$\|\Delta x_{i+1}\| \leq (1 - C_i)\|\Delta x_i\| + C_i\|\Delta f(x_i)\|$$

$$\leq (1 - C_i)\|\Delta x_i\| + C_i\|\Delta x_i\| = \|\Delta x_i\| \; ,$$

and this implies, $\|\Delta x_i\| = \|\Delta f(x_i)\|$. Assume for contradiction that,

$$\|\Delta x_i\| = \|\Delta f(x_i)\| = \beta > 0, \; i = 1, 2, \ldots. \tag{6.14}$$

Choose $N, K \in \mathbb{N}^+$ sufficiently large. From (6.14) we have, for $i = N + K$,

$$\|\Delta x_{N+K}\| = \|\Delta f(x_{N+K})\| = \beta > 0 \; . \tag{6.15}$$

Let $f^* \in X^*$ such that $\|f^*\| = 1$ and $f^*(\Delta x_{N+K}) = \|\Delta x_{N+K}\|$. Then for $j = 0, 1, 2, \ldots$,

$$f^*(\Delta f(x_{N+K-j})) \leq \|f^*\| \cdot \|\Delta f(x_{N+K-j})\| = \|\Delta f(x_{N+K-j})\| = s \cdot \tag{6.16}$$

From $x_{N+K-j+1} = (1 - C_{N+K-j})x_{N+K-j} + C_{N+K-j}f(x_{N+K-j})$ we obtain, using $C_i = C_{i-1}$ for all i,

$$\Delta x_{N+K-j+1} = (1 - C_{N+K-j})\Delta x_{N+K-j} + C_{N+K-j}\Delta f(x_{N+K-j}). \tag{6.17}$$

We will show that applying f^* to this equation yields:

$$f^*(\Delta x_{N+K-j}) \geq \beta \text{ for } j = 0, 1, \ldots. \tag{6.18}$$

We establish (6.18) by induction. Observe that $f^*(\Delta x_{N+K}) = \|\Delta x_{N+K}\| = \beta$ satisfies (6.18) with $j = 0$. Now if $j = 1$, applying f^* to (6.17) and using (6.16) yield:

$$f^*(\Delta x_{N+K-1}) = \left(\frac{1}{1 - C_{N+K-1}}\right) f^*(\Delta x_{N+K}) - \left(\frac{C_{N+K-1}}{1 - C_{N+K-1}}\right) \times$$
$$f^*(\Delta f(x_{N+K-1}))$$

$$\geq \left(\frac{1}{1 - C_{N+K-1}}\right) \beta - \left(\frac{C_{N+K-1}}{1 - C_{N+K-1}}\right) \beta = \beta \,,$$

and (6.18) holds for $j = 1$. Assume it holds for $j = 0, 1, \ldots, t$. Then using (6.16), (6.17), and the inductive hypothesis we have,

$$f^*(\Delta x_{N+K-t-1}) = \left(\frac{1}{1 - C_{N+K-t-1}}\right) f^*(\Delta x_{N+K-t})$$
$$- \left(\frac{C_{N+K-t-1}}{1 - C_{N+K-t-1}}\right) f^*(\Delta f(x_{N+K-t-1}))$$
$$\geq \left(\frac{1}{1 - C_{N+K-t-1}}\right) \beta - \left(\frac{C_{N+K-t-1}}{1 - C_{N+K-t-1}}\right) \beta = \beta \,,$$

which completes the induction. Using the technique of the proof of Theorem 6.16 and summing (6.18) from $j = 0$ to $K - 1$ yields:

$$\|x_{N+K} - x_N\| \geq f^*(x_{N+K} - x_N) \geq K\beta \,, \tag{6.19}$$

and this implies that the sequence $\{x_i\}_{i=0}^{\infty}$ cannot have a convergent subsequence, a contradiction of the fact that $\{x_n\}_{n=0}^{\infty}$ has a cluster point. Hence $\beta = 0$ and $f(q) = q$. That $x_n \to q$ now follows readily from the nonexpansiveness of f. $\qquad \square$

For our next result the following definition is needed.

Definition 6.18. (Petryshyn, [381]). Let C be a subset of a real normed linear space X. A mapping $f : C \to X$ is said to be *demicompact* at $h \in X$ if, for any bounded sequence $\{x_n\}_{n=0}^{\infty}$ in C such that $x_n - f(x_n) \to h$ as $n \to \infty$, there exist a subsequence $\{x_{n_j}\}_{j=0}^{\infty}$ and an $x \in C$ such that $x_{n_j} \to x$ as $j \to \infty$ and $x - f(x) = h$.

Corollary 6.19 *Suppose X is a real normed linear space, C is a closed bounded convex subset of X and f is a nonexpansive mapping of C into C. Suppose further that either, (i) f is demicompact at 0, or (ii) $(I - f)$ maps closed bounded subsets of X into closed subsets of X. For $x_0 \in C$ let $\{x_n\}_{n=0}^{\infty} \subseteq C$ be an admissible sequence where the real sequence $\{C_n\}_{n=0}^{\infty}$*

also satisfies $0 < a \leq C_n \leq b < 1$ for all $n \geq 1$. Then $\{x_n\}_{n=0}^{\infty}$ converges strongly to a fixed point of f in C.

Proof. (i) From $x_{n+1} = (1 - C_n)x_n + C_n f(x_n)$ we obtain $x_n - f(x_n) = \frac{1}{C_n}\{x_n - x_{n-1}\}$. Since C is bounded, $\{x_n\}_{n=0}^{\infty}$ is a bounded sequence and also $\{C_n\}_{n=1}^{\infty}$ bounded away from 0 implies (by Theorem 6.15) that $\{x_n - f(x_n)\}$ is convergent to 0 so that by the demicompactness of f at 0, $\{x_n\}_{n=0}^{\infty}$ has a cluster point in C. The result follows by Theorem 6.17.

(ii) If q is a fixed point of f, $\{\|x_n - q\|\}_{n=0}^{\infty}$ does not increase with n. It suffices, therefore, to show that there exists a subsequence of $\{x_n\}_{n=0}^{\infty}$ which converges strongly to a fixed point of f. For $x_0 \in C$, let K be the strong closure of the set $\{x_n\}_{n=0}^{\infty}$. By Theorem 6.15, $\{(I - f)(x_n)\}$ converges strongly to 0 as $n \to \infty$. Hence, 0 lies in the strong closure of $(I - f)(K)$ and since the latter is closed by hypothesis (since K is closed and bounded), 0 lies in $(I - f)(K)$. Hence, there is a subsequence $\{x_{n_j}\}_{j=0}^{\infty}$ such that $x_{n_j} \to \mu \in C$, where μ is a point such that $(I - f)\mu = 0$. Hence $x_n \to \mu$. $\square$

6.5 Weak Convergence

Theorem 6.20. *Let X be an Opial space and $f : K \to K$ be a nonexpansive self-mapping of a weakly compact convex subset K of X. For any x_0 in K, let $\{x_n\}_{n=0}^{\infty} \subseteq K$ be the corresponding admissible sequence which is such that the non-increasing sequence $\{C_n\}_{n=1}^{\infty}$ also satisfies $0 < a \leq C_n < 1$ for all $n \geq 1$. Then $\{x_n\}_{n=0}^{\infty}$ converges weakly to a fixed point of f.*

We shall need the following definition.

Definition 6.21. A mapping $T : K \to X$ is called *demiclosed* at y if, for any sequence $\{x_n\}_{n=0}^{\infty} \subseteq K$ which converges weakly to an x in K, the strong convergence of the sequence $\{T(x_n)\}_{n=0}^{\infty}$ to y in K implies $Tx = y$.

The technique of Edelstein and O'Brian [218] together with Theorem 6.15 yields the following proof.

Proof of Theorem 6.20. Since X is an Opial space and f is nonexpansive, $(I - f)$ is demiclosed (see, e.g. Opial, [366]). Furthermore, by Theorem 6.15, f is asymptotically regular. Hence, by a result of Browder and Petryshyn [51], any weak cluster point of $\{x_n\}_{n=0}^{\infty} \subseteq K$ is a fixed point of f. We claim that $\{x_n\}_{n=0}^{\infty} \subseteq K$ has a unique weak cluster point. Suppose there exist two distinct weak cluster points of $\{x_n\}_{n=0}^{\infty}$, say q_1 and q_2, and two subsequences $\{x_{n_i}\}_{i=1}^{\infty}$ and $\{x_{n_j}\}_{j=1}^{\infty}$ such that $\{x_{n_i}\}_{i=1}^{\infty}$ converges weakly to q_1 and $\{x_{n_j}\}_{j=1}^{\infty}$ converges weakly to q_2. Let $p \in F(f)$ where $F(f)$ denotes the fixed point set of f. Then, it is easy to see that $\|x_{n+1} - p\| \leq \|x_n - p\|$ for each $n \geq 0$ so that $\lim_{n \to \infty} \|x_n - p\|$ exists for every $p \in F(f)$. Thus, since X is an Opial space, it follows that

$$\lim_{n \to \infty} \|x_n - q_1\| = \lim_{i \to \infty} \|x_{n_i} - q_1\| < \lim_{i \to \infty} \|x_{n_i} - q_2\|$$

$$= \lim_{n \to \infty} \|x_n - q_2\|$$

$$\lim_{n \to \infty} \|x_n - q_2\| = \lim_{j \to \infty} \|x_{n_j} - q_2\| < \lim_{j \to \infty} \|x_{n_j} - q_1\|$$

$$= \lim_{n \to \infty} \|x_n - q_1\|,$$

this contradiction implies there exists exactly one weak cluster point q of $\{x_n\}_{n=0}^{\infty} \subseteq K$. By weak compactness of K, $\{x_n\}_{n=0}^{\infty}$ converges weakly to q. $\square$

Theorem 6.22. *Let K be a closed convex subset of a reflexive Banach space X, and T a continuous mapping of K into X such that (i) $F(T) \neq \emptyset$, where $F(T)$ denotes the fixed point set of T in K, (ii) If $Tp = p$, then $\|Tx - p\| \leq \|x - p\|$ for all x in K, (iii) There exist an x_0 in K and a corresponding admissible sequence $\{x_n\}_{n=0}^{\infty} \subseteq K$, (iv) T is asymptotically regular at x_0, (v) If $\{x_{n_j}\}_{j=1}^{\infty}$ is a subsequence of $\{x_n\}_{n=0}^{\infty}$ such that $\{x_{n_j}\}_{j=1}^{\infty}$ converges weakly to $\tilde{x}$ in K and $\{x_{n_j} - Tx_{n_j}\}$ converges strongly to zero then $\tilde{x} - T\tilde{x} = 0$, (vi) X is an Opial space. Then the sequence $\{x_n\}_{n=0}^{\infty}$ converges weakly to a fixed point of T in K.*

Proof. Theorem 4.2 of Petryshyn and Williamson [382] implies the sequence $\{x_n\}_{n=0}^{\infty}$ contains a weakly convergent subsequence with its limit in $F(T)$ and, furthermore, that every weakly convergent subsequence of $\{x_n\}_{n=0}^{\infty}$ has a point q in $F(T)$ for its limit. As in the proof of Theorem 6.20, the sequence $\{x_n\}_{n=0}^{\infty}$ has a unique weak cluster point q in K. By weak compactness of K, $\{x_n\}_{n=0}^{\infty}$ converges weakly to q. $\square$

6.6 Some Examples

Let K be a nonempty subset of a real normed linear space X. Recall that a mapping $T : K \to X$ is called *quasi-nonexpansive* provided T has a fixed point in K and that if $Tp = p$, p in K, then $\|Tx - p\| \leq \|x - p\|$ for all x in K. In this section, we shall exhibit large classes of quasi-nonexpansive mappings and these classes, in particular, properly contain the class of nonexpansive mappings with fixed points.

The concept of quasi-nonexpansive mappings was essentially introduced by Diaz and Metcalf [205]. A nonexpansive map $T : K \to K$ with at least one fixed point in K is quasi-nonexpansive. Also, a linear quasi-nonexpansive mapping on a subspace is nonexpansive on that subspace; but there exist continuous and discontinuous nonlinear quasi-nonexpansive mappings which are not nonexpansive (see, e.g., Example 6.23 below). We proved in Section 6.3 that if K is a nonempty closed convex bounded subset of X and T is a *nonexpansive* mapping of K into a bounded subset of X, the iterates of the map $S_\lambda = (1-\lambda)I + \lambda T$, $\lambda \in (0,1)$, are *uniformly asymptotically regular* on K. In this section, we shall show by means of an example that this result does not extend to the class of quasi-nonexpansive maps.

We start with the following example which shows that the class of quasi-nonexpansive mappings properly includes that of nonexpansive maps with fixed points.

Example 6.23. Let $X = \ell_\infty$ and $K := \{x \in \ell_\infty : \|x\|_\infty \leq 1\}$. Define $f : K \to K$ by $f(x) = (0, x_1^2, x_2^2, x_3^2, \dots)$ for $x = (x_1, x_2, x_3, \dots)$ in K. Then it is clear that f is continuous and maps K into K. Moreover $f(p) = p$ if and only if $p = 0$. Furthermore,

$$\|f(x) - p\|_\infty = \|f(x)\|_\infty = \|(0, x_1^2, x_2^2, x_3^2, \dots)\|_\infty$$
$$\leq \|(0, x_1, x_2, x_3, \dots)\|_\infty = \|x\|_\infty = \|x - p\|_\infty$$

for all x in K. Therefore, f is quasi-nonexpansive. However, f is not nonexpansive.

For, if $x = (\frac{3}{4}, \frac{3}{4}, \dots)$ and $y = (\frac{1}{2}, \frac{1}{2}, \dots)$, it is clear that x and y belong to K. Furthermore, $\|x - y\|_\infty = \|(\frac{1}{4}, \frac{1}{4}, \dots)\| = \frac{1}{4}$, and $\|f(x) - f(y)\|_\infty = \|(0, \frac{5}{16}, \frac{5}{16}, \dots)\|_\infty = \frac{5}{16} > \frac{1}{4} = \|x - y\|_\infty$. □

Before we exhibit a large class of quasi-nonexpansive mappings we need the following preliminaries.

Suppose X is a Banach space and K is a bounded closed and convex subset of X. Within the past thirty years or so numerous papers have appeared concerning variants of the following contractive condition for mappings $T : K \to K$ introduced by Kannan [271]:

$$\|Tx - Ty\| \leq \frac{1}{2}(\|x - Tx\| + \|y - Ty\|), x, y \in K \qquad (*)$$

(see e.g., Bianchini [31], Ciric [190], Hardy and Rogers [248], Ray [398], Reich [399]-[402], Rhoades [415]-[420], Shimi [448], Soardi [457]). These mappings are neither stronger nor weaker than the nonexpansive mappings. Nevertheless, it appears that most of the fixed point Theorems for nonexpansive mappings also hold for mappings which are continuous and satisfy $(*)$. A more general class of mappings was introduced in Hardy and Rogers [248] and the following result was proved.

Theorem 6.24. *(Hardy and Rogers, [248]) Let (M, d) be a complete metric space and $T : M \to M$ a continuous mapping satisfying for $x, y \in M$:*

$$d(Tx, Ty) \leq a_1 d(x, y) + a_2 d(x, Tx) + a_3 d(y, Ty)$$
$$+ a_4 d(x, Ty) + a_5 d(y, Tx), \qquad (6.20)$$

where $a_i \geq 0$, $\sum_{i=1}^{5} a_i < 1$. Then T has a unique fixed point in M.

Condition (6.20), of course, implies T is a strict contraction if $a_i = 0$, $i = 2,\ldots,5$; it reduces to the condition studied by Reich [399] if $a_4 = a_5 = 0$; and to a condition of Kannan [271] if $a_1 = a_4 = a_5 = 0$, $a_2 = a_3$.

In the case that M is replaced by a uniformly convex Banach space, inequality (6.20) has been weakened to allow $\sum_{i=1}^{5} a_i = 1$.

Theorem 6.25. *(Goebel, Kirk and Shimi, [233]) Let X be a uniformly convex Banach space, K a nonempty bounded closed and convex subset of X, $T : K \to K$ a continuous mapping satisfying for all x, y in K:*

$$\|Tx - Ty\| \le a_1\|x - y\| + a_2\|x - Tx\| + a_3\|y - Ty\|$$
$$+a_4\|x - Ty\| + a_5\|y - Tx\| \tag{6.21}$$

where $a_i \ge 0$, $\sum_{i=1}^{5} a_i \le 1$. Then T has a fixed point in K.

Remark 6.26. Since the four points $\{x, y, Tx, Ty\}$ determine six distances in M, inequality (6.21) amounts to saying that the image distance $d(Tx, Ty)$ never exceeds a fixed convex combination of the remaining five distances. Geometrically, this type of condition is quite natural.

We now have the following proposition.

Proposition 6.27. *Let $T : K \to K$ be a map satisfying inequality (6.21). Then T is quasi-nonexpansive.*

Iterative methods for approximating fixed points of quasi-nonexpansive mappings have been studied by various authors (e.g., Chidume [93, 96, 101, 105], Dotson [211, 212, 213], Hardy and Rogers [248], Johnson [263], Outlaw [377], Outlaw and Groetsch [378], Reich [399]-[402], Rhoades [419], Senter and Dotson [441], Shimi [448]) and a host of other authors.

We now turn to the main example of this section which shows that a quasi-nonexpansive mapping in an arbitrary real Banach space need not be uniformly asymptotically regular.

Example 6.28. Let $X = \ell_\infty$ and $B(0,1) = \{x \in \ell_\infty : \|x\|_\infty \le 1\}$. The example is the construction of an $f : \ell_\infty \to \ell_\infty$ such that, (i) f is continuous; (ii) $f : \ell_\infty \to B(0,1)$; (iii) $f(p) = p$ if and only if $p = 0$; (iv) $\|f(x) - p\|_\infty \le \|x - p\|_\infty$ for all $x \in \ell_\infty$ and the fixed point p; (v) for all $n \in \mathbb{N}^+$, there exists $x \in B(0,1)$ such that

$$\|S_\lambda^{n+1}x - S_\lambda^n x\| > \lambda^2(1 - \lambda)^2 \ ,$$

for arbitrary $\lambda \in (0,1)$ where $S_\lambda^n(x) = S_\lambda(S_\lambda^{n-1}x)$ and $S_\lambda x = \lambda x + (1-\lambda)f(x)$. Define $f : \ell_\infty \to B(0,1) \subset \ell_\infty$ by

$$f(x) = \begin{cases} (0, x_1^2, x_2^2, x_3^2, \ldots, \ldots); & \text{if } \|x\|_\infty \le 1 \\ \|x\|_\infty^{-2}(0, x_1^2, x_2^2, x_3^2, \ldots); & \text{if } \|x\|_\infty > 1 \ , \end{cases}$$

where $x = (x_1, x_2, x_3, \dots) \in \ell_\infty$. Then it is clear that f satisfies $(i) - (iii)$ above. For (iv) we have,

$$\|f(x) - p\|_\infty = \|f(x)\|_\infty \leq \begin{cases} \|x\|_\infty^2 \leq \|x\|_\infty; & \text{if } \|x\|_\infty \leq 1 \\ 1; & \text{if } \|x\|_\infty > 1 \end{cases},$$

i.e., $\|f(x) - p\|_\infty \leq \|x - p\|_\infty$ and (iv) is satisfied. For (v), we examine f and S_λ more closely. Now,

$$S_\lambda(x) = \{\lambda x_1, \lambda x_2 + (1 - \lambda)x_1^2, \lambda x_3 + (1 - \lambda)x_2^2, \dots\}.$$

Thus, if $x = (x_1, x_2, x_3, \dots)$ and $x_j = a$, where a is a constant for $j \geq k$, then

$$(S_\lambda(x))_j = \lambda a + (1 - \lambda)a^2 \text{ for } j \geq k + 1. \tag{6.22}$$

More generally, by induction, if $x_j = a$ for $j \geq k$ then $(S_\lambda^n(x))_j = a_n$, a constant for $j \geq k + n$, where the $\{a_n\}$ satisfies the recurrence relation,

$$a_n = \lambda a_{n-1} + (1 - \lambda)a_{n-1}^2, \; n \geq 1, \; a_0 = a. \tag{6.23}$$

Suppose we have chosen $x = (a, a, a, \dots)$. Then

$$\begin{aligned}
\|S_\lambda^{n+1}x - S_\lambda^n x\|_\infty &= \|\lambda(S_\lambda^n(x)) + (1 - \lambda)f(S_\lambda^n(x)) - S_\lambda^n(x)\|_\infty \\
&= (1 - \lambda)\|S_\lambda^n(x) - f(S_\lambda^n(x))\|_\infty \\
&= (1 - \lambda)\sup_{j \geq 1}|(S_\lambda^n(x))_{j-1} - (S_\lambda^n(x))_{j-1}^2| \\
&\geq (1 - \lambda)\sup_{j > n}|(S_\lambda^n(x))_{j-1} - (S_\lambda^n(x))_{j-1}^2| \\
&\geq (1 - \lambda)\lambda|a_n - a_{(n-1)}^2|.
\end{aligned}$$

If we could choose $0 < a < 1$ such that $a_n = \lambda$ then $x \in B(0, 1)$ and condition (v) would be satisfied, completing the example. To do this, we note first that from (6.23), if $a < 1$, then $a_k < a$, inductively, and so

$$a_k = \lambda a_{k-1} + (1 - \lambda)a_{k-1}^2 < \lambda a_{k-1} + (1 - \lambda)a_{k-1} = a_{k-1}.$$

Thus, $1 > a > a_1 > a_2 > \dots > a_k > \dots$. Also, if we know a_k then we can find a_{k-1} from (6.23) (i.e., solving) as

$$a_{k-1} = 2^{-1}(1 - \lambda)^{-1}\{-\lambda + [\lambda^2 + 4(1 - \lambda)a_k]^{\frac{1}{2}}\}. \tag{6.24}$$

Note that if $a_k < 1$, then $a_{k-1} < 1$ for, from (6.24), using $a_k < 1$ we have,

$$\begin{aligned}
a_{k-1} &< 2^{-1}(1 - \lambda)^{-1}\{-\lambda + [\lambda^2 + 4(1 - \lambda)]^{\frac{1}{2}}\} \\
&= 2^{-1}(1 - \lambda)^{-1}\{-\lambda + (2 - \lambda)\} = 1.
\end{aligned}$$

We can now show that (v) is satisfied. Choose $n \in N^+$, and put $a_n = \lambda$. Use (6.24) to compute $a_{n-1}, a_{n-2}, \ldots, a_1, a_0$; $a_0 < 1$. Then starting with $x = (a_0, a_0, a_0, \ldots)$ we have $x \in B(0,1)$, $a_n = \lambda$ and so, $\|S_\lambda^{n+1} x - S_\lambda^n x\| > \lambda^2 (1-\lambda)^2$, as required. $\square$

6.7 Halpern-type Iteration Method

Let E be a real Banach space, K a closed convex subset of E and $T : K \to K$ a nonexpansive mapping. For fixed $t \in (0,1)$ and arbitrary $u \in K$, let $z_t \in K$ denote the unique fixed point of T_t defined by $T_t x := tu + (1-t)Tx, x \in K$. Assume $F(T) := \{x \in K : Tx = x\} \neq \emptyset$. Browder [44] proved that if $E = H$, a Hilbert space, then $\lim_{t \to 0} z_t$ exists and is a fixed point of T. Reich [412] extended this result to uniformly smooth Banach spaces. Kirk [284] obtained the same result in arbitrary Banach spaces under the additional assumption that T has pre-compact range.

For a sequence $\{\alpha_n\}$ in $[0,1]$ and an arbitrary $u \in K$, let the sequence $\{x_n\}$ in K be iteratively defined by $x_0 \in K$,

$$x_{n+1} := \alpha_n u + (1 - \alpha_n)Tx_n, n \geq 0. \qquad (6.25)$$

Concerning this process, Reich [412] posed the following question.

Question. *Let E be a Banach space. Is there a sequence $\{\alpha_n\}$ such that whenever a weakly compact convex subset K of E has the fixed point property for nonexpansive mappings, then the sequence $\{x_n\}$ defined by (6.25) converges to a fixed point of T for arbitrary fixed $u \in K$ and all nonexpansive $T : K \to K$?*

Halpern [245] was the first to study the convergence of the algorithm (6.25) in the framework of Hilbert spaces. He proved the following Theorem.

Theorem H (Halpern, [245]) *Let K be a bounded closed convex subset of a Hilbert space H and $T : K \to K$ be a nonexpansive mapping. Let $u \in K$ be arbitrary. Define a real sequence $\{\alpha_n\}$ in $[0,1]$ by $\alpha_n = n^{-\theta}$, $\theta \in (0,1)$. Define a sequence $\{x_n\}$ in K by $x_1 \in K$, $x_{n+1} = \alpha_n u + (1-\alpha_n)Tx_n$, $n \geq 1$. Then, $\{x_n\}$ converges strongly to the element of $F(T) := \{x \in K : Tx = x\}$ nearest to u.*

An iteration method with recursion formula of the form (6.25) is now referred to as a *Halpern-type iteration method*.

Lions [313] improved Theorem H, still in Hilbert spaces, by proving strong convergence of $\{x_n\}$ to a fixed point of T if the real sequence $\{\alpha_n\}$

satisfies the following conditions: (i) $\lim\limits_{n\to\infty} \alpha_n = 0$; (ii) $\sum\limits_{n=1}^{\infty} \alpha_n = \infty$; and (iii) $\lim\limits_{n\to\infty} \frac{|\alpha_n - \alpha_{n-1}|}{\alpha_n^2} = 0$.

Reich [412] gave an affirmative answer to the above question in the case when E is uniformly smooth and $\alpha_n = n^{-a}$ with $0 < a < 1$. It was observed that both Halpern's and Lions' conditions on the real sequence $\{\alpha_n\}$ excluded the natural choice, $\alpha_n := (n+1)^{-1}$. This was overcome by Wittmann [505] who proved, still in Hilbert spaces, the strong convergence of $\{x_n\}$ if $\{\alpha_n\}$ satisfies the following conditions:

$$(i)\ \lim_{n\to\infty} \alpha_n = 0;\ (ii)\ \sum_{n=1}^{\infty} \alpha_n = \infty;\ \text{and}\ ;\ (iii)\ \sum_{n=1}^{\infty} |\alpha_{n+1} - \alpha_n| < \infty. \quad (6.26)$$

Reich [413] extended this result of Wittmann to the class of Banach spaces which are uniformly smooth and have weakly sequentially continuous duality maps (e.g., $l_p(1 < p < \infty)$), where the sequence $\{\alpha_n\}$ is required to satisfy conditions (i) and (ii) of (6.26) and to be decreasing (and hence also satisfies (iii) of (6.26)). Shioji and Takahashi [450] extended Wittmann's result to Banach spaces with uniformly Gâteaux differentiable norms and in which each nonempty closed convex bounded subset of K has the fixed point property for nonexpansive mappings (e.g., L_p spaces $(1 < p < \infty)$). They proved the following theorem.

Theorem ST. *Let E be a Banach space whose norm is uniformly Gâteaux differentiable and let K be a closed convex subset of E. Let T be a nonexpansive mapping from K into K such that the set $F(T)$ of fixed points of T is nonempty. Let $\{\alpha_n\}$ be a sequence which satisfies the following conditions: $0 \le \alpha_n \le 1$, $\lim \alpha_n = 0$, $\sum \alpha_n = \infty$, $\sum |\alpha_{n+1} - \alpha_n| < \infty$. Let $u \in K$ and let $\{x_n\}$ be the sequence defined by $x_0 \in K$, $x_{n+1} = \alpha_n u + (1 - \alpha_n)Tx_n, n \ge 0$. Assume that $\{z_t\}$ converges strongly to $z \in F(T)$ as $t \downarrow 0$, where for $0 < t < 1, z_t$ is the unique element of K which satisfies $z_t = tu + (1 - t)Tz_t$. Then, $\{x_n\}$ converges strongly to z.*

A result of Reich [409] and that of Takahashi and Ueda [478] show that if K satisfies some additional assumption, then $\{z_t\}$ defined above converges strongly to a fixed point of T. In particular, the following is true.

Let E be a Banach space whose norm is uniformly Gâteaux differentiable and let K be a weakly compact convex subset of E. Let T be a nonexpansive mapping from K into K. Let $u \in K$ and let z_t be the unique element of K which satisfies $z_t = tu + (1-t)Tz_t$ for $0 < t < 1$. Assume that each nonempty T-invariant closed convex subset of K contains a fixed point of T. Then, $\{z_t\}$ converges strongly to a fixed point of T.

Morales and Jung [341] established the following result.

Theorem MJ (Morales and Jung, [341]) *Let K be a nonempty closed convex subset of a reflexive Banach space E which has uniformly Gâteaux differentiable norm and $T : K \to K$ be a nonexpansive mapping with $F(T) \neq \emptyset$. Suppose that every nonempty closed convex bounded subset of K has the fixed point property for nonexpansive mappings. Then there exists a continuous path $t \to z_t$, $0 < t < 1$ satisfying $z_t = tu + (1-t)Tz_t$, for arbitrary but fixed $u \in K$, which converges strongly to a fixed point of T.*

Xu [511] (see, also [510]) showed that the result of Halpern holds in uniformly smooth Banach spaces if condition (iii) of Lions is replaced with the condition $(iii)^*$ $\lim\limits_{n\to\infty} \frac{|\alpha_n - \alpha_{n-1}|}{\alpha_n} = 0$. He proved the following theorem.

Theorem 6.29. *(Xu, [511]) Let E be a uniformly smooth Banach space, K be a closed convex nonempty subset of E, $T : K \to K$ be a nonexpansive mapping with $F(T) \neq \emptyset$. Let $u, x_0 \in K$ be given and let $\{\alpha_n\} \subset [0, 1]$ satisfy the conditions: (a) $\lim \alpha_n = 0$; (b) $\sum \alpha_n = \infty$; and (c) $\lim \frac{|\alpha_{n-1} - \alpha_n|}{\alpha_n} = 0$. Then the sequence $\{x_n\}$ generated by $x_0 \in K$, $x_{n+1} := (1 - \alpha_n)Tx_n + \alpha_n u$, $n \geq 0$, converges strongly to some $x^* \in F(T)$.*

Remark 6.30. Wittman [505] had earlier proved Theorem 6.29 with condition (c) replaced by: $(c)^*$ $\sum |\alpha_{n+1} - \alpha_n| < \infty$. The conditions (c) and $(c)^*$ are not comparable. For instance, the sequence $\{\alpha_n\}$ defined by

$$\alpha_n := \begin{cases} n^{-\frac{1}{2}}, & \text{if } n \text{ is odd,} \\ (n^{-\frac{1}{2}} - 1)^{-1}, & \text{if } n \text{ is even,} \end{cases}$$

satisfies (c) but fails to satisfy $(c)^*$.

Remark 6.31. Halpern showed that the conditions (i) $\lim\limits_{n\to\infty} \alpha_n = 0$ and (ii) $\sum\limits_{n=0}^{\infty} \alpha_n = \infty$ are necessary for the convergence of the sequence $\{x_n\}$ defined by (6.25). It is not known if generally they are sufficient. Some authors have established that if in the recursion formula (6.25), Tx_n is replaced with $T_n x_n := \left(\frac{1}{n}\right) \sum\limits_{k=0}^{n-1} T^k x_n$, then conditions (i) and (ii) are sufficient.

In order to prove the main theorems of this section, we shall make use of the following lemmas.

Lemma 6.32. *(Tan and Xu, [487]) Let $\{a_n\}$ be a sequence of nonnegative real numbers satisfying the following relation:*

$$a_{n+1} \leq a_n + \sigma_n, \ n \geq 0,$$

such that $\sum\limits_{n=1}^{\infty} \sigma_n < \infty$. Then, $\lim a_n$ exists. If, in addition, $\{a_n\}$ has a subsequence that converges to 0, then a_n converges to 0 as $n \to \infty$.

Proof. From $0 \leq a_{n+1} \leq a_n + \sigma_n$, $n \geq 0$, we obtain that $0 \leq a_{n+1} \leq a_1 + \sum_1^n \sigma_n \leq a_1 + \sum_1^\infty \sigma_n < \infty$, $n \geq 0$, and so $\{a_n\}$ is bounded. Furthermore, for fixed $m \in \mathbb{N}$, we have

$$0 \leq a_{n+m} \leq a_{n+m-1} + \sigma_{n+m-1}$$

$$\leq a_{n+m-2} + \sigma_{n+m-2} + \sigma_{n+m-1}$$

$$\vdots$$

$$\leq \alpha_n + \sum_{i=n}^{n+m-1} \sigma_i.$$

Taking "lim sup" as $m \to \infty$, we obtain that $\limsup_{m\to\infty} a_n \leq a_n + \sum_{i=n}^\infty \sigma_i$. Now, taking "lim inf" as $n \to \infty$, we get $\limsup_{n\to\infty} a_n \leq \liminf a_n$. Thus, $\liminf_{n\to\infty} a_n = \limsup_{n\to\infty} a_n$, and the limit exists. If, in addition, $\{a_n\}$ has a subsequence that converges to 0, since the limit of $\{a_n\}$ exists, then $\{a_n\}$ converges to 0 as $n \to \infty$. $\qquad\square$

Aliter. We give another proof of the first part of Lemma 6.32.

Define $\rho_n := \sum_{k=1}^{n-1} \sigma_k$. Recall that $\lim_{n\to\infty} \rho_n := \sum_{k=1}^\infty \sigma_k < \infty$. Let $\lim_{n\to\infty} \rho_n = L < \infty$, for some $L \in \mathbb{R}$. Now, $a_{n+1} + \rho_n \leq a_n + \sigma_n + \rho_n = a_n + \rho_{n+1}$ so that $a_{n+1} - \rho_{n+1} \leq a_n - \rho_n \; \forall \; n \geq 1$. This implies that $\{a_n - \rho_n\}$ is non-increasing. Two cases arise.

Case 1: $\lim_{n\to\infty} (a_n - \rho_n) = -\infty$, or, **Case 2:** $\lim_{n\to\infty} (a_n - \rho_n) = M$, for some $M \in \mathbb{R}$. We show Case 1 is impossible. Suppose, for contradiction, case 1 holds. Then, $\lim_{n\to\infty} a_n = \lim_{n\to\infty} \left(a_n - \rho_n + \rho_n \right) = -\infty + L = -\infty$, contradicting the hypothesis that $a_n \geq 0$. So, case 2 holds. Hence, $\lim_{n\to\infty} a_n = \lim_{n\to\infty} \left(a_n - \rho_n + \rho_n \right) = M + L$, completing the proof.

Lemma 6.33. *(Suzuki, [463]).* *Let $\{x_n\}$ and $\{y_n\}$ be bounded sequences in a Banach space E and let $\{\beta_n\}$ be a sequence in $[0,1]$ with $0 < \liminf_{n\to\infty} \beta_n \leq \limsup_{n\to\infty} \beta_n < 1$. Suppose $x_{n+1} = \beta_n y_n + (1 - \beta_n)x_n$ for all integers $n \geq 0$ and $\limsup_{n\to\infty}(\|y_{n+1} - y_n\| - \|x_{n+1} - x_n\|) \leq 0$. Then, $\lim_{n\to\infty} \|y_n - x_n\| = 0$.*

Lemma 6.34. *(Xu, [511]).* *Let $\{a_n\}$ be a sequence of nonnegative real numbers satisfying the following relation:*

$$a_{n+1} \leq (1 - \alpha_n)a_n + \alpha_n \sigma_n + \gamma_n, n \geq 0,$$

where,

(i) $\{\alpha_n\} \subset [0,1]$, $\sum \alpha_n = \infty$;
(ii) $\limsup\limits_{n\to\infty} \sigma_n \leq 0$;
(iii) $\gamma_n \geq 0$; $(n \geq 0)$, $\sum \gamma_n < \infty$. *Then,* $a_n \to 0$ *as* $n \to \infty$.

6.7.1 Convergence Theorems

In the sequel, $F(T) := \{x \in K : Tx = x\}$. In the next theorem, we shall assume that $\{z_t\}$ converges strongly to a fixed point z of T as $t \to 0$, where z_t is the unique element of K which satisfies $z_t = tu + (1-t)Tz_t$ for arbitrary $u \in K$.

Theorem 6.35. *Let K be a nonempty closed convex subset of a real Banach space E which has a uniformly Gâteaux differentiable norm and $T : K \to K$ be a nonexpansive mapping with $F(T) \neq \emptyset$. For a fixed $\delta \in (0,1)$, define $S : K \to K$ by $Sx := (1-\delta)x + \delta Tx \ \forall \ x \in K$. Let $\{\alpha_n\}$ be a real sequence in $(0,1)$ which satisfies the following conditions: $C1 : \lim \alpha_n = 0$; $C2 : \sum \alpha_n = \infty$. For arbitrary $u, x_0 \in K$, let the sequence $\{x_n\}$ be defined iteratively by*

$$x_{n+1} = \alpha_n u + (1 - \alpha_n)Sx_n, n \geq 0. \tag{6.27}$$

Then, $\{x_n\}$ converges strongly to a fixed point of T.

Proof. Observe first that S is nonexpansive and has the same set of fixed points as T. Define

$$\beta_n := (1 - \delta)\alpha_n + \delta \ \forall \ n \geq 0; \ y_n := \frac{x_{n+1} - x_n + \beta_n x_n}{\beta_n}, \ n \geq 0. \tag{6.28}$$

Observe that $\beta_n \to \delta$ as $n \to \infty$, and that if $\{x_n\}$ is bounded, then $\{y_n\}$ is bounded. Let $x^* \in F(T) = F(S)$. One easily shows by induction that $||x_n - x^*|| \leq \max\{||x_0 - x^*||, ||u - x^*||\}$ for all integers $n \geq 0$, and so, $\{x_n\}, \{y_n\}, \{Tx_n\}$ and $\{Sx_n\}$ are all bounded. Also,

$$||x_{n+1} - Sx_n|| = \alpha_n||u - Sx_n|| \to 0, \tag{6.29}$$

as $n \to \infty$. Observe also that from the definitions of β_n and S, we obtain that $y_n = \frac{1}{\beta_n}(\alpha_n u + (1 - \alpha_n)\delta Tx_n)$ so that

$$||y_{n+1} - y_n|| - ||x_{n+1} - x_n|| \leq \left|\frac{\alpha_{n+1}}{\beta_{n+1}} - \frac{\alpha_n}{\beta_n}\right| \cdot ||u||$$

$$+ \frac{(1 - \alpha_{n+1})}{\beta_{n+1}} \delta \, ||Tx_{n+1} - Tx_n||$$

$$+ \left|\frac{1 - \alpha_{n+1}}{\beta_{n+1}} - \frac{1 - \alpha_n}{\beta_n}\right| \delta \, ||Tx_n|| - ||x_{n+1} - x_n||,$$

so that, since $\{x_n\}$ and $\{Tx_n\}$ are bounded, we obtain that, for some constants $M_1 > 0$, and $M_2 > 0$,

$$\limsup_{n \to \infty}(\|y_{n+1} - y_n\| - \|x_{n+1} - x_n\|) \leq \limsup_{n \to \infty}\left\{ \left| \frac{\alpha_{n+1}}{\beta_{n+1}} - \frac{\alpha_n}{\beta_n} \right| \cdot \|u\| \right.$$
$$+ \left| \frac{(1 - \alpha_{n+1})}{\beta_{n+1}}\delta - 1 \right| M_1$$
$$\left. + \left| \frac{1 - \alpha_{n+1}}{\beta_{n+1}} - \frac{1 - \alpha_n}{\beta_n} \right| \delta M_2 \right\} \leq 0.$$

Hence, by Lemma 6.33, $\|y_n - x_n\| \to 0$ as $n \to \infty$. Consequently, $\lim_{n\to\infty}\|x_{n+1} - x_n\| = \lim_{n \to \infty}\beta_n \|y_n - x_n\| = 0$. Combining this with (6.29) yields that

$$\|x_n - Sx_n\| \to 0 \ as \ n \to \infty. \tag{6.30}$$

We now show that

$$\limsup_{n \to \infty}\langle u - z, j(x_n - z)\rangle \leq 0. \tag{6.31}$$

For each integer $n \geq 1$, let $t_n \in (0, 1)$ be such that

$$t_n \to 0, \ and \ \frac{\|x_n - Sx_n\|}{t_n} \to 0, \ n \to \infty. \tag{6.32}$$

Let $z_{t_n} \in K$ be the unique fixed point of the contraction mapping S_{t_n} given by $S_{t_n}x := t_n u + (1 - t_n)Sx$, $x \in K$. Then, $z_{t_n} - x_n = t_n(u - x_n) + (1 - t_n)(Sz_{t_n} - x_n)$. Using inequality (4.4), we compute as follows:

$$\|z_{t_n} - x_n\|^2 \leq (1 - t_n)^2\|Sz_{t_n} - x_n\|^2 + 2t_n\langle u - x_n, j(z_{t_n} - x_n)\rangle$$
$$\leq (1 - t_n)^2(\|Sz_{t_n} - Sx_n\| + \|Sx_n - x_n\|)^2 + 2t_n(\|z_{t_n} - x_n\|^2$$
$$+ \langle u - z_{t_n}, j(z_{t_n} - x_n)\rangle)$$
$$\leq (1 + t_n^2)\|z_{t_n} - x_n\|^2 + \|Sx_n - x_n\| \times$$
$$(2\|z_{t_n} - x_n\| + \|Sx_n - x_n\|)$$
$$+ 2t_n\langle u - z_{t_n}, j(z_{t_n} - x_n)\rangle,$$

and hence,

$$\langle u - z_{t_n}, j(x_n - z_{t_n})\rangle \leq \frac{t_n}{2}\|z_{t_n} - x_n\|^2 + \frac{\|Sx_n - x_n\|}{2t_n}$$
$$\times (2\|z_{t_n} - x_n\| + \|Sx_n - x_n\|).$$

Since $\{x_n\}, \{z_{t_n}\}$ and $\{Sx_n\}$ are bounded and $\frac{\|Sx_n - x_n\|}{2t_n} \to 0, n \to \infty$, it follows from the last inequality that

$$\limsup_{n \to \infty}\langle u - z_{t_n}, j(x_n - z_{t_n})\rangle \leq 0. \tag{6.33}$$

Moreover, we have that

$$\langle u - z_{t_n}, j(x_n - z_{t_n}) \rangle = \langle u - z, j(x_n - z) \rangle$$
$$+ \langle u - z, j(x_n - z_{t_n}) - j(x_n - z) \rangle$$
$$+ \langle z - z_{t_n}, j(x_n - z_{t_n}) \rangle. \tag{6.34}$$

But, by hypothesis, $z_{t_n} \to z \in F(S)$, $n \to \infty$. Thus, using the boundedness of $\{x_n\}$ we obtain that $\langle z - z_{t_n}, j(x_n - z_{t_n}) \rangle \to 0$, $n \to \infty$. Also, $\langle u - z, j(x_n - z_{t_n}) - j(x_n - z) \rangle \to 0$, $n \to \infty$, since j is norm-to-weak* uniformly continuous on bounded subsets of E. Hence, we obtain from (6.33) and (6.34) that $\limsup_{n \to \infty} \langle u - z, j(x_n - z) \leq 0$. Furthermore, from (6.27) we get that $x_{n+1} - z = \alpha_n(u - z) + (1 - \alpha_n)(Sx_n - z)$. It then follows that

$$||x_{n+1} - z||^2 \leq (1 - \alpha_n)^2 ||Sx_n - z||^2 + 2\alpha_n \langle u - z, j(x_{n+1} - z) \rangle$$
$$\leq (1 - \alpha_n)||x_n - z||^2 + \alpha_n \sigma_n,$$

where $\sigma_n := 2\langle u - z, j(x_{n+1} - z) \rangle$; $\gamma_n \equiv 0 \ \forall \ n \geq 0$. Thus, by Lemma 6.34, $\{x_n\}$ converges strongly to a fixed point of T. $\qquad \square$

Remark 6.36. We note that every uniformly smooth Banach space has a uniformly Gâteaux differentiable norm and is such that every nonempty closed convex and bounded subset of E has the fixed point property for nonexpansive maps (see e.g., [189]).

Let $S_n(x) := \frac{1}{n} \sum_{k=0}^{n-1} S^k x$, where $S : K \to K$ is a nonexpansive map. With this definition, Xu proved the following theorem.

Theorem HKX (Xu, [511], Theorem 3.2) *Assume that E is a real uniformly convex and uniformly smooth Banach space. For given $u, x_0 \in K$, let $\{x_n\}$ be generated by the algorithm:*

$$x_{n+1} = \alpha_n u + (1 - \alpha_n) S_n x_n, n \geq 0. \tag{6.35}$$

Assume that (i) $\lim \alpha_n = 0$; (i) $\sum \alpha_n = \infty$. Then, $\{x_n\}$ converges strongly to a fixed point of S.

Remark 6.37. Theorem 6.35 is a significant improvement of Theorem HKX in the sense that the recursion formula (6.27) is simpler and requires less computation at each stage than the recursion formula (6.35). Moreover, the requirement that E be also uniformly convex imposed in Theorem HKX is dispensed with in Theorem 6.35. Furthermore, Theorem 6.35 is proved in the framework of the more general real Banach spaces with uniformly Gâteaux differentiable norms.

6.7.2 The Case of Non-self Mappings

Definition 6.38. Let K be a nonempty subset of a Banach space E. For $x \in K$, the *inward set* of x, $I_K x$, is defined by $I_K x := \{x + \alpha(u - x) : u \in K, \alpha \geq 1\}$. A mapping $T : K \to E$ is called *weakly inward* if $Tx \in cl[I_K(x)]$ for all $x \in K$, where $cl[I_K(x)]$ denotes the closure of the inward set. Every self-map is trivially weakly inward.

Definition 6.39. Let $K \subseteq E$ be closed convex and Q be a mapping of E onto K. Then Q is said to be *sunny* if $Q(Qx + t(x - Qx)) = Qx$ for all $x \in E$ and $t \geq 0$. A mapping Q of E into E is said to be a *retraction* if $Q^2 = Q$. If a mapping Q is a retraction, then $Qz = z$ for every $z \in R(Q)$, *range of Q*. A subset K of E is said to be a *sunny nonexpansive retract* of E if there exists a sunny nonexpansive retraction of E onto K and it is said to be a *nonexpansive retract* of E if there exists a nonexpansive retraction of E onto K. If $E = H$, the metric projection P_K is a sunny nonexpansive retraction from H to any closed convex subset of H.

Remark 6.40. We note that, if $T : K \to E$ is weakly inward, then $F(T) = F(QT)$, where Q is a sunny nonexpansive retraction of E onto K. In fact, clearly, $F(T) \subseteq F(QT)$. We show $F(QT) \subseteq F(T)$. Suppose this is not the case. Then there exists $x \in F(QT)$ such that $x \notin F(T)$. But, since T is weakly inward there exists $u \in K$, such that $Tx = x + \lambda(u - x)$ for some $\lambda > 1$ and $u \neq x$. Observe that if $u = x$ then $Tx = x$, a contradiction, since $x \notin F(T)$. As Q is sunny nonexpansive, we have $Q[QTx + t(Tx - QTx)] = x$ for all $t \geq 0$. But $QTx = x$ so that $Q[tTx + (1 - t)x] = x$ for all $t \geq 0$. Since T is weakly inward, there exists $t_0 \in (0, 1)$ such that $u := t_0 Tx + (1 - t_0)x$, and since $u \in K$, $Qu = u$. This implies $u = Qu = x$, a contradiction, since $x \neq u$. Therefore, $F(QT) \subseteq F(T)$, which implies that $F(QT) = F(T)$.

We now prove the following convergence theorem.

Theorem 6.41. *(Chidume et al., [184]) Let K be a nonempty closed convex subset of a real Banach space E which has a uniformly Gâteaux differentiable norm, and $T : K \to E$ be a nonexpansive mapping satisfying weakly inward condition with $F(T) \neq \emptyset$. Assume K is a sunny nonexpansive retract of E with Q as the sunny nonexpansive retraction. Assume that $\{z_t\}$ converges strongly to a fixed point z of QT as $t \to 0$, where for $0 < t < 1$, z_t is the unique element of K which satisfies $z_t = tx + (1 - t)QT z_t$. Let $\{\alpha_n\}$ be a real sequence in $(0, 1)$ which satisfies the following conditions: (i) $\lim_{n \to \infty} \alpha_n = 0$, (ii) $\sum_{n=1}^{\infty} \alpha_n = \infty$, and either (iii) $\sum_{n=1}^{\infty} |\alpha_n - \alpha_{n-1}| < \infty$, or (iii)* $\lim_{n \to \infty} \frac{|\alpha_n - \alpha_{n-1}|}{\alpha_n} = 0$. For fixed $u, x_0 \in K$, let the sequence $\{x_n\}$ be defined iteratively by*

$$x_{n+1} := \alpha_n u + (1 - \alpha_n)QT x_n, n \geq 0. \tag{6.36}$$

Then, $\{x_n\}_{n=0}^{\infty}$ converges strongly to a fixed point of T.

Proof. Let $x^* \in F(T)$. One easily shows by induction that $||x_n - x^*|| \leq max\{||x_0 - x^*||, ||u - x^*||\}$ for all integers $n \geq 0$, and hence $\{x_n\}$ and $\{QTx_n\}$ are bounded. But this implies from (6.36) that

$$||x_{n+1} - QTx_n|| = \alpha_n ||u - QTx_n|| \to 0 \quad \text{as } n \to \infty. \quad (6.37)$$

Furthermore, for some constant $M > 0$,

$$||x_{n+1} - x_n|| = ||(\alpha_n - \alpha_{n-1})(u - QTx_{n-1}) + (1 - \alpha_n)(QTx_n - QTx_{n-1})||$$
$$\leq M|\alpha_{n-1} - \alpha_n| + (1 - \alpha_n)||x_n - x_{n-1}||.$$

We consider two cases.

Case 1. Condition $(iii)^*$ is satisfied. Then, $||x_{n+1} - x_n|| \leq (1 - \alpha_n)||x_n - x_{n-1}|| + \sigma_n$, where $\sigma_n := \alpha_n \beta_n$ and $\beta_n := (|\alpha_n - \alpha_{n-1}|M/\alpha_n)$ so that $\sigma_n = o(\alpha_n)$.

Case 2. Condition (iii) is satisfied. Then, $||x_{n+1} - x_n|| \leq (1 - \alpha_n)||x_n - x_{n-1}|| + \sigma_n$, where $\sigma_n := M|\alpha_n - \alpha_{n-1}|$ so that $\sum_{n=1}^{\infty} \sigma_n < \infty$. In either case, a lemma of [511] (see, Exercises 6.1, Problem 8) yields that $||x_{n+1} - x_n|| \to 0$ as $n \to \infty$. Combining this with (6.36), we obtain that $||x_n - QTx_n|| \to 0$ as $n \to \infty$. For each integer $n \geq 0$, let $t_n \in (0, 1)$ be such that $t_n \to 0$ and $\frac{||x_n - QTx_n||}{t_n} \to 0$. Let $z_{t_n} \in K$ be the unique fixed point of the contraction mapping T_{t_n} given by $T_{t_n} x := t_n u + (1 - t_n)QTx, x \in K$. Then, $z_{t_n} - x_n = t_n(u - x_n.) + (1 - t_n)(QTz_{t_n} - x_n)$. Moreover, using inequality (4.4), we have

$$||z_{t_n} - x_n||^2$$
$$\leq (1 - t_n)^2 ||QTz_{t_n} - x_n||^2 + 2t_n \langle u - x_n, j(z_{t_n} - x_n) \rangle$$
$$\leq (1 - t_n)^2 (||QTz_{t_n} - QTx_n|| + ||QTx_n - x_n||)^2 + 2t_n(||z_{t_n} - x_n||^2$$
$$+ \langle u - z_{t_n}, j(z_{t_n} - x_n) \rangle)$$
$$\leq (1 + t_n^2)||z_{t_n} - x_n||^2 + ||QTx_n - x_n||(2||z_{t_n} - x_n|| + ||QTx_n - x_n||)$$
$$+ 2t_n \langle u - z_{t_n}, j(z_{t_n} - x_n) \rangle,$$

and hence,

$$\langle u - z_{t_n}, j(x_n - z_{t_n}) \rangle \leq \frac{t_n}{2}||z_{t_n} - x_n||^2 + \frac{||QTx_n - x_n||}{2t_n}$$
$$\times (2||z_{t_n} - x_n|| + ||QTx_n - x_n||).$$

Since $\{x_n\}$, $\{z_{t_n}\}$ and $\{Tx_n\}$ are bounded and $\frac{||x_n - QTx_n||}{t_n} \to 0$ as $n \to \infty$, it follows from the last inequality that

$$\limsup_{n \to \infty} \langle u - z_{t_n}, j(x_n - z_{t_n}) \rangle \leq 0. \quad (6.38)$$

Moreover, we have that

$$\langle u - z_{t_n}, j(x_n - z_{t_n})\rangle = \langle u - z, j(x_n - z)\rangle + \langle u - z, j(x_n - z_{t_n})$$
$$-j(x_n - z)\rangle + \langle z - z_{t_n}, j(x_n - z_{t_n})\rangle. \quad (6.39)$$

But, by hypothesis, $z_{t_n} \to z \in F(QT)$ as $n \to 0$ and by Remark 6.40 we have that $QTz = z = Tz$. Thus, $\langle z - z_{t_n}, j(x_n - z_{t_n})\rangle \to 0$ as $n \to \infty$ (since $\{x_n\}$ is bounded). Also, $\langle u - z, j(x_n - z_{t_n}) - j(x_n - z)\rangle \to 0$ as $n \to \infty$ (since j is norm-to-w^* uniformly continuous on bounded subsets of E). Therefore, we obtain from (6.38) and (6.39) that $\limsup_{n\to\infty}\langle u - z, j(x_n - z)\rangle \leq 0$. Now from (6.36), we get $x_{n+1} - z = \alpha_n(u - z) + (1 - \alpha_n)(QTx_n - z)$. It follows that

$$||x_{n+1} - z||^2 \leq (1 - \alpha_n)^2||QTx_n - z||^2 + 2\alpha_n\langle u - z, j(x_{n+1} - z)\rangle$$
$$\leq (1 - \alpha_n)||x_n - z||^2 + \sigma_n,$$

where $\sigma_n := 2\alpha_n\langle u - z, j(x_{n+1} - z)\rangle$ so that $\limsup_{n\to\infty}\sigma_n \leq 0$. Thus, (see Exercise 6.1, Problem 8), $\{x_n\}$ converges strongly to a fixed point z of T. □

Remark 6.42. In [453], Shioji and Takahashi proved that if E is *a uniformly convex Banach space* whose norm is uniformly Gâteaux differentiable, and T is self-map of $K \subseteq E$ with $F(T) := \{x \in K : Tx = x\} \neq \emptyset$, and T satisfies the following condition: $||T^n x - T^n y|| \leq k_n||x - y|| \ \forall \ x, y \in K, n \in \mathbb{N}$ for some sequence $\{k_n\}, k_n \geq 1, \ \lim k_n = 1$, then the sequence $\{x_n\}$ defined iteratively by $u, x_0 \in K$ arbitrary,

$$x_n = \alpha_n u + (1 - \alpha_n)\frac{1}{n+1}\sum_{j=0}^{n} T^j x_n, n \geq N_0,$$

for N_0 sufficiently large, converges strongly to Pu, where P is the *sunny nonexpansive retract* from K onto $F(T)$ and $\{\alpha_n\}$ satisfies the following conditions: $0 \leq \alpha_n \leq 1, \ \lim \alpha_n = 0, \sum \alpha_n = \infty$ and

$$\sum\left(\left(1 - \alpha_n\right)\left(\frac{1}{n+1}\sum_{j=0}^{n}k_j\right)^2 - 1\right) < \infty.$$

The two authors [450] had earlier established the same result in Hilbert space for the iterative scheme, $x_0, u \in K$ arbitrary,

$$y_n = \beta_n u + \frac{(1 - \beta_n)}{n+1}\sum_{j=0}^{n}T^j x_n; \ x_{n+1} = \alpha_n u + \frac{(1 - \alpha_n)}{n+1}\sum_{j=0}^{n}T^j y_n.$$

In these results, α_n and β_n are real sequences satisfying appropriate conditions.

Finally, we have the following Theorem which holds in *uniformly convex Banach spaces*.

Theorem 6.43. *(Xu, [511]). Let E be a uniformly convex Banach space, K a nonempty closed convex subset of E, $T : K \to K$ a nonexpansive mapping with $F(T) \neq \emptyset$. Assume that E has a Fréchet differentiable norm or satisfies Opial's condition. With an initial $x_0 \in K$, let $\{x_n\}$ be defined by $x_{n+1} := (1 - \alpha_n)T_n x_n + \alpha_n u$, $n \geq 0$, where $\{\alpha_n\} \subset [0, 1]$ satisfies the following conditions: (i) $\lim \alpha_n = 0$; (ii)$\sum \alpha_n = \infty$, and T_n is defined by $T_n x := \left(\frac{1}{n}\right) \sum_{j=1}^{n-1} T^j x$.*

Then, $\{x_n\}$ converges weakly to some $x^* \in F(T)$.

EXERCISES 6.1

1. Verify the assertions made in Example 6.4.
2. Prove Theorem 6.7, i.e., suppose M is a metric space and $T : M \to M$ is continuous and asymptotically regular at x_0 in M. Then, any cluster point of $\{T^n(x_0)\}$ is a fixed point of T.
3. Prove Proposition 6.27.
4. (a) Find an example of a complete metric space (E, ρ) and a mapping $f : E \to E$ such that $\rho(f(x), f(y)) < \rho(x, y) \ \forall \ x, y \in E$, and f has no fixed points.
   Hint: Consider $(E, \rho) \equiv (\mathbb{R}, \rho)$ where ρ is the usual metric and $f(x) = \ell n \, (1 + e^x)$.
   (b) Prove that if the Contraction Mapping Principle applies to f^n where n is a positive integer, then f has a unique fixed point.
   (c) Let (E, ρ) be a complete metric space and $f, g : E \to E$ be functions. Suppose f is a contraction and $f(g(x)) = g(f(x)) \ \forall \ x \in E$. Prove that g has a fixed point but that such a fixed point need not be unique.
5. Let $E = C[0, 1]$ with "sup" norm and let $K = \{f \in C[0, 1] : f(0) = 0, f(1) = 1, 0 \leq f(x) \leq 1\}$. For each $f \in K$ define $\varphi : K \to C[0, 1]$ by $(\varphi f)(x) = x f(x)$. Prove: (a) K is nonempty, closed, convex and bounded; (b) φ maps K into K; (c) φ is nonexpansive; (d) φ has no fixed points.
6. (a) State the Contraction Mapping Principle.
   (b) Let (E, ρ) be a complete metric space and $T : E \to E$ a contraction map with constant $k < 1$. Define the sequence $\{x_n\}$ inductively by $x_{n+1} = T x_n$, $n = 1, 2, \ldots$, $x_0 \in E$. If x^* is the unique fixed point of T, prove: (i) $x_n \to x^*$ as $n \to \infty$; (ii) $\rho(x_n, x^*) \leq \dfrac{k^n}{1 - k} \rho(x_1, x_0)$.
7. Let $C[0, 1]$ be endowed with the "sup" metric. Define $T : C[0, 1] \to C[0, 1]$ by $(Tf)(t) = \int_0^t f(s)ds$, $f \in C[0, 1]$, $t \in [0, 1]$. Prove:
   (a) T is *not* a contraction map;
   (b) T^2 is a contraction map.

(Note: "sup" metric ρ is given by $\rho(f,g) = \sup\limits_{0 \le t \le 1} |f(t) - g(t)|$).

(c) Does T have a fixed point?

8. Let $\{a_n\}$ be a sequence of nonnegative real numbers satisfying the following relation:

$$a_{n+1} \le (1 - \alpha_n)a_n + \sigma_n, n \ge 0,$$

where (i) $0 < \alpha_n < 1$; (ii) $\sum\limits_{n=1}^{\infty} \alpha_n = \infty$. Suppose, $\sum\limits_{n=1}^{\infty} \sigma_n < \infty$. Prove that $a_n \to 0$ as $n \to \infty$.

6.8 Historical Remarks

Remark 6.44. If the constant vector u in the Halpern-type recursion formula (6.25) is replaced with $f(x_n)$, where $f : K \to K$ is a strict contraction, an iteration method involving the resulting formula is called the *viscosity method*. We make the following remarks concerning this method.

- The recursion formula with $f(x_n)$ involves more computation at each stage of the iteration than that with u and does not result in any improvement in the speed or rate of convergence of the scheme. Consequently, from the practical point of view, it is undesirable.
- When a Theorem has been proved using a Halpern-type recursion formula with a constant vector, say u, the proof of the same Theorem with u replaced by $f(x_n)$, the so-called viscosity method, generally does not involve any new ideas or method. Such a proof is generally an unnecessary repetition of the proof when the vector u is used.
- The so-called viscosity method may be useful in other iteration processes. But for the approximation of fixed points of nonexpansive and related operators, there seems to be no justification for studying it.

Let $I = [a, b]$ and let T be a self-map of I and suppose T has a unique fixed point in I. Mann [319] proved that the iteration process: $x_0 \in I$,

$$x_{n+1} = (1 - c_n)x_n + c_n T x_n, \text{ with } c_n = \frac{1}{n+1} \tag{6.40}$$

converges to the fixed point. Franks and Marzec [225] proved that the uniqueness assumption was unnecessary. Rhoades ([415], Theorem 1) extended the Franks and Marzec result to: $0 \le c_n \le 1, \sum c_n = \infty$. Outlaw and Groetcsh [378] obtained convergence for a nonexpansive mapping T of a convex compact subset of the complex plane. Groetsch [242] generalized the method for nonexpansive mappings on uniformly convex Banach spaces. Dotson [212] also used the method for quasi-nonexpansive mappings on strictly convex Banach spaces.

The concept of *uniform* asymptotic regularity was introduced by Edelstein and O'Brian in [218] where they also proved that, on any normed linear space E, and on any bounded convex subset $K \subset E$, S_λ is uniformly asymptotically regular. Results on the asymptotic regularity of S_λ were first obtained by Browder and Petryshyn [52]. They proved that if E is uniformly convex and $T : K \to K$ is a nonexpansive self-mapping of K, where K is a nonempty closed convex and bounded subset of E, then S_λ is asymptotically regular. As is easily seen, $S_\lambda(x^*) = x^*$ is equivalent to $Tx^* = x^*$, so that problems pertaining to the existence and location of fixed points for T reduce to similar problems for S_λ, where, by the result of Edelstein and O'Brian cited above, S_λ can be assumed to be uniformly asymptotically regular. Theorem 6.3 is the well known Browder-Göhde-Kirk theorem [42, 238, 283]. Theorem 6.7 and Proposition 6.8 are due to Edelstein and O'Brian [218]; Example 6.10 is due to Genel and Lindenstrauss [228]. Part (i) of Corollary 6.19 was originally proved by Petryshyn [381] for uniformly convex Banach spaces and part (ii) was first proved by Browder and Petryshyn [51] again, for uniformly convex Banach spaces. Edelstein and O'Brian [218] extended these results to arbitrary normed linear spaces for the sequence $\{S_\lambda^n\}$, defined by (6.2).

A consequence of a result of Browder and Petryshyn [51] shows that if T is asymptotically regular and $(I - T)$ is demiclosed, then any weak cluster point of $\{T^n(x_0)\}$ is a fixed point of T. It is also known that *in an Opial space*, $(I - T)$ is always demiclosed for any nonexpansive self-map T of a nonempty closed convex and bounded subset K. Edelstein and O'Brian [218] then proved that in an Opial space E, if $K \subset E$ is weakly compact and convex and T is a self-mapping of K, then for any $x_0 \in K$, the sequence $\{S_\lambda^n(x_0)\}$ converges weakly to a fixed point of T. This result is a generalization of an earlier result of Opial [366] who had proved the same result under the assumption that E is uniformly convex and has a weakly continuous duality map. It is pertinent to mention here that Gossez and Lami Dozo [241] have shown that for any normed linear space E, the existence of a weakly continuous duality map implies that E satisfies Opial's condition which in turn implies that E has *normal structure*, (see e.g., Brodskii and Mil'man, [38] for definition), but that none of the converse implications hold.

Theorems 6.14, 6.15 and 6.16 were proved by Edelstein and O'Brian [218] where the sequence $\{x_n\}$ is defined by (6.2). Theorem 6.15 was proved by Ishikawa for the more general sequence defined by (6.3). But then, while this result is somewhat stronger than the result of Edelstein and O'Brian in the sense that it involves the more general Mann iterates, the theorems of Edelstein and O'Brian are stronger in the sense that *uniform* asymptotic regularity is proved. Theorems 6.14, 6.15 and 6.16 unify these results of Ishikawa and those of Edelstein and O'Brian. The theorems are due to Chidume [87] who used a method which seems simpler and totally different from those of Ishikawa and, Edelstein and O'Brian. Finally, the results of Sections 6.3 to

6.5 together with Example 6.23 and Example 6.28 are also due to Chidume [87, 93]. Theorem 6.41 is due to Chidume *et al.* [184].

Strong convergence theorems for a generalization of nonexpansive mappings (relatively weak nonexpansive mappings) can be found in Zegeye and Shahzad [548].

Chapter 7
Hybrid Steepest Descent Method for Variational Inequalities

7.1 Introduction

Let (E, ρ) be a metric space and K be a nonempty subset of E. For every $x \in E$, the distance between the point x and K is denoted by $\rho(x, K)$ and is defined by the following minimum problem:

$$\rho(x, K) := \inf_{y \in K} \rho(x, y).$$

The *metric projection operator* (also called the *nearest point mapping*) P_K defined on E is a mapping from E to 2^K such that

$$P_K(x) := \{z \in K : \rho(x, z) = \rho(x, K)\} \; \forall \; x \in E.$$

If $P_K(x)$ is singleton for every $x \in E$, then K is said to be a *Chebyshev set*. It is well known that if E is a uniformly convex and uniformly smooth Banach space, then any closed convex nonempty subset K of E is a Chebyshev set (see e.g., Johnson [262]).

We state a property of P_K as it applies to real Banach spaces which are uniformly convex and uniformly smooth. For properties of P_K in more general Banach spaces, the reader may consult any of the following references Isac [257]; Li [298]; Wen and Zhengjun [504]; Takahashi, [475]).

Lemma 7.1. *Let E be a uniformly convex and uniformly smooth real Banach space and $K \subseteq E$ be a nonempty closed and convex subset. For any given $x \in E$, P_K is the metric projection of x onto K if and only if*

$$\langle j_q(x - P_K x), P_K x - y \rangle \geq 0 \; \forall \; y \in K,$$

where for $q > 1$, j_q is the generalized normalized duality map on E.

Remark 7.2. The above lemma is easily proved in a $q-$uniformly smooth space E.

C. Chidume, *Geometric Properties of Banach Spaces and Nonlinear Iterations,*
Lecture Notes in Mathematics 1965,
© Springer-Verlag London Limited 2009

From the definition of P_K, we have the following inequality: for arbitrary $x \in E$, it is clear that

$$||x - P_K x|| \leq ||x - \omega|| \ \forall \ \omega \in K.$$

For $\lambda \in (0,1)$, we have that if $u \in K$, then $v := (1 - \lambda)P_K x + \lambda u$ is in K. Substituting this for ω, and using inequality (5.10), we obtain the following estimates:

$$\begin{aligned}
||x - P_K x||^q &\leq ||x - P_K x + \lambda(P_K x - u)||^q \\
&\leq ||x - P_K x||^q + q\lambda\langle P_K x - u, j_q(x - P_K x)\rangle \\
&\quad + d_q \lambda^q ||P_K x - x||^q.
\end{aligned}$$

This implies,

$$0 \leq q\langle P_K x - u, j_q(x - P_K x)\rangle + d_q \lambda^{q-1} ||P_K x - x||^q,$$

so that, letting $\lambda \to 0$, we obtain the desired result.

Let K be a nonempty closed convex subset of a real normed space E, and $S : K \to E$ be a nonlinear operator. The *variational inequality problem* is formulated as follows: Find a point $x^* \in K$ such that

$$VIP(S,K): \ \langle j_q(Sx^*), y - x^*\rangle \geq 0 \ \ \forall y \in K, \qquad (7.1)$$

where $j_q \in J(q)$ and J_q is the generalized duality map on E. If $E = H$, a real Hilbert space, the variational inequality problem reduces to the following: Find a point $x^* \in K$ such that

$$VIP(S,K): \ \langle Sx^*, y - x^*\rangle \geq 0 \ \ \forall y \in K. \qquad (7.2)$$

A mapping $G : K \to E$ is said to be *accretive* if $\forall x, y \in E$, there exists $j_q(x - y) \in J_q(x - y)$ such that

$$\langle Gx - Gy, j_q(x - y)\rangle \geq 0. \qquad (7.3)$$

In Hilbert spaces, accretive operators are called *monotone*. For some real number $\eta > 0$, G is called $\eta-strongly\ accretive$ if $\forall x, y \in E$, there exists $j_q(x - y) \in J_q(x - y)$ such that

$$\langle Gx - Gy, j_q(x - y)\rangle \geq \eta ||x - y||^q. \qquad (7.4)$$

Applications of variational inequalities span as diverse disciplines as differential equations, time-optimal control, optimization, mathematical programming, mechanics, finance and so on (see, for example, [281, 358] for more details).

By using Lemma 7.1, the following lemma is easily proved.

Lemma 7.3. *Let E be a uniformly convex and uniformly smooth real Banach space and $K \subseteq E$ be a nonempty closed and convex subset. Let $S : K \to E$ be a mapping. Then, an element $x^* \in K$ is a solution of the variational inequality problem, VIP(S,K) if and only if $x^* = P_K(I - \delta S)x^*$ for some suitable $\delta > 0$.*

Remark 7.4. It then follows from Lemma 7.3 that solving the variational inequality problem, VIP(S,K), is equivalent to finding a fixed point of the mapping $P_K(I - \delta S) : K \to K$.

It is known that if S is Lipschitz and $\eta-$*strongly accretive*, then $VIP(S, K)$ has a unique solution. An important problem is how to find a solution of $VIP(S, K)$ whenever it exists. Considerable efforts have been devoted to this problem (see, e.g. [518, 531] and the references contained therein).

Lemma 7.3 asserts that in a real Hilbert space, the $VIP(S, K)$ is equivalent to the following fixed point equation

$$x^* = P_K(x^* - \delta Sx^*), \qquad (7.5)$$

where $\delta > 0$ is a suitable fixed constant and P_K is the *nearest point projection map* from H onto K, i.e., $P_K x = y$ where $\|x - y\| = \inf_{u \in K} \|x - u\|$ for $x \in H$.

Consequently, under appropriate conditions on S and δ, fixed point methods can be used to find or approximate a solution of $VIP(S, K)$. For instance, if S is $\eta-$strongly monotone and Lipschitz, then a mapping $G : H \to H$ defined by $Gx = P_K(x - \delta Sx)$, $x \in H$ with $\delta > 0$ sufficiently small is a strict contraction. Hence, the *Picard iteration*, $x_0 \in H$, $x_{n+1} = Gx_n$, $n \geq 0$, of the classical *Banach contraction mapping principle* converges to the unique solution of the $VIP(S, K)$.

In applications, however, the projection operator P_K in the fixed point formulation (7.5) may make the computation of the iterates difficult due to possible complexity of the convex set K where K is an *arbitrary* closed convex subset of H. In order to reduce the possible difficulty with the use of P_K, Yamada [531] introduced a *hybrid steepest descent method* for solving $VIP(S, K)$. Let $T : H \to H$ be a map and let $K := \{x \in H : Tx = x\} \neq \emptyset$. Let S be $\eta-$strongly monotone and $\kappa-$Lipschitz on H. Let $\delta \in (0, \frac{2\eta}{\kappa^2})$ be an arbitrary but fixed real number and let a sequence $\{\lambda_n\}$ in $(0, 1)$ satisfy the following conditions: $C1 : \lim_{n \to \infty} \lambda_n = 0$; $C2 : \sum \lambda_n = \infty$; and $C3 : \lim_{n \to \infty} \frac{\lambda_n - \lambda_{n+1}}{\lambda_n^2} = 0$. Starting with an arbitrary initial guess $x_0 \in H$, let a sequence $\{x_n\}$ be generated by the following algorithm

$$x_{n+1} = Tx_n - \lambda_{n+1}\delta S(Tx_n), \quad n \geq 0. \qquad (7.6)$$

Then, Yamada [531] proved that $\{x_n\}$ converges strongly to the unique solution of $VIP(S, K)$.

In the case that $K = \bigcap_{i=1}^{r} F(T_i) \neq \emptyset$, where $\{T_i\}_{i=1}^{r}$ is a finite family of nonexpansive mappings, Yamada [531] studied the following algorithm,

$$x_{n+1} = T_{[n+1]}x_n - \lambda_{n+1}\delta S(T_{[n+1]}x_n), \quad n \geq 0, \qquad (7.7)$$

where $T_{[k]} = T_{k \bmod r}$, for $k \geq 1$, with the *mod function* taking values in the set $\{1, 2, ..., r\}$, where the sequence $\{\lambda_n\}$ satisfies the conditions $C1$, $C2$ and the following condition: $C4: \sum |\lambda_n - \lambda_{n+N}| < \infty$. Under these conditions, he proved the strong convergence of $\{x_n\}$ to the unique solution of the $VIP(S, K)$.

Xu and Kim [518] studied the convergence of the algorithms (7.6) and (7.7), still in the framework of Hilbert spaces, and proved strong convergence theorems with condition $C3$ replaced by $C5: \lim_{n \to \infty} \frac{\lambda_n - \lambda_{n+1}}{\lambda_{n+1}} = 0$ and with condition $C4$ replaced by $C6: \lim_{n \to \infty} \frac{\lambda_n - \lambda_{n+r}}{\lambda_{n+r}} = 0$. They proved the following theorems.

Theorem 7.5. *Let H be a real Hilbert space, $T : H \to H$ be a nonexpansive mapping, $K := \{x \in H : Tx = x\} = F(T) \neq \emptyset$. Let G be an $\eta-strongly$ monotone and $\kappa-Lipschitzian$ map on H. Let $\mu \in (0, \frac{2\eta}{\kappa^2})$ and a sequence $\{\lambda_n\}$ in $(0, 1)$ satisfy the following conditions:*

$$C1: \lim \lambda_n = 0, \quad C2: \sum \lambda_n = \infty, \quad C5: \lim \frac{\lambda_n - \lambda_{n+1}}{\lambda_{n+1}} = 0.$$

Then, the sequence $\{x_n\}$ defined by (7.6) converges strongly to the unique solution x^* of the $VI(G, K)$.

Theorem 7.6. *Let H be a real Hilbert space, $T_i : H \to H$, $i = 1, 2, ...r$ be a finite family of nonexpansive mappings, $K := \bigcap_{i=1}^{r}(T_i) = F(T_1 T_2 ... T_r) = F(T_r T_1 ... T_{r-1}) = ... = F(T_r T_{r-1} ... T_1) \neq \emptyset$. Let G be an $\eta-strongly$ monotone and $\kappa-Lipscitzian$ map on H. Let $\mu \in (0, \frac{2\eta}{\kappa^2})$ and a sequence $\{\lambda_n\}$ in $(0, 1)$ satisfy the following conditions:*

$$C1: \lim \lambda_n = 0, \quad C2: \sum \lambda_n = \infty, \quad C6: \lim \frac{\lambda_n - \lambda_{n+r}}{\lambda_{n+r}} = 0.$$

Then, the sequence $\{x_n\}$ defined by (7.7) converges strongly to the unique solution x^* of the $VI(G, K)$ where $T_{[k]} = T_{k \bmod r}$, for $k \geq 1$, with the mod function taking values in the set $\{1, 2, ..., r\}$.

These theorems are improvements on the results of Yamada. In particular, the canonical choice $\lambda_n := \frac{1}{n+1}$ is applicable in the results of Xu and Kim but is not in the result of Yamada [531] with condition C3. For further recent results on the schemes (7.6) and (7.7), the reader my consult Wang [499], Zeng and Yao [553] and the references contained in them.

In the next section, we present theorems which extend the results of Xu and Kim [518] to q-uniformly smooth real Banach spaces, $q \geq 2$. In particular, the theorems to be presented will be applicable in L_p spaces, $2 \leq p < \infty$. In section 7.4, we present an iteration process which again extends the results of Xu and Kim [518] to q-uniformly smooth Banach spaces, $q \geq 2$ but where the iteration parameter $\{\lambda_n\}$ is now required to satisfy only conditions C1 and C2.

We remark that q-uniformly smooth Banach spaces, $q \geq 2$, include the L_p space, $2 \leq p < \infty$. They do not include L_p spaces, $1 < p < 2$. In section 7.5, we use a different tool to extend all the theorems of sections 7.3 and 7.4 to L_p spaces, $1 < p \leq 2$.

7.2 Preliminaries

Let K be a nonempty closed convex and bounded subset of a Banach space E and let the *diameter* of K be defined by $d(K) := \sup\{\|x - y\| : x, y \in K\}$. For each $x \in K$, let $r(x, K) := \sup\{\|x - y\| : y \in K\}$ and let $r(K) := \inf\{r(x, K) : x \in K\}$ denote the *Chebyshev radius* of K relative to itself. The *normal structure coefficient* $N(E)$ of E (see e.g. Bynum, [62]) is defined by $N(E) := \inf\left\{\frac{d(K)}{r(K)} : K \text{ is a closed convex and bounded subset of E}\right.$ with $d(K) > 0\Big\}$. A space E such that $N(E) > 1$ is said to have *uniform normal structure*. It is known that all uniformly convex and uniformly smooth Banach spaces have uniform normal structure (see e.g., Lim and Xu, [307]).

We shall denote a *Banach limit* by μ. Recall that μ is an element of $(l^\infty)^*$ such that $\|\mu\| = 1$, $\liminf\limits_{n \to \infty} a_n \leq \mu_n a_n \leq \limsup\limits_{n \to \infty} a_n$ and $\mu_n a_n = \mu_{n+1} a_n$ for all $\{a_n\}_{n \geq 0} \in l^\infty$.

In the sequel, we shall also make use of the following lemmas.

Lemma 7.7. (Shioji and Takahashi, [450]) *Let* $(a_0, a_1, ...) \in l^\infty$ *be such that* $\mu_n(a_n) \leq 0$ *for all Banach limits* μ_n *and* $\limsup\limits_{n \to \infty} (a_{n+1} - a_n) \leq 0$. *Then,* $\limsup\limits_{n \to \infty} a_n \leq 0$.

Lemma 7.8. *(Lim and Xu, [307], Theorem 1) Suppose E is a Banach space with uniform normal structure, K is a nonempty bounded subset of E, and $T : K \to K$ is uniformly k-Lipschitzian mapping with $k < N(E)^{\frac{1}{2}}$. Suppose also there exists a nonempty bounded closed convex subset C of K with the property (P):*
$$(P) \quad x \in C \quad \text{implies} \quad \omega_w(x) \subset C,$$
where $\omega_w(x)$ is the ω-limi set of T at x, i.e., the set
$$\{y \in E : y = \text{weak} - \lim_j T^{n_j} x \text{ for some } n_j \to \infty\}.$$

Then, T has a fixed point in C.

Lemma 7.9. *Let X be a reflexive real Banach space and $f : X \longrightarrow \mathbb{R} \cup \{+\infty\}$ be a convex* $\boxed{proper}$ *lower semi-continuous function. Suppose*

$$\lim_{\|x\| \to \infty} f(x) = +\infty.$$

Then, $\exists \bar{x} \in X$ such that $f(\bar{x}) \le f(x)$, $x \in X$, i.e.,

$$f(\bar{x}) = \inf_{x \in X} f(x).$$

7.3 Convergence Theorems

We first prove the following lemma which will be central in the sequel.

Lemma 7.10. *Let E be a $q-$uniformly smooth real Banach space with constant d_q, $q \ge 2$. Let $T : E \to E$ be a nonexpansive mapping and $G : E \to E$ be an $\eta-$ strongly accretive and $\kappa-$Lipschitzian map. For $\lambda \in \left(0, \frac{2}{q(q-1)}\right)$ and $\delta \in \left(0, \min\left\{\frac{q}{4\eta}, (\frac{q\eta}{d_q \kappa^q})^{\frac{1}{(q-1)}}\right\}\right)$, define a map $T^\lambda : E \to E$ by $T^\lambda x = Tx - \lambda \delta G(Tx)$, $x \in E$. Then, T^λ is a strict contraction. Furthermore,*

$$\|T^\lambda x - T^\lambda y\| \le (1 - \lambda\alpha)\|x - y\| \ \forall \ x, y \in E, \tag{7.8}$$

where $\alpha := \frac{q}{2} - \sqrt{\frac{q^2}{4} - \delta(q\eta - \delta^{q-1} d_q \kappa^q)} \in (0, 1)$.

Proof. For $x, y \in E$, using inequality (5.10), we have:

$$
\begin{aligned}
\|T^\lambda x - T^\lambda y\|^q &= \|Tx - Ty - \lambda\delta(G(Tx) - G(Ty))\|^q \\
&\le \|Tx - Ty\|^q - q\lambda\delta\langle G(Tx) - G(Ty), j_q(Tx - Ty)\rangle \\
&\quad + d_q \lambda^q \delta^q \|G(Tx) - G(Ty)\|^q \\
&\le \|Tx - Ty\|^q - q\lambda\delta\eta\|Tx - Ty\|^q + d_q \lambda^q \delta^q \kappa^q \|Tx - Ty\|^q \\
&\le \left[1 - \lambda\delta\big(q\eta - d_q \lambda^{q-1}\delta^{q-1}\kappa^q\big)\right]\|x - y\|^q \\
&\le \left[1 - \lambda\delta\big(q\eta - d_q \delta^{q-1}\kappa^q\big)\right]\|x - y\|^q.
\end{aligned}
$$

Define

$$f(\lambda) := 1 - \lambda\delta(q\eta - d_q \delta^{q-1}\kappa^q) = (1 - \lambda\tau)^q, \text{ for some } \tau \in (0, 1), \text{ say.}$$

By Taylor development, there exists $\xi \in (0, \lambda)$ such that

$$1 - \lambda\delta(q\eta - d_q \delta^{q-1}\kappa^q) = 1 - q\tau\lambda + \frac{1}{2}q(q - 1)(1 - \xi\tau)^{q-2}\lambda^2\tau^2.$$

Using $\lambda \in \left(0, \frac{2}{q(q-1)}\right)$ which implies $\frac{1}{2}q(q-1)\lambda < 1$, we obtain that

$$1 - \lambda\delta(q\eta - d_q\delta^{q-1}\kappa^q) < 1 - q\tau\lambda + \frac{1}{2}q(q-1)\lambda^2\tau^2 < 1 - q\tau\lambda + \lambda\tau^2,$$

so that

$$\tau^2 - q\tau + \delta(q\eta - d_q\delta^{q-1}\kappa^q) > 0.$$

Solving this quadratic inequality in τ, we obtain, $\tau < \frac{q}{2} - \sqrt{\frac{q^2}{4} - \delta(q\eta - d_q\delta^{q-1}\kappa^q)}$. Now, set

$$\alpha := \frac{q}{2} - \sqrt{\frac{q^2}{4} - \delta(q\eta - d_q\delta^{q-1}\kappa^q)}.$$

Observe that

$$\frac{q^2}{4} - \delta(q\eta - d_q\delta^{q-1}\kappa^q) = \left(\frac{q^2}{4} - \delta q\eta\right) + d_q\delta^{q-1}\kappa^q > 0,$$

since $\delta < \frac{q}{4\eta}$. Moreover, since $q \geq 2$ and $\lambda < \frac{2}{q(q-1)} < \frac{2}{q}$, we have

$$1 - \lambda\alpha = 1 - \frac{\lambda q}{2} + \sqrt{\frac{q^2\lambda^2}{4} - \lambda^2\delta(q\eta - d_q\delta^{q-1}\kappa^q)} \in (0,1).$$

The proof is complete. $\qquad\qquad\qquad\qquad\qquad\qquad\qquad\qquad\square$

We note that L_p spaces, $2 \leq p < \infty$, are 2–uniformly smooth and the following inequality holds (see e.g., inequality (5.10)): For each $x, y \in L_p$, $2 \leq p < \infty$,

$$\|x + y\|^2 \leq \|x\|^2 + 2\langle y, j(x)\rangle + (p-1)\|y\|^2.$$

It then follows that by setting $q = 2$, $d_q = (p-1)$ in Lemma 7.10, we obtain the following corollary.

Corollary 7.11 *Let $E = L_p$, $2 \leq p < \infty$. Let $T : E \to E$ be a nonexpansive map and $G : E \to E$ be an η–strongly accretive and κ–Lipschitzian map. For $\lambda \in \left(0,1\right)$ and $\delta \in \left(0, \min\left\{\frac{1}{2\eta}, \frac{2\eta}{(p-1)\kappa^2}\right\}\right)$, define a map $T^\lambda : E \to E$ by $T^\lambda x = Tx - \lambda\delta G(Tx)$, $x \in E$. Then, T^λ is a strict contraction. In particular,*

$$\|T^\lambda x - T^\lambda y\| \leq (1 - \lambda\alpha)\|x - y\|, \quad x, y \in E, \tag{7.9}$$

where $\alpha := 1 - \sqrt{1 - \delta(2\eta - (p-1)\delta\kappa^2)} \in (0,1)$.

By setting $p = 2$ in Corollary 7.11, we obtain the following corollary.

Corollary 7.12 *Let H be a real Hilbert space, $T : H \to H$ be a nonexpansive map and $G : H \to H$ be an η–strongly monotone and κ–Lipschitzian map. For $\lambda \in \left(0,1\right)$ and $\delta \in \left(0, \min\left\{\frac{1}{2\eta}, \frac{2\eta}{\kappa^2}\right\}\right)$, define a map $T^\lambda : H \to H$ by $T^\lambda x = Tx - \lambda\delta G(Tx)$, $x \in H$. Then, T^λ is a strict contraction. In particular,*

$$\|T^\lambda x - T^\lambda y\| \le (1 - \lambda\alpha)\|x - y\|, \quad x, y \in H, \qquad (7.10)$$

where $\alpha := 1 - \sqrt{1 - \delta(2\eta - \delta\kappa^2)} \in (0, 1)$.

Remark 7.13. Corollary 7.12 is a result of Yamada [531] and is the main tool used in Wang [499], Xu and Kim [518], Yamada [531], Zheng and Yao [553]. Lemma 7.10 and Corollary 7.11 extend this result to q-uniformly smooth spaces, $q \ge 2$, and L_p spaces, $2 \le p < \infty$, respectively.

We prove the following theorem for family of nonexpansive maps. In the theorem, d_q is the constant which appears in inequality (5.10).

Theorem 7.14. *Let E be a q-uniformly smooth real Banach space with constant d_q, $q \ge 2$. Let $T_i : E \to E$, $i = 1, 2, ..., r$ be a finite family of nonexpansive mappings with $K := \bigcap\limits_{i=1}^{r} Fix(T_i) \ne \emptyset$. Let $G : E \to E$ be an η-strongly accretive map which is also κ-Lipschitzian. Let $\{\lambda_n\}$ be a sequence in $[0, 1]$ satisfying*

$$C1 : lim\lambda_n = 0; \ C2 : \sum \lambda_n = \infty; \ C6 : lim\frac{\lambda_n - \lambda_{n+r}}{\lambda_{n+r}} = 0.$$

For $\delta \in \left(0, \min\left\{\frac{q}{4\eta}, (\frac{q\eta}{d_q\kappa^q})^{\frac{1}{(q-1)}}\right\}\right)$, define a sequence $\{x_n\}$ iteratively in E by $x_0 \in E$,

$$x_{n+1} = T_{[n+1]}^{\lambda_{n+1}}x_n = T_{[n+1]}x_n - \delta\lambda_{n+1}G(T_{[n+1]}x_n), \quad n \ge 0, \qquad (7.11)$$

where $T_{[n]} = T_{n \bmod r}$. Assume also that

$$K = Fix(T_rT_{r-1}...T_1) = Fix(T_1T_r...T_2) = ... = Fix(T_{r-1}T_{r-2}...T_r).$$

Then, $\{x_n\}$ converges strongly to the unique solution x^* of the variational inequality $VI(G, K)$.

Proof. Let $x^* \in K$, then the sequence $\{x_n\}$ satisfies

$$\|x_n - x^*\| \le \max\left\{\|x_0 - x^*\|, \frac{\delta}{\alpha}\|G(x^*)\|\right\}, \ n \ge 0.$$

It is obvious that this is true for $n = 0$. Assume it is true for $n = k$ for some $k \in \mathbb{N}$. $\qquad \Box$

From the recursion formula (7.11) and condition C1, we have

$$\begin{aligned}
\|x_{k+1} - x^*\| &= \|T_{[k+1]}^{\lambda_{k+1}}x_k - x^*\| \\
&\le \|T_{[k+1]}^{\lambda_{k+1}}x_k - T_{[k+1]}^{\lambda_{k+1}}x^*\| + \|T_{[k+1]}^{\lambda_{k+1}}x^* - x^*\| \\
&\le (1 - \lambda_{k+1}\alpha)\|x_k - x^*\| + \lambda_{k+1}\delta\|G(x^*)\| \\
&\le \max\left\{\|x_0 - x^*\|, \frac{\delta}{\alpha}\|G(x^*)\|\right\},
\end{aligned}$$

and the claim follows by induction. Thus the sequence $\{x_n\}$ is bounded and so are $\{T_{[n+1]}x_n\}$ and $\{G(T_{[n+1]}x_n)\}$. Using the recursion formula (7.11) we get,

$$\|x_{n+1} - T_{[n+1]}x_n\| = \lambda_{n+1}\delta\|G(T_{[n+1]}x_n)\| \to 0 \quad \text{as} \quad n \to \infty.$$

Also,

$$\begin{aligned}
\|x_{n+r} - x_n\| &= \|T_{[n+r]}^{\lambda_{n+r}}x_{n+r-1} - T_{[n]}^{\lambda_n}x_{n-1}\| \\
&\leq \|T_{[n+r]}^{\lambda_{n+r}}x_{n+r-1} - T_{[n+r]}^{\lambda_{n+r}}x_{n-1}\| + \|T_{[n+r]}^{\lambda_{n+r}}x_{n-1} - T_{[n]}^{\lambda_n}x_{n-1}\| \\
&\leq (1 - \lambda_{n+r}\alpha)\|x_{n+r-1} - x_{n-1}\| \\
&\quad + \alpha\lambda_{n+r}\left(\frac{|\lambda_{n+r} - \lambda_n|}{\alpha\lambda_{n+r}}\delta\|G(T_{[n]}x_{n-1})\|\right).
\end{aligned}$$

By Lemma 6.34 and condition $C6$, we have

$$\|x_{n+r} - x_n\| \to 0 \text{ as } n \to \infty. \tag{7.12}$$

In particular,

$$\|x_{n+1} - x_n\| \to 0 \text{ as } n \to \infty. \tag{7.13}$$

Using the recursion formula (7.11), replacing n by $(n+r-1)$ in this formula and denoting $[n+r]$ by $n+r$, we have,

$$\|x_{n+r} - T_{n+r}x_{n+r-1}\| = \delta\lambda_{n+r}\|G(T_{n+r}x_{n+r-1})\| \to 0, \ n \to \infty.$$

Using the fact that T_i is nonexpansive for each i, we obtain the following finite table:

$$x_{n+r} - T_{n+r}x_{n+r-1} \to 0 \quad as \quad n \to \infty;$$

$$T_{n+r}x_{n+r-1} - T_{n+r}T_{n+r-1}x_{n+r-2} \to 0 \quad as \quad n \to \infty;$$

$$\vdots$$

$$T_{n+r}T_{n+r-1}...T_{n+2}x_{n+1} - T_{n+r}T_{n+r-1}...T_{n+2}T_{n+1}x_n \to 0 \quad as \quad n \to \infty;$$

and adding up the table yields

$$x_{n+r} - T_{n+r}T_{n+r-1}...T_{n+1}x_n \to 0 \quad as \quad n \to \infty.$$

Using this and (7.12) we get that

$$\lim_{n\to\infty} \|x_n - T_{n+r}T_{n+r-1}...T_{n+1}x_n\| = 0. \tag{7.14}$$

Define a map $\varphi : E \to \mathbb{R}$ by $\varphi(y) = \mu_n\|x_{n+1} - y\|^2$, where μ_n denotes a Banach limit. Then, φ is continuous, convex and $\varphi(y) \to +\infty$ as $\|y\| \to +\infty$. Thus, since E is a reflexive Banach space, Lemma 7.9 implies that there exists $y^* \in E$ such that $\varphi(y^*) = \min\limits_{u \in E} \varphi(u)$. So, the set $K^* := \{x \in E : \varphi(x) = \min\limits_{u \in E} \varphi(u)\} \neq \emptyset$. We now show T_i has a fixed point in K^* for each $i = 1, 2, ..., r$. We shall assume, from equation (7.14), that $\forall i$,

$$\lim_{n \to \infty} ||x_n - T_i x_n|| = 0. \tag{7.15}$$

We shall make use of Lemma 7.8. If x is in K^* and $y := \omega - \lim_j T_i^{m_j} x$ belongs to the weak $\omega - limit$ set $\omega_w(x)$ of T_i at x, then, from the w-l.s.c. of φ and equation (7.15), we have, (since equation (7.15) implies $||x_n - T_i^m x_n|| \to 0$ as $n \to \infty$, this is easily proved by induction),

$$\varphi(y) \leq \liminf_{j} \varphi\left(T_i^{m_j} x\right) \leq \limsup_{m} \varphi\left(T_i^m x\right)$$

$$= \limsup_{m} \left(\mu_n ||x_n - T_i^m x||^2\right)$$

$$= \limsup_{m} \left(\mu_n ||x_n - T_i^m x_n + T_i^m x_n - T_i^m x||^2\right)$$

$$\leq \limsup_{m} \left(\mu_n ||T_i^m x_n - T_i^m x||^2\right) \leq \limsup_{m} \left(\mu_n ||x_n - x||^2\right) = \varphi(x)$$

$$= \inf_{u \in E} \varphi(u).$$

So, $y \in K^*$. By Lemma 7.8, T_i has a fixed point in $K^* \forall i$ and so $K^* \cap K \neq \emptyset$.

Let $x^* \in K^* \cap K$ and $t \in (0, 1)$. It then follows that $\varphi(x^*) \leq \varphi(x^* - tG(x^*))$. Using inequality (4.4), we have that

$$||x_n - x^* + tG(x^*)||^2 \leq ||x_n - x^*||^2 + 2t\langle G(x^*), j(x_n - x^* + tG(x^*))\rangle.$$

Thus, taking Banach limits over $n \geq 1$ gives

$$\mu_n ||x_n - x^* + tG(x^*)||^2 \leq \mu_n ||x_n - x^*||^2$$
$$+ 2t\mu_n \langle G(x^*), j(x_n - x^* + tG(x^*))\rangle.$$

This implies,

$$\mu_n \langle -G(x^*), j(x_n - x^* + tG(x^*))\rangle \leq \varphi(x^*) - \varphi(x^* - tG(x^*)) \leq 0.$$

This therefore implies that

$$\mu_n \langle -G(x^*), j(x_n - x^* + tG(x^*))\rangle \leq 0 \,\forall\, n \geq 1.$$

Since the normalized duality mapping is norm-to-norm uniformly continuous on bounded subsets of E, we obtain, as $t \to 0$, that

$$\langle -G(x^*), j(x_n - x^*)\rangle - \langle -G(x^*), j(x_n - x^* + tG(x^*))\rangle \to 0.$$

Hence, for all $\varepsilon > 0$, there exists $\delta > 0$ such that $\forall t \in (0, \delta)$ and for all $n \geq 1$,

$$\langle -G(x^*), j(x_n - x^*) \rangle < \langle -G(x^*), j(x_n - x^* + tG(x^*)) \rangle + \varepsilon.$$

Consequently,

$$\mu_n \langle -G(x^*), j(x_n - x^*) \rangle \leq \mu_n \langle -G(x^*), j(x_n - x^* + tG(x^*)) \rangle + \varepsilon \leq \varepsilon.$$

Since ε is arbitrary, we have

$$\mu_n \langle -G(x^*), j(x_n - x^*) \rangle \leq 0.$$

Moreover, from the norm-to-norm uniform continuity of j on bounded sets, we obtain, that

$$\lim_{n \to \infty} \left(\langle -G(x^*), j(x_{n+1} - x^*) \rangle - \langle -G(x^*), j(x_n - x^*) \rangle \right) = 0.$$

Thus, the sequence $\{ \langle -G(x^*), j(x_n - x^*) \rangle \}$ satisfies the conditions of Lemma 7.7. Hence, we obtain that

$$\limsup_{n \to \infty} \langle -G(x^*), j(x_n - x^*) \rangle \leq 0.$$

Define

$$\varepsilon_n := \max \left\{ \langle -G(x^*), j(x_{n+1} - x^*) \rangle, 0 \right\}.$$

Then, $\lim \varepsilon_n = 0$, and $\langle -G(x^*), j(x_{n+1} - x^*) \rangle \leq \varepsilon_n$. From the recursion formula (7.11), and inequality (4.4), we have,

$$
\begin{aligned}
\|x_{n+1} - x^*\|^2 &= \|T_{[n+1]}^{\lambda_{n+1}} x_n - T_{[n+1]}^{\lambda_{n+1}} x^* + T_{[n+1]}^{\lambda_{n+1}} x^* - x^*\|^2 \\
&\leq \|T_{[n+1]}^{\lambda_{n+1}} x_n - T_{[n+1]}^{\lambda_{n+1}} x^*\|^2 + 2\lambda_{n+1} \delta \langle -G(x^*), j(x_{n+1} - x^*) \rangle \\
&\leq (1 - \lambda_{n+1} \alpha) \|x_n - x^*\|^2 + 2\lambda_{n+1} \delta \langle -G(x^*), j(x_{n+1} - x^*) \rangle
\end{aligned}
$$

and by Lemma 6.34, we have that $x_n \to x^*$ as $n \to \infty$. This completes the proof. $\qquad \square$

The following corollaries follow from Theorem 7.14.

Corollary 7.15 *Let $E = L_p$, $2 \leq p < \infty$. Let $T_i : E \to E$, $i = 1, 2, ..., r$ be a finite family of nonexpansive mappings with $K := \bigcap_{i=1}^{r} Fix(T_i) \neq \emptyset$. Let $G : E \to E$ be an $\eta-$strongly accretive map which is also $\kappa-$Lipschitzian. Let $\{\lambda_n\}$ be a sequence in $[0, 1]$ that satisfies conditions $C1$, $C2$ and $C6$ as in Theorem 7.14. For $\delta \in \left(0, \min \left\{ \frac{1}{2\eta}, \frac{2\eta}{(p-1)\kappa^2} \right\} \right)$, define a sequence $\{x_n\}$ iteratively in E by (7.11). Then, $\{x_n\}$ converges strongly to the unique solution x^* of the variational inequality $VI(G, K)$.*

Corollary 7.16 *Let H be a real Hilbert space. Let $T_i : H \to H$, $i = 1, 2, ..., r$ be a finite family of nonexpansive mappings with $K = \overset{r}{\underset{i=1}{\cap}} Fix(T_i) \neq \emptyset$. Let $G : H \to H$ be an $\eta-$strongly monotone map which is also $\kappa-$Lipschitzian. Let $\{\lambda_n\}$ be a sequence in $[0, 1]$ that satisfies conditions C1, C2 and C6 as in Theorem 7.15. For $\delta \in \left(0, \min \left\{ \frac{1}{2\eta}, \frac{2\eta}{\kappa^2} \right\} \right)$, define a sequence $\{x_n\}$ iteratively in H by (7.11). Then, $\{x_n\}$ converges strongly to the unique solution x^* of the variational inequality $VI(G, K)$.*

Theorem 7.17. *Let E be a $q-$uniformly smooth real Banach space with constant d_q, $q \geq 2$. Let $T : E \to E$ be a nonexpansive map. Assume that $K := F(T) = \{x \in E : Tx = x\} \neq \emptyset$. Let $G : E \to E$ be an $\eta-$strongly accretive and $\kappa-$Lipschitzian map. Let $\{\lambda_n\}$ be a sequence in $[0, 1]$ satisfying the following conditions:*

$$C1 : lim\lambda_n = 0; \quad C2 : \sum \lambda_n = \infty; \quad C5 : lim\frac{|\lambda_n - \lambda_{n+1}|}{\lambda_{n+1}} = 0.$$

For $\delta \in \left(0, \min \left\{ \frac{q}{4\eta}, \left(\frac{q\eta}{d_q \kappa^q} \right)^{\frac{1}{(q-1)}} \right\} \right)$, define a sequence $\{x_n\}$ iteratively in E by $x_0 \in E$,

$$x_{n+1} = T^{\lambda_{n+1}} x_n = Tx_n - \delta\lambda_{n+1}G(Tx_n), \quad n \geq 0. \qquad (7.16)$$

Then, $\{x_n\}$ converges strongly to the unique solution x^* of the variational inequality $VI(G, K)$.

Proof. Take $T_1 = T_2 = ... = T_r = T$ in Theorem 7.14 and the result follows.

The following corollaries follow from Theorem 7.17.

Corollary 7.18 *Let $E = L_p$, $2 \leq p < \infty$. Let $T : E \to E$ be a nonexpansive map. Assume that $K := F(T) = \{x \in E : Tx = x\} \neq \emptyset$. Let $G : E \to E$ be an $\eta-$strongly accretive and $\kappa-$Lipschitzian map. Let $\{\lambda_n\}$ be a sequence in $[0, 1]$ that satisfies conditions C1, C2 and C5 as in Theorem 7.17. For $\delta \in \left(0, \min \left\{ \frac{1}{2\eta}, \frac{2\eta}{(p-1)\kappa^2} \right\} \right)$, define a sequence $\{x_n\}$ iteratively in E by (7.16). Then, $\{x_n\}$ converges strongly to the unique solution x^* of the variational inequality $VI(G, K)$.*

Corollary 7.19 *Let H be a real Hilbert space. Let $T : H \to H$ be a nonexpansive map. Assume that $K := F(T) = \{x \in E : Tx = x\} \neq \emptyset$. Let $G : H \to H$ be an $\eta-$strongly monotone and $\kappa-$Lipschitzian map. Let $\{\lambda_n\}$ be a sequence in $[0, 1]$ that satisfies conditions C1, C2 and C5 as in Theorem 7.17. For $\delta \in \left(0, \min \left\{ \frac{1}{2\eta}, \frac{2\eta}{\kappa^2} \right\} \right)$, define a sequence $\{x_n\}$ iteratively in H by (7.16). Then, $\{x_n\}$ converges strongly to the unique solution x^* of the variational inequality $VI(G, K)$.*

7.4 Further Convergence Theorems

In the last section, we extended the results of Xu and Kim [518] to $q-$uniformly smooth Banach spaces, $q \geq 2$ under conditions $C1, C2$ and $C5$ or $C6$ (as in the result of Xu and Kim).

In this section, we introduce *new recursion formulas* and prove strong convergence theorems for the unique solution of the variational inequality problem $VI(K, S)$, *requiring only conditions C1 and C2 on the parameter sequence* $\{\lambda_n\}$. Furthermore in the case $T_i : E \to E \quad i = 1, 2, ..., r$ is a family of nonexpansive mappings with $K := \underset{i=1}{\overset{r}{\cap}} F(T_i) \neq \emptyset$, we prove a convergence theorem where condition $C6$ is replaced with $\underset{n \to \infty}{lim} \|T_{[n+2]}x_n - T_{[n+1]}x_n\| = 0$. An example satisfying this condition is given in Chidume and Ali [125]. All the theorems of this section are proved in $q-$uniformly smooth Banach spaces, $q \geq 2$. In particular, the theorems are applicable in L_p spaces, $2 \leq p < \infty$.

7.4.1 Convergence Theorems

We first prove the following lemma which will be central in the sequel.

Lemma 7.20. *Let E be a $q-$uniformly smooth real Banach space with constant d_q, $q \geq 2$, $T : E \to E$ be a nonexpansive mapping and $G : E \to E$ be an $\eta-$strongly accretive and $\kappa-$Lipschitzian mapping. For*

$$\delta \in \left(0, \min\left\{\frac{q}{4\sigma\eta}, (\frac{q\eta}{d_q\kappa^q})^{\frac{1}{(q-1)}}\right\}\right), \quad \sigma, \lambda \in (0, 1),$$

define a mapping $T^\lambda : E \to E$ by:

$$T^\lambda x := (1 - \sigma)x + \sigma[Tx - \lambda\delta G(Tx)], \quad x \in E.$$

Then, T^λ is a strict contraction. Furthermore,

$$\|T^\lambda x - T^\lambda y\| \leq (1 - \lambda\alpha)\|x - y\| \quad x, y \in E, \tag{7.17}$$

where

$$\alpha := \frac{q}{2} - \sqrt{\frac{q^2}{4} - \sigma\delta(q\eta - \delta^{q-1}d_q\kappa^q)} \quad \in (0, 1).$$

Proof. For $x, y \in E$, using the convexity of $\|.\|^q$ and inequality (5.10), we have,

$$\|T^\lambda x - T^\lambda y\|^q = \|(1 - \sigma)(x - y) + \sigma[Tx - Ty - \lambda\delta(G(Tx) - G(Ty))]\|^q$$

$$\leq (1 - \sigma)\|x - y\|^q + \sigma\Big[\|Tx - Ty\|^q - q\lambda\delta\langle G(Tx) - G(Ty), j_q(Tx - Ty)\rangle$$

$$+ d_q\lambda^q\delta^q\|G(Tx) - G(Ty)\|^q\Big]$$

$$\leq (1 - \sigma)\|x - y\|^q + \sigma\Big[\|Tx - Ty\|^q - q\lambda\delta\eta\|Tx - Ty\|^q$$

$$+ d_q\lambda^q\delta^q\kappa^q\|Tx - Ty\|^q\Big]$$

$$\leq \Big[1 - \sigma\lambda\delta(q\eta - d_q\lambda^{q-1}\delta^{q-1}\kappa^q)\Big]\|x - y\|^q$$

$$\leq \Big[1 - \sigma\lambda\delta(q\eta - d_q\delta^{q-1}\kappa^q)\Big]\|x - y\|^q.$$

Define

$$f(\lambda) := 1 - \sigma\lambda\delta(q\eta - d_q\delta^{q-1}\kappa^q) = (1 - \lambda\tau)^q, \quad \text{for some} \quad \tau \in (0,1), \quad \text{say.}$$

Then, there exists $\xi \in (0, \lambda)$ such that

$$1 - \sigma\lambda\delta(q\eta - d_q\delta^{q-1}\kappa^q) = 1 - q\tau\lambda + \frac{1}{2}q(q-1)(1 - \xi\tau)^{q-2}\lambda^2\tau^2,$$

and since $q \geq 2$, this implies

$$1 - \sigma\lambda\delta(q\eta - d_q\delta^{q-1}\kappa^q) \leq 1 - q\tau\lambda + \frac{1}{2}q(q-1)\lambda^2\tau^2,$$

which yields,

$$\tau^2 - q\tau + \sigma\delta\Big(q\eta - d_q\delta^{q-1}\kappa^q\Big) > 0,$$

since $\lambda \in \Big(0, \frac{2}{q(q-1)}\Big)$. Thus we have,

$$\tau \leq \frac{q}{2} - \sqrt{\frac{q^2}{4} - \sigma\delta(q\eta - \delta^{q-1}d_q\kappa^q)} \quad \in (0,1).$$

Set

$$\alpha := \frac{q}{2} - \sqrt{\frac{q^2}{4} - \sigma\delta(q\eta - \delta^{q-1}d_q\kappa^q)} \quad \in (0,1).$$

and the proof is complete. $\square$

Recall (Exercises 5.1, Problem 3) that in L_p spaces, $2 \leq p < \infty$, the following inequality holds: For each $x, y \in L_p$, $2 \leq p < \infty$,

$$\|x + y\|^2 \leq \|x\|^2 + 2\langle y, j(x)\rangle + (p - 1)\|y\|^2.$$

It then follows that by setting $q = 2$, $d_q = p - 1$ in Lemma 7.20, the following corollary is easily proved.

Corollary 7.21 Let $E = L_p$, $2 \leq p < \infty$, $T : E \to E$ be a nonexpansive mapping and $G : E \to E$ be an η−strongly accretive and κ−Lipschitzian *mapping. For* λ, $\sigma \in (0,1)$, *and* $\delta \in \left(0, \min\left\{\frac{1}{2\sigma\eta}, \frac{2\eta}{(p-1)\kappa^2}\right\}\right)$, *define a mapping* $T^\lambda : E \to E$ *by:*

$$T^\lambda x := (1 - \sigma)x + \sigma[Tx - \lambda\delta G(Tx)] \ \forall \ x \in E.$$

Then, T^λ *is a strict contraction. In particular,*

$$\|T^\lambda x - T^\lambda y\| \le (1 - \lambda\alpha)\|x - y\| \quad x, y \in H, \tag{7.18}$$

where $\alpha := 1 - \sqrt{1 - \sigma\delta(2\eta - (p-1)\delta\kappa^2)} \ \in (0,1)$.

We also have the following corollary.

Corollary 7.22 *Let* H *be a real Hilbert space,* $T : H \to H$ *be a nonexpansive mapping,* $G : H \to H$ *be an* $\eta-$*strongly monotone and* $\kappa-$*Lipschitzian mapping. For* $\lambda, \sigma \in (0,1)$ *and* $\delta \in \left(0, \min\left\{\frac{1}{2\sigma\eta}, \frac{2\eta}{\kappa^2}\right\}\right)$, *define a mapping* $T^\lambda : H \to H$ *by:*

$$T^\lambda x = (1 - \sigma)x + \sigma[Tx - \lambda\delta G(Tx)] \ \forall \ x \in H.$$

Then, T^λ *is a strict contraction. In particular,*

$$\|T^\lambda x - T^\lambda y\| \le (1 - \lambda\alpha)\|x - y\| \quad \forall \ x, y \in H, \tag{7.19}$$

where $\alpha := 1 - \sqrt{1 - \sigma\delta(2\eta - \delta\kappa^2)} \ \in (0,1)$.

Proof. Set $p = 2$ in Corollary 7.21 and the result follows.

We now prove the following convergence theorem.

Theorem 7.23. *Let* E *be a* $q-$*uniformly smooth real Banach space with constant* d_q, $q \ge 2$ *and* $T : E \to E$ *be a nonexpansive mapping. Assume* $K := \{x \in E : Tx = x\} \ne \emptyset$. *Let* $G : E \to E$ *be an* $\eta-$*strongly accretive and* $\kappa-$*Lipschitzian mapping. Let* $\{\lambda_n\}$ *be a sequence in* $[0,1]$ *satisfying the conditions:*

$$C1 : \lim\lambda_n = 0; \quad C2 : \sum \lambda_n = \infty.$$

For $\delta \in \left(0, \min\left\{\frac{q}{4\sigma\eta}, \left(\frac{q\eta}{d_q\kappa^q}\right)^{\frac{1}{(q-1)}}\right\}\right)$, $\sigma \in (0,1)$, *define a sequence* $\{x_n\}$ *iteratively in* E *by* $x_0 \in E$,

$$x_{n+1} = T^{\lambda_{n+1}}x_n = (1 - \sigma)x_n + \sigma[Tx_n - \delta\lambda_{n+1}G(Tx_n)], \quad n \ge 0. \tag{7.20}$$

Then, $\{x_n\}$ *converges strongly to the unique solution* x^* *of the variational inequality* $VI(G, K)$.

Proof. Let $x^* \in K := F(T)$, then the sequence $\{x_n\}$ satisfies

$$\|x_n - x^*\| \le \max\left\{\|x_0 - x^*\|, \frac{\delta}{\alpha}\|G(x^*)\|\right\}, \quad n \ge 0.$$

It is obvious that this is true for $n = 0$. Assume it is true for $n = k$ for some $k \in \mathbb{N}$. From the recursion formula (7.20), we have

$$
\begin{aligned}
\|x_{k+1} - x^*\| &= \|T^{\lambda_{k+1}} x_k - x^*\| \\
&\leq \|T^{\lambda_{k+1}} x_k - T^{\lambda_{k+1}} x^*\| + \|T^{\lambda_{k+1}} x^* - x^*\| \\
&\leq (1 - \lambda_{k+1}\alpha)\|x_k - x^*\| + \lambda_{k+1}\delta\|G(x^*)\| \\
&\leq max\left\{ \|x_0 - x^*\|, \frac{\delta}{\alpha}\|G(x^*)\| \right\},
\end{aligned}
$$

and the claim follows by induction. Thus the sequence $\{x_n\}$ is bounded and so are the sequences $\{Tx_n\}$ and $\{G(Tx_n)\}$.

Define two sequences $\{\beta_n\}$ and $\{y_n\}$ by $\beta_n := (1 - \sigma)\lambda_{n+1} + \sigma$ and $y_n := \frac{x_{n+1} - x_n + \beta_n x_n}{\beta_n}$. Then,

$$
y_n = \frac{(1 - \sigma)\lambda_{n+1}x_n + \sigma[Tx_n - \lambda_{n+1}\delta G(Tx_n)]}{\beta_n}.
$$

Observe that $\{y_n\}$ is bounded and that

$$
\begin{aligned}
\|y_{n+1} - y_n\| - \|x_{n+1} - x_n\| &\leq \left| \frac{\sigma}{\beta_{n+1}} - 1 \right| \|x_{n+1} - x_n\| \\
&+ \left| \frac{\sigma}{\beta_{n+1}} - \frac{\sigma}{\beta_n} \right| \|Tx_n\| + \frac{\lambda_{n+2}(1 - \sigma)}{\beta_{n+1}}\|x_{n+1} - x_n\| \\
&+ (1 - \sigma)\left| \frac{\lambda_{n+2}}{\beta_{n+1}} - \frac{\lambda_{n+1}}{\beta_n} \right| \|x_n\| + \frac{\lambda_{n+1}\sigma\delta}{\beta_n}\|G(Tx_n) - G(Tx_{n+1})\| \\
&+ \sigma\delta\left| \frac{\lambda_{n+1}}{\beta_n} - \frac{\lambda_{n+2}}{\beta_{n+1}} \right| \|G(Tx_{n+1})\|.
\end{aligned}
$$

This implies, $\underset{n\to\infty}{limsup}\ (\|y_{n+1} - y_n\| - \|x_{n+1} - x_n\|) \leq 0,$, and therefore by Lemma 6.33,

$$
\lim_{n\to\infty} \|y_n - x_n\| = 0.
$$

Hence,

$$
\|x_{n+1} - x_n\| = \beta_n\|y_n - x_n\| \to 0 \quad \text{as} \quad n \to \infty. \tag{7.21}
$$

From the recursion formula (7.20), we have that

$$
\sigma\|x_{n+1} - Tx_n\| \leq (1 - \sigma)\|x_{n+1} - x_n\| + \lambda_{n+1}\sigma\delta\|G(Tx_n)\| \to 0 \quad \text{as} \quad n \to \infty.
$$

which implies,

$$
\|x_{n+1} - Tx_n\| \to 0 \quad \text{as} \quad n \to \infty. \tag{7.22}
$$

From (7.21) and (7.22) we have

$$
\|x_n - Tx_n\| \leq \|x_n - x_{n+1}\| + \|x_{n+1} - Tx_n\| \to 0 \quad \text{as} \quad n \to \infty. \tag{7.23}
$$

We now prove that

$$\limsup_{n\to\infty}\langle -G(x^*), j(x_{n+1} - x^*)\rangle \leq 0.$$

Define a map $\phi : E \to \mathbb{R}$ by

$$\phi(x) = \mu_n\|x_n - x\|^2 \quad \forall\, x \in E,$$

where μ_n is a Banach limit for each n. Then, $\phi(x) \to \infty$ as $\|x\| \to \infty$, ϕ is continuous and convex, so as E is reflexive, it follows from Lemma 7.9 that there exists $y^* \in E$ such that $\phi(y^*) = \min_{u\in E} \phi(u)$. Hence, the set

$$K^* := \left\{x \in E : \phi(x) = \min_{u\in E}\phi(u)\right\} \neq \emptyset.$$

We now show T has a fixed point in K^*. We know

$$\lim_{n\to\infty}\|x_n - Tx_n\| = 0. \tag{7.24}$$

We shall make use of Lemma 7.8. If x is in K^* and $y := w-\lim_j T^{m_j} x$ belongs to the weak $\omega - limit$ set $\omega_w(x)$ of T at x, then, from the w-l.s.c. (since φ is l.s.c. and convex) of φ and equation (7.24), we have,

$$\varphi(y) \leq \liminf_j \varphi\left(T^{m_j} x\right) \leq \limsup_m \varphi\left(T^m x\right)$$
$$= \limsup_m \left(\mu_n\|x_n - T^m x\|^2\right)$$
$$= \limsup_m \left(\mu_n\|x_n - T^m x_n + T^m x_n - T^m x\|^2\right)$$
$$\leq \limsup_m \left(\mu_n\|T^m x_n - T^m x\|^2\right) \leq \limsup_m \left(\mu_n\|x_n - x\|^2\right) = \varphi(x)$$
$$= \inf_{u\in E}\varphi(u).$$

So, $y \in K^*$.

By Lemma 7.8, $K^* \cap K \neq \emptyset$. Let $x^* \in K^* \cap K$ and let $t \in (0,1)$. Then, it follows that $\phi(x^*) \leq \phi(x^* - tG(x^*))$ and using inequality (4.4), we obtain that

$$\|x_n - x^* + tG(x^*)\|^2 \leq \|x_n - x^*\|^2 + 2t\langle G(x^*), j(x_n - x^* + tG(x^*))\rangle$$

which implies,

$$\mu_n\langle -G(x^*), j(x_n - x^* + tG(x^*))\rangle \leq 0.$$

The rest now follows exactly as in the proof of Theorem 7.14 to yield that $x_n \to x^*$ as $n \to \infty$. This completes the proof. $\qquad\square$

The following corollaries follow from Theorem 7.23.

Corollary 7.24 *Let* $E = L_p$, $2 \leq p < \infty$, $T : E \to E$ *be a nonexpansive mapping. Assume* $K := \{x \in E : Tx = x\} \neq \emptyset$. *Let* $G : E \to E$ *be an* $\eta-$*strongly accretive and* $\kappa-$*Lipschitzian mapping. Let* $\{\lambda_n\}$ *be a sequence in* $[0,1]$ *that satisfies conditions C1 and C2 as in Theorem 7.23. For* $\delta \in \left(0, \min\left\{\frac{1}{2\sigma\eta}, \frac{2\eta}{(p-1)\kappa^2}\right\}\right)$, $\sigma \in (0,1)$, *define a sequence* $\{x_n\}$ *iteratively in* E *by* (7.20). *Then,* $\{x_n\}$ *converges strongly to the unique solution* x^* *of the variational inequality problem* $VI(G, K)$.

Corollary 7.25 *Let* H *be a real Hilbert space,* $T : H \to H$ *be a nonexpansive mapping. Assume* $K := \{x \in H : Tx = x\} \neq \emptyset$. *Let* $G : H \to H$ *be an* $\eta-$*strongly monotone* $\kappa-$*Lipschitzian mapping. Further, let* $\{\lambda_n\}$ *be a sequence in* $[0,1]$ *that satisfies conditions C1 and C2 as in Theorem 7.23. For* $\delta \in \left(0, \min\left\{\frac{1}{2\sigma\eta}, \frac{2\eta}{\kappa^2}\right\}\right)$, $\sigma \in (0,1)$, *define a sequence* $\{x_n\}$ *iteratively in* H *by* (7.20). *Then,* $\{x_n\}$ *converges strongly to the unique solution* x^* *of the variational inequality problem* $VI(G, K)$.

Finally, we prove the following theorem for a finite family of nonexpansive mappings.

Theorem 7.26. *Let* E *be a* $q-$*uniformly smooth real Banach space with constant* d_q, $q \geq 2$, $T_i : E \to E$, $i = 1, 2, ..., r$ *be a finite family of nonexpansive mappings with* $K := \bigcap_{i=1}^{r} F(T_i) \neq \emptyset$. *Let* $G : E \to E$ *be an* $\eta-$*strongly accretive and* $\kappa-$*Lipschitzian mapping, and* $\{\lambda_n\}$ *a sequence in* $[0,1]$ *satisfying the conditions:*

$$C1 : \lim \lambda_n = 0; \quad C2 : \sum \lambda_n = \infty.$$

For $\delta \in \left(0, \min\left\{\frac{q}{4\sigma\eta}, \left(\frac{q\eta}{d_q\kappa^q}\right)^{\frac{1}{(q-1)}}\right\}\right)$, $\sigma \in (0,1)$, *define a sequence* $\{x_n\}$ *iteratively in* E *by* $x_0 \in E$,

$$x_{n+1} = T_{[n+1]}^{\lambda_{n+1}} x_n = (1-\sigma)x_n + \sigma[T_{[n+1]}x_n - \delta\lambda_n G(T_{[n+1]}x_n)], \quad n \geq 0, \quad (7.25)$$

where $T_{[n]} = T_{n \bmod r}$. *Assume also that*

$$K = F(T_r T_{r-1}...T_1) = F(T_1 T_r...T_2) = ... = F(T_{r-1}T_{r-2}...T_r)$$

and $\lim_{n \to \infty} \|T_{[n+2]}x_n - T_{[n+1]}x_n\| = 0$. *Then,* $\{x_n\}$ *converges strongly to the unique solution* x^* *of the variational inequality problem* $VI(G, K)$.

Proof. Let $x^* \in K$, then the sequence $\{x_n\}$ satisfies

$$\|x_n - x^*\| \leq \max\left\{\|x_0 - x^*\|, \frac{\delta}{\alpha}\|G(x^*)\|\right\}, \quad n \geq 0.$$

It is obvious that this is true for $n = 0$. Assume it is true for $n = k$ for some $k \in \mathbb{N}$.

From the recursion formula (7.25), we have

$$\begin{aligned}
\|x_{k+1} - x^*\| &= \|T_{[k+1]}^{\lambda_{k+1}} x_k - x^*\| \\
&\leq \|T_{[k+1]}^{\lambda_{k+1}} x_k - T_{[k+1]}^{\lambda_{k+1}} x^*\| + \|T_{[k+1]}^{\lambda_{k+1}} x^* - x^*\| \\
&\leq (1 - \lambda_{k+1}\alpha)\|x_k - x^*\| + \lambda_{k+1}\delta\|G(x^*)\| \\
&\leq max\left\{ \|x_0 - x^*\|, \frac{\delta}{\alpha}\|G(x^*)\| \right\},
\end{aligned}$$

and the claim follows by induction. Thus the sequence $\{x_n\}$ is bounded and so are $\{T_{[n]}x_n\}$ and $\{G(T_{[n]}x_n)\}$.

Define two sequences $\{\beta_n\}$ and $\{y_n\}$ by $\beta_n := (1-\sigma)\lambda_{n+1} + \sigma$ and $y_n := \frac{x_{n+1} - x_n + \beta_n x_n}{\beta_n}$. Then,

$$y_n = \frac{(1-\sigma)\lambda_{n+1}x_n + \sigma[T_{[n+1]}x_n - \lambda_{n+1}\delta G(T_{[n+1]}x_n)]}{\beta_n}.$$

Observe that $\{y_n\}$ is bounded and that

$$\begin{aligned}
\|y_{n+1} - y_n\| - \|x_{n+1} - x_n\| &\leq \left| \frac{\sigma}{\beta_{n+1}} - 1 \right| \|x_{n+1} - x_n\| \\
&+ \frac{\sigma}{\beta_{n+1}}\|T_{[n+2]}x_n - T_{[n+1]}x_n\| + \left| \frac{\sigma}{\beta_{n+1}} - \frac{\sigma}{\beta_n} \right| \|T_{[n+1]}x_n\| \\
&+ \frac{\lambda_{n+2}(1-\sigma)}{\beta_{n+1}}\|x_{n+1} - x_n\| + (1-\sigma)\left| \frac{\lambda_{n+2}}{\beta_{n+1}} - \frac{\lambda_{n+1}}{\beta_n} \right| \|x_n\| \\
&+ \frac{\lambda_{n+1}\sigma\delta}{\beta_n}\|G(T_{[n+1]}x_n) - G(T_{[n+2]}x_{n+1})\| \\
&+ \sigma\delta\left| \frac{\lambda_{n+1}}{\beta_n} - \frac{\lambda_{n+2}}{\beta_{n+1}} \right| \|G(T_{[n+2]}x_{n+1})\|.
\end{aligned}$$

This implies,

$$\limsup_{n \to \infty} (\|y_{n+1} - y_n\| - \|x_{n+1} - x_n\|) \leq 0,$$

and by Lemma 6.33, $\lim_{n \to \infty} \|y_n - x_n\| = 0$. Hence,

$$\|x_{n+1} - x_n\| = \beta_n\|y_n - x_n\| \to 0 \tag{7.26}$$

as $n \to \infty$. From the recursion formula (7.25), we have that

$$\sigma\|x_{n+1} - T_{[n+1]}x_n\| \leq (1-\sigma)\|x_{n+1} - x_n\| + \lambda_{n+1}\sigma\delta\|G(T_{[n+1]}x_n)\| \to 0,$$

as $n \to \infty$, which implies,

$$\|x_{n+1} - T_{[n+1]}x_n\| \to 0 \quad \text{as} \quad n \to \infty. \tag{7.27}$$

Note that from (7.26) and (7.27) we have

$$\|x_n - T_{[n+1]}x_n\| \le \|x_n - x_{n+1}\| + \|x_{n+1} - T_{[n+1]}x_n\| \to 0 \text{ as } n \to \infty. \quad (7.28)$$

Also,

$$\|x_{n+r} - x_n\| \le \|x_{n+r} - x_{n+r-1}\| + \|x_{n+r-1} - x_{n+r-2}\| + \cdots + \|x_{n+1} - x_n\|$$

and so,

$$\|x_{n+r} - x_n\| \to 0 \text{ as } n \to \infty. \quad (7.29)$$

Using the fact that T_i is nonexpansive for each i, we obtain the following finite table:

$$x_{n+r} - T_{n+r}x_{n+r-1} \to 0 \quad as \quad n \to \infty;$$

$$T_{n+r}x_{n+r-1} - T_{n+r}T_{n+r-1}x_{n+r-2} \to 0 \quad as \quad n \to \infty;$$

$$\vdots$$

$$T_{n+r}T_{n+r-1}\cdots T_{n+2}x_{n+1} - T_{n+r}T_{n+r-1}\cdots T_{n+2}T_{n+1}x_n \to 0 \quad as \quad n \to \infty;$$

and adding up the table yields

$$x_{n+r} - T_{n+r}T_{n+r-1}\cdots T_{n+1}x_n \to 0 \quad as \quad n \to \infty.$$

Using this and (7.29) we get that $\lim_{n\to\infty} \|x_n - T_{n+r}T_{n+r-1}\cdots T_{n+1}x_n\| = 0$.

Carrying out similar arguments as in the proof of Theorem 7.23, we easily get that

$$\limsup_{n\to\infty} \langle -G(x^*), j(x_{n+1} - x^*) \rangle \le 0.$$

From the recursion formula (7.25), and inequality (4.4), we have

$$\|x_{n+1} - x^*\|^2 = \|T_{[n+1]}^{\lambda_{n+1}}x_n - T_{[n]}^{\lambda_{n+1}}x^* + T_{[n+1]}^{\lambda_{n+1}}x^* - x^*\|^2$$

$$\le \|T_{[n+1]}^{\lambda_{n+1}}x_n - T_{[n+1]}^{\lambda_{n+1}}x^*\|^2 + 2\lambda_{n+1}\sigma\delta\langle -G(x^*), j(x_{n+1} - x^*)\rangle$$

$$\le (1 - \lambda_{n+1}\alpha)\|x_n - x^*\|^2 + 2\lambda_{n+1}\sigma\delta\langle -G(x^*), j(x_{n+1} - x^*)\rangle$$

which, by Lemma 6.34, gives that $x_n \to x^*$ as $n \to \infty$, completing the proof. $\square$

The following corollaries follow from Theorem 7.26.

Corollary 7.27 *Let* $E = L_p$, $2 \le p < \infty$, $T_i : E \to E$, $i = 1, 2, ..., r$ *be a finite family of nonexpansive mappings with* $K := \bigcap_{i=1}^{r} F(T_i) \ne \emptyset$. *Let* $G : E \to E$ *be an* $\eta-$*strongly accretive and* $\kappa-$*Lipschitzian mapping. Let* $\{\lambda_n\}$ *be a sequence in* $[0, 1]$ *that satisfies conditions C1 and C2 as in Theorem 7.26 and let* $\lim_{n\to\infty} \|T_{[n+2]}x_n - T_{[n+1]}x_n\| = 0$. *For* $\delta \in \left(0, \min\left\{\frac{1}{2\sigma\eta}, \frac{2\eta}{(p-1)\kappa^2}\right\}\right)$,

$\sigma \in (0, 1)$, *define a sequence* $\{x_n\}$ *iteratively in* E *by* (7.25). *Then,* $\{x_n\}$ *converges strongly to the unique solution* x^* *of the variational inequality problem* $VI(G, K)$.

Corollary 7.28 *Let* H *be a real Hilbert space,* $T_i : H \rightarrow H$, $i = 1, 2, ..., r$ *be a finite family of nonexpansive mappings with* $K := \overset{r}{\underset{i=1}{\cap}} F(T_i) \neq \emptyset$. *Let* $G : H \rightarrow H$ *be an* $\eta-$*strongly monotone and* $\kappa-$*Lipschitzian mapping. Let* $\{\lambda_n\}$ *be a sequence in* $[0, 1]$ *that satisfies conditions* $C1$ *and* $C2$ *as in Theorem 7.26 and let* $\lim\limits_{n \to \infty} \|T_{[n+2]}x_n - T_{[n+1]}x_n\| = 0$. *For* $\delta \in \left(0, \min\left\{\frac{1}{2\sigma\eta}, \frac{2\eta}{\kappa^2}\right\}\right)$, $\sigma \in (0, 1)$, *define a sequence* $\{x_n\}$ *iteratively in* H *by* (7.25). *Then,* $\{x_n\}$ *converges strongly to the variational inequality problem* $VI(G, K)$.

7.5 The Case of L_p spaces, $1 < p \leq 2$

We begin with the following definition. A Banach space E is called a *lower weak parallelogram space with constant* $b \geq 0$ or, briefly, E is $LWP(b)$, in the terminology of Bynum [61] if

$$\|x + y\|^2 + b\|x - y\|^2 \leq 2(\|x\|^2 + \|y\|^2) \tag{7.30}$$

holds for all $x, y \in E$. It is proved in [61] that l_p space, $1 < p \leq 2$, is a lower weak parallelogram space with $(p - 1)$ as the largest number b for which (7.30) holds. Furthermore, if L_p, $(1 < p \leq 2)$, has at least two disjoint sets of positive finite measure, then it is a lower weak parallelogram space with $(p - 1)$ as the largest number b for which (7.30) holds. We shall assume, without loss of generality, that L_p, $(1 < p \leq 2)$, has at least two disjoint sets of positive finite measure. In the sequel, we shall state all theorems and lemmas only for L_p spaces, $1 < p \leq 2$, with the understanding that they also hold for l_p spaces, $1 < p \leq 2$.

In terms of the normalized duality mapping, Bynum [61] proved that a real Banach space is a lower weak parallelogram space if and only if for each $x, y \in E$ and $f \in J(x)$, the following inequality holds:

$$\|x + y\|^2 \geq \|x\|^2 + b\|y\|^2 + 2\langle y, f \rangle. \tag{7.31}$$

In particular, for $E = L_p$, $1 < p \leq 2$, the following inequality holds:

$$\|x + y\|^2 \geq \|x\|^2 + (p - 1)\|y\|^2 + 2\langle y, j(x) \rangle \ \forall \ x, y \in E. \tag{7.32}$$

We now obtain the following lemmas which will be central in the sequel.

Lemma 7.29. *Let* $E = L_p, 1 < p \leq 2$. *Then, for all* $x, y \in E$, *the following inequality holds:*

$$(p-1)\|x+y\|^2 \leq \|x\|^2 + 2\langle y, j(x)\rangle + \|y\|^2. \tag{7.33}$$

Proof. Observe first that E is smooth so that the normalized duality map on E is single-valued. Now, replacing x by $(-x)$ and y by $(x+y)$ in inequality (7.32), we obtain $\|y\|^2 \geq \|x\|^2 + 2\langle x+y, j(-x)\rangle + (p-1)\|x+y\|^2$, so that

$$(p-1)\|x+y\|^2 \leq \|y\|^2 - \|x\|^2 + 2\langle x+y, j(x)\rangle$$
$$= \|x\|^2 + 2\langle y, j(x)\rangle + \|y\|^2,$$

establishing the lemma.　　　　　　　　　　　　　　　　　　　　　□

Lemma 7.30. *Let $E = L_p$, $1 < p \leq 2$, $T : E \to E$ be a nonexpansive mapping and $G : E \to E$ be an $\eta-$strongly accretive and κ- Lipschitzian mapping. For,*

$$\lambda \in (0,1), \quad \sigma \in (0,1), \quad \delta \in \left(0, \; min\left\{\frac{2\eta}{\kappa^2}, \; \frac{(p-1)^2}{2\eta\sigma}\right\}\right),$$

define a map $T^\lambda : E \to E$ by

$$T^\lambda x := (1-\sigma)x + \sigma\Big[Tx - \lambda\delta G(Tx)\Big], \quad x \in E.$$

Then, T^λ is a strict contraction. Furthermore,

$$\|T^\lambda x - T^\lambda y\| \leq (1 - \lambda\alpha)\|x - y\| \quad \forall \; x,y \in E, \tag{7.34}$$

where
$$\alpha := (p-1) - \sqrt{(p-1)^2 - \sigma\delta(2\eta - \delta\kappa^2)} \quad \in (0,1).$$

Proof. For $x, y \in E$, using the convexity of $\|.\|^2$, and Lemma 7.29, we have,

$$\|T^\lambda x - T^\lambda y\|^2 = \|(1-\sigma)(x-y) + \sigma[Tx - Ty - \lambda\delta(G(Tx) - G(Ty))]\|^2$$

$$\leq (1-\sigma)\|x-y\|^2 + \frac{\sigma}{(p-1)}\Big[\|Tx - Ty\|^2$$

$$- 2\lambda\delta\langle G(Tx) - G(Ty), j(Tx - Ty)\rangle + \lambda^2\delta^2\|G(Tx) - G(Ty)\|^2\Big]$$

$$\leq (1-\sigma)\|x-y\|^2 + \frac{\sigma}{(p-1)}\Big[\|Tx - Ty\|^2 - 2\lambda\delta\eta\|Tx - Ty\|^2$$

$$+ \lambda^2\delta^2\kappa^2\|Tx - Ty\|^2\Big]$$

$$\leq \left[1 + \sigma\left(\frac{1}{p-1} - 1\right) - \frac{2\sigma\lambda\delta\eta}{(p-1)} + \frac{\sigma\lambda\delta^2\kappa^2}{(p-1)}\right]\|x-y\|^2, \quad (\lambda<1).$$

Define

$$f(\lambda) := 1 + \sigma\Big(\frac{1}{p-1} - 1\Big) - \frac{2\sigma\lambda\delta\eta}{(p-1)} + \frac{\sigma\lambda\delta^2\kappa^2}{(p-1)} = (1 - \lambda\tau)^2,$$

for some $\tau \in (0,1)$, say. Since $\Big(\frac{1}{p-1} - 1\Big) > 0$, and $\lambda(p-1) \le 1$, this implies,

$$-\frac{2\sigma\delta\eta}{(p-1)} + \frac{\sigma\delta^2\kappa^2}{(p-1)} \le -2\tau + \tau^2,$$

which yields

$$\tau^2 - 2(p-1)\tau + 2\sigma\delta\eta - \sigma\delta^2\kappa^2 \ge 0,$$

implying that

$$\tau \le (p-1) - \sqrt{(p-1)^2 - \sigma\delta(2\eta - \delta\kappa^2)} \quad \in (0,1).$$

Now set

$$\alpha := (p-1) - \sqrt{(p-1)^2 - \sigma\delta(2\eta - \delta\kappa^2)} \quad \in (0,1),$$

and the proof is complete. $\qquad\square$

Corollary 7.31 *Let H be a real Hilbert space, $T : H \to H$ be a nonexpansive mapping, $G : H \to H$ be an η-strongly κ-Lipschitzian mapping. For $\lambda \in (0,1)$ and $\delta \in \Big(0, \min\Big\{\frac{1}{2\sigma\eta}, \frac{2\eta}{\kappa^2}\Big\}\Big)$, $\sigma \in (0,1)$, define a mapping $T^\lambda : H \to H$ by: $T^\lambda x = (1 - \sigma)x + \sigma[Tx - \lambda\delta G(Tx)]$, $x \in H$. Then, T^λ is a strict contraction. In particular,*

$$\|T^\lambda x - T^\lambda y\| \le (1 - \lambda\alpha)\|x - y\| \quad \forall \ x, y \in H, \qquad (7.35)$$

where $\alpha := 1 - \sqrt{1 - \sigma\delta(2\eta - \delta\kappa^2)} \quad \in (0,1)$.

Proof. Set $p = 2$ in Lemma 7.30 and the result follows.

We now prove the following theorem.

Theorem 7.32. *Let $E = L_p$, $1 < p \le 2$, $T : E \to E$ be a nonexpansive mapping. Assume $K := \{x \in E : Tx = x\} \ne \emptyset$. Let $G : E \to E$ be an η-strongly accretive and κ-Lipschitzian mapping. Let $\{\lambda_n\}$ be a sequence in $[0,1]$ satisfying the conditions:*

$$C1 : \lim\lambda_n = 0; \quad C2 : \sum\lambda_n = \infty.$$

For $\sigma \in (0,1)$, and $\delta \in \Big(0, \min\Big\{\frac{2\eta}{\kappa^2}, \frac{(p-1)^2}{2\eta\sigma}\Big\}\Big)$, define a sequence $\{x_n\}$ iteratively in E by $x_0 \in E$,

$$x_{n+1} = T^{\lambda_{n+1}}x_n = (1 - \sigma)x_n + \sigma[Tx_n - \delta\lambda_{n+1}G(Tx_n)], \quad n \ge 0. \quad (7.36)$$

Then, $\{x_n\}$ converges strongly to the unique solution x^* of the variational inequality problem $VI(G, K)$.

Proof. This follows exactly as in the proof of Theorem 7.23, using Lemma 7.30.

The following corollary follows from Theorem 7.32.

Corollary 7.33 *Let H be a real Hilbert space, $T : H \to H$ be a nonexpansive mapping. Assume $K := \{x \in E : Tx = x\} \neq \emptyset$. Let $G : H \to H$ be an η-strongly monotone κ-Lipschitzian mapping. Let $\{\lambda_n\}$ be a sequence in $[0, 1]$ that satisfies conditions $C1$ and $C2$ as in Theorem 7.23. For $\delta \in \left(0, \min\left\{\frac{1}{2\sigma\eta}, \frac{2\eta}{\kappa^2}\right\}\right)$, $\sigma \in (0, 1)$, define a sequence $\{x_n\}$ iteratively in H by (7.20). Then, $\{x_n\}$ converges strongly to the unique solution x^* of the variational inequality $VI(G, K)$.*

Following the method of Section 7.4, the following theorem and corollary are easily proved.

Theorem 7.34. *Let $E = L_p$, $1 < p \leq 2$, and $T_i : E \to E, i = 1, 2, ..., r$ be a finite family of nonexpansive mappings with $K := \overset{r}{\underset{i=1}{\cap}} F(T_i) \neq \emptyset$. Let $G : E \to E$ be an η-strongly accretive and κ-Lipschitzian mapping. Let $\{\lambda_n\}$ be a sequence in $[0, 1]$ satisfying the conditions: $C1 : \lim \lambda_n = 0$;　$C2 : \sum \lambda_n = \infty$. For $\sigma \in (0, 1)$, and $\delta \in \left(0, \min\left\{\frac{2\eta}{\kappa^2}, \frac{(p-1)^2}{2\eta\sigma}\right\}\right)$, define a sequence $\{x_n\}$ iteratively in E by $x_0 \in E$,*

$$x_{n+1} = T_{[n+1]}^{\lambda_{n+1}}x_n = (1-\sigma)x_n + \sigma[T_{[n+1]}x_n - \delta\lambda_n G(T_{[n+1]}x_n)], \quad n \geq 0, \quad (7.37)$$

where $T_{[n]} = T_{n \bmod r}$. Assume also that

$$K = F(T_r T_{r-1}...T_1) = F(T_1 T_r...T_2) = ... = F(T_{r-1}T_{r-2}...T_r)$$

and $\lim\limits_{n \to \infty} \|T_{[n+2]}x_n - T_{[n+1]}x_n\| = 0$. Then, $\{x_n\}$ converges strongly to the unique solution x^* of the variational inequality problem $VI(G, K)$.

Corollary 7.35 *Let H be a real Hilbert space, $T_i : H \to H, \quad i = 1, 2, ..., r$ be a finite family of nonexpansive mappings with $K := \overset{r}{\underset{i=1}{\cap}} F(T_i) \neq \emptyset$. Let $G : H \to H$ be an η-strongly monotone and κ-Lipschitzian mapping. Let $\{\lambda_n\}$ be a sequence in $[0, 1]$ that satisfies conditions $C1$ and $C2$ as in Theorem 7.34 and let $\lim\limits_{n \to \infty} \|T_{[n+2]}x_n - T_{[n+1]}x_n\| = 0$. For $\delta \in \left(0, \min\left\{\frac{1}{2\sigma\eta}, \frac{2\eta}{\kappa^2}\right\}\right)$, $\sigma \in (0, 1)$, define a sequence $\{x_n\}$ iteratively in H by (7.37). Then, $\{x_n\}$ converges strongly to the unique solution x^* of the variational inequality problem $VI(G, K)$.*

Remark 7.36. The theorems of this section are extensions of the results of Yamada [531], Wang [499], Xu and Kim [518], Zeng and Yao [553] from real Hilbert spaces to L_p spaces. $1 < p \le 2$. Moreover, in this general setting, the iteration parameter is required to satisfy only conditions $C1$ and $C2$.

7.6 Historical Remarks

All the theorems of this chapter are due to Chidume *et al.* [133], [134].

Chapter 8
Iterative Methods for Zeros
of $\Phi-$Accretive-Type Operators

8.1 Introduction and Preliminaries

In this chapter, we continue to apply the Mann iteration method introduced in Chapter 6. Here, we use it to approximate the zeros of Φ-*strongly accretive* operators (and to approximate fixed points of Φ-*strong pseudo-contractions*).

The pseudo-contractions are important generalizations of the nonexpansive maps and are intimately connected with the important class of nonlinear *accretive* operators defined earlier (Chapter 7). This connection and the importance of accretive operators will become clear in what follows. Let E be a real normed space with dual E^*. A mapping $T : E \to E$ is called *strongly pseudo-contractive* if for all $x, y \in E$, the following inequality holds:

$$||x - y|| \le ||(1 + r)(x - y) - rt(Tx - Ty)|| \qquad (8.1)$$

for *all* $r > 0$ and *some* $t > 1$. If $t = 1$ in inequality (8.1), then T is called *pseudo-contractive*. As a consequence of Proposition 3.11, it follows from inequality (8.1) that T is strongly pseudo-contractive if and only if

$$\langle (I - T)x - (I - T)y, j(x - y) \rangle \ge k||x - y||^2 \qquad (8.2)$$

holds for all $x, y \in E$ and for some $j(x - y) \in J(x - y)$, where $k = \frac{1}{t}(t - 1) \in (0, 1)$. Consequently, it follows easily (again from Proposition 3.11 and inequality (8.2)) that T is strongly pseudo-contractive if and only if the following inequality holds:

$$||x - y|| \le ||x - y + s[(I - T - kI)x - (I - T - kI)y]|| \qquad (8.3)$$

for all $x, y \in E$ and for *all* $s > 0$. Again, it follows from inequality (8.2) that T is strongly pseudo-contractive if and only if $\langle Tx - Ty, j(x - y) \rangle \le \gamma ||x - y||^2$

C. Chidume, *Geometric Properties of Banach Spaces and Nonlinear Iterations*,
Lecture Notes in Mathematics 1965,
© Springer-Verlag London Limited 2009

holds for all $x, y \in E$ and $\gamma = \frac{1}{t} \in (0, 1)$, and is pseudo-contractive (t=1) if and only if

$$\langle Tx - Ty, j(x - y) \rangle \leq ||x - y||^2. \tag{8.4}$$

It is easy to see that every nonexpansive map is pseudo-contractive. We note immediately that pseudo-contractive maps are not necessarily continuous as can be seen from the following example.

Example 8.1. Let $T : [0, 1] \to \mathbb{R}$ be defined by

$$Tx = \begin{cases} x - \frac{1}{2}, & \text{if } x \in [0, \frac{1}{2}), \\ x - 1, & \text{if } x \in (\frac{1}{2}, 1]. \end{cases}$$

Then T is pseudo-contractive, but is neither nonexpansive nor continuous.

We also have the following example.

Example 8.2. Let $T : [0, 1] \to \mathbb{R}$ be defined by $T(x) = 1 - x^{\frac{2}{3}}$. Then T is a *continuous* pseudo-contraction which is not nonexpansive.

It is now clear that the class of pseudo-contractive maps properly contains the class of nonexpansive maps. For Hilbert spaces, the form of pseudo-contractive maps is given in the following proposition.

Proposition 8.3. *Let H be a real Hilbert space. Then, T is strongly pseudo-contractive if it satisfies the following inequality*

$$||Tx - Ty||^2 \leq ||x - y||^2 + k||(I - T)x - (I - T)y||^2, \tag{8.5}$$

for all $x, y \in H$, and for some $k \in (0, 1)$; and T is called pseudo-contractive if $k = 1$.

Observe that if $F(T) := \{x \in H : Tx = x\} \neq \emptyset$, then inequality (8.5) reduces, for the case $k = 1$, to the following one: $||Tx - x^*||^2 \leq ||x - x^*||^2 + ||x - Tx||^2$, for all $x \in H, x^* \in F(T)$. This now leads to our next definition.

Definition 8.4. A mapping $T : D(T) \subseteq H \to H$ is called *hemi-contractive* (see e.g., Qihou, [392]) if $F(T) \neq \emptyset$ and

$$||Tx - x^*||^2 \leq ||x - x^*||^2 + ||x - Tx||^2 \tag{8.6}$$

holds for all $x \in D(T), x^* \in F(T)$. Here H is a real Hilbert space.

For arbitrary real normed spaces, we have the following definition.

Definition 8.5. Let E be an arbitrary real normed linear space. A mapping $T : D(T) \subseteq E \to E$ is called *strongly hemi-contractive* if $F(T) \neq \emptyset$, and there exists $t > 1$ such that for all $r > 0$,

$$||x - x^*|| \leq ||(1 + r)(x - x^*) - rt(Tx - x^*)|| \tag{8.7}$$

holds for all $x \in D(T), x^* \in F(T)$. If $t = 1$, T is called *hemi-contractive*.

If $E = H$, a Hilbert space, and $t = 1$, it is easy to verify that inequality (8.7) is equivalent to inequality (8.6). It is clear that pseudo-contractive maps with nonempty fixed point sets are hemi-contractive, and that any strongly pseudo-contractive mapping T with $F(T) \neq \emptyset$ is strongly hemi-contractive. Clearly, the class of hemi-contractive maps includes the class of *quasi-nonexpansive* maps. The following example shows that the class of strong hemi-contractions properly includes the class of strong pseudo-contractions with fixed points.

Example 8.6. Take $E = \mathbb{R}$ with the usual norm, $K = [0, 2\pi]$. Define $T : K \to \mathbb{R}$ by

$$Tx = \frac{2}{3}x \cos x, \tag{8.8}$$

for each $x \in K$. Then, T is a Lipschitz strongly hemi-contractive map which is not strongly pseudo-contractive (Exercises 8.1, Problem 7).

The pseudo-contractive maps, apart from being a generalization of the nonexpansive maps, are intimately connected with an important class of nonlinear maps (the *accretive operators*) as is shown below.

We begin with the following definitions.

Definition 8.7. Let E be a real normed linear space with dual E^*. A mapping A with domain $D(A)$ and range $R(A)$ in E is called *accretive* if and only if for all $x, y \in D(A)$, the following inequality is satisfied:

$$||x - y|| \leq ||x - y + s(Ax - Ay)|| \; \forall \; s > 0. \tag{8.9}$$

As a consequence of Proposition 3.11, it follows that A is accretive if and only if for each $x, y \in D(A)$, there exists $j(x - y) \in J(x - y)$ such that

$$\langle Ax - Ay, j(x - y) \rangle \geq 0, \tag{8.10}$$

where $J : E \to 2^{E^*}$ is the normalized duality map on E. An operator A is called *dissipative* if $(-A)$ is accretive. It follows from inequality (8.9) that A is accretive if and only if $(I + sA)$ is *expansive*, and consequently, its inverse, $(I + sA)^{-1}$ exists and is nonexpansive as a mapping from $R(I + sA)$ into $D(A)$, where $R(I + sA)$ denotes the range of $(I + sA)$. The range of $(I + sA)$ need not be all of E. This leads to the following definition.

Definition 8.8. An operator A is said to be $m-accretive$ if A is accretive and the range of $(I + sA)$ is all of E for some $s > 0$. It can be shown that if $R(I + sA) = E$ for *some* $s > 0$, then this holds for *all* $s > 0$.

Example 8.9. The operator $-\Delta$, where Δ denotes the Laplacian, is an $m-$accretive operator.

Definition 8.10. Recall that an operator $A : D(A) \subseteq E \to E$ is called *strongly accretive* if there exists some $k > 0$ such that for each $x, y \in D(A)$, there exists $j(x - y) \in J(x - y)$ such that

$$\langle Ax - Ay, j(x - y) \rangle \geq k||x - y||^2. \tag{8.11}$$

8.2 Some Remarks on Accretive Operators

The accretive operators were introduced independently in 1967 by Browder [45] and Kato [275]. Interest in such mappings stems mainly from their firm connection with the existence theory for nonlinear equations of evolution in Banach spaces. It is well known that many physically significant problems can be modelled in terms of an initial value problem of the form

$$\frac{du}{dt} + Au = 0, \ u(0) = u_0, \tag{8.12}$$

where A is an accretive map on an appropriate Banach space. Typical examples of such evolution equations are found in models involving the heat, wave or Schrödinger equation (see e.g., Browder [47]). An early fundamental result in the theory of accretive operators, due to Browder [45], states that the initial value problem (8.12) is solvable if A is locally Lipschitzian and accretive on E. Utilizing the existence result for equation (8.12), Browder [45] proved that if A is locally Lipschitzian and accretive on E, then A is $m-$accretive. Clearly, a consequence of this is that the equation

$$x + Tx = f, \tag{8.13}$$

for a given $f \in E$, where $T := I - A$, has a solution. In [397], Ray gave an elementary elegant proof of this result of Browder by using a fixed point theorem of Caristi [65]. Martin [322, 323] proved that equation (8.12) is solvable if A is *continuous* and accretive on E, and utilizing this result, he further proved that if A is continuous and accretive, then A is $m-$accretive. In [45], Browder proved that if $A : E \to E$ is a Lipschitz and strongly accretive mapping, then A is surjective. Consequently, for each $f \in E$, the equation

$$Ax = f \tag{8.14}$$

has a solution in E. This result was subsequently generalized (see e.g., Deimling, [199], Theorem 13.1) to the *continuous strongly* accretive operators. Other existence theorems for zeros of accretive operators can be found in Browder ([39, 40, 43, 46]).

Definition 8.11. A mapping A with domain $D(A)$ and range $R(A)$ in E is said to be *strongly $\phi-$accretive* if, for any $x, y \in D(A)$ there exist $j(x - y) \in J(x - y)$ and a strictly increasing function $\phi : [0, \infty) \to [0, \infty)$ with $\phi(0) = 0$ such that $\langle Ax - Ay, j(x - y) \rangle \geq \phi(||x - y||)||x - y||$. The mapping A is called *generalized $\Phi-$accretive* if, for any $x, y \in D(A)$, there exist $j(x-y) \in J(x-y)$ and a strictly increasing function $\Phi : [0, \infty) \to [0, \infty)$ with $\Phi(0) = 0$ such that $\langle Ax-Ay, j(x-y) \rangle \geq \Phi(||x-y||)$. It is well known that the class of generalized $\Phi-$accretive mappings includes the class of strongly $\phi-$accretive operators as a special case (one sets $\Phi(s) = s\phi(s)$ for all $s \in [0, \infty)$).

Let $N(A) := \{x \in E : Ax = 0\} \neq \emptyset$. The mapping A is called *strongly quasi-accretive* if for all $x \in E, x^* \in N(A)$, there exists $k \in (0,1)$ such that $\langle Ax - Ax^*, j(x - x^*) \rangle \geq k||x - x^*||^2$; A is called *strongly $\phi-$quasi-accretive* if, for all $x \in E, x^* \in N(A)$, there exists ϕ such that $\langle Ax - Ax^*, j(x - x^*) \rangle \geq \phi(||x - x^*||)||x - x^*||$. Finally, A is called *generalized $\Phi-$quasi-accretive* if, for all $x \in E, x^* \in N(A)$, there exists $j(x - x^*) \in J(x - x^*)$ such that $\langle Ax - Ax^*, j(x - x^*) \rangle \geq \Phi(||x - x^*||)$.

The relation between the class of accretive-type mappings and those of *pseudo-contractive-type* is contained in the following proposition.

Proposition 8.12. *A mapping $T : E \to E$ is strongly pseudo-contractive if and only if $(I - T)$ is strongly accretive, and is strongly $\phi-$pseudo-contractive if and only if $(I - T)$ is strongly $\phi-$accretive. The mapping T is generalized $\Phi-$pseudo-contractive if and only if $(I - T)$ is generalized $\Phi-$accretive.*

Proposition 8.13. *If $F(T) := \{x \in E : Tx = x\} \neq \emptyset$, the mapping T is strongly hemi-contractive if and only if $(I - T)$ is strongly quasi-accretive; it is $\phi-$hemi-contractive if and only if $(I - T)$ is strongly $\phi-$quasi-accretive; and T is generalized $\Phi-$ hemi-contractive if and only if $(I - T)$ is generalized $\Phi-$quasi-accretive.*

The class of generalized $\Phi-$hemi-contractive mappings is the most general (among those defined above) for which T has a unique fixed point.

We note that accretive operators can also be defined in terms of their *graphs*. An operator $A : E \to 2^E$ is said to be (a): *accretive* if for each $(x, u), (y, v) \in G(A)$, (graph of A), there exists $j(x - y) \in J(x - y)$ such that $\langle u - v, j(x - y) \rangle \geq 0$; (b) *maximal accretive* if it is accretive and the inclusion $G(A) \subseteq G(B)$, with B accretive, implies $G(A) = G(B)$; (c) *m−accretive* (hypermaximal accretive in Browder's terminology) if A is accretive and $R(A + I) = E$. The map T is called *maximal pseudo-contractive* if and only if $(I - T)$ is maximal accretive.

We remark that in the evolution equation (8.12), if u is independent of t, then $\frac{du}{dt} = 0$ and the equation reduces to $Au = 0$ whose solutions correspond to the *equilibrium points* of the system described by the equation. If we set $T = I - A$, and if A is accretive, then T is pseudo-contractive and the zeros of A are the fixed points of T. Thus, in the last 30 years or so, considerable research efforts have been devoted to iterative methods for approximating fixed points of pseudo-contractive maps (or, equivalently, zeros of accretive maps).

8.3 Lipschitz Strongly Accretive Maps

Let E be *an arbitrary real Banach space* and $A : E \to E$ be a *Lipschitz* and *strongly* accretive map. In this section, we construct a *Picard-type sequence* which converges strongly to the solution of the equation $Au = 0$. Furthermore, if K is a nonempty convex (not necessarily bounded) subset of E, $T : K \to K$ is a Lipschitz strong pseudo-contraction, and if T has a fixed point in K, we construct a *Picard-type sequence* which converges strongly to the fixed point. In the sequel, $L > 0$ will denote the Lipschitz constant of an operator A which is *strongly accretive*. Furthermore, $\varepsilon > 0$ is defined by $\varepsilon := \frac{1}{2} \left[\frac{k}{1+L(3+L-k)} \right]$, where k is the strong accretivity constant given in inequality (8.11). With these notations, we prove the following theorem.

Theorem 8.14. *Let E be an arbitrary real Banach space, $A : E \to E$ be a Lipschitz and strongly accretive map. Let x^* denote a solution of the equation $Ax = 0$. Define $A_\varepsilon : E \to E$ by $A_\varepsilon x := x - \varepsilon Ax$ for each $x \in E$. For arbitrary $x_0 \in E$, define the sequence $\{x_n\}_{n=0}^{\infty}$ in E by*

$$x_{n+1} = A_\varepsilon x_n, n \geq 0. \tag{8.15}$$

Then, $\{x_n\}_{n=0}^{\infty}$ converges strongly to x^* with $||x_n - x^*|| \leq \delta^n ||x_0 - x^*||$, where $\delta = (1 - \frac{1}{2} k\varepsilon) \in (0, 1)$. Moreover, x^* is unique.

Proof. Existence of x^* follows from Theorem 13.1 of Deimling, [199]. Define $T := (I - A)$ where I denotes the identity map on E. Observe that $Ax^* = 0$ if and only if x^* is a fixed point of T. Moreover, T is strongly pseudo-contractive, i.e., (from inequality (8.11)) T satisfies the following inequality:

$$\langle (I - T)x - (I - T)y, j(x - y) \rangle \geq k||x - y||^2 \tag{8.16}$$

for all $x, y \in E$ and for some $j(x - y) \in J(x - y)$. Consequently, from Proposition 3.11, this implies that

$$||x - y|| \leq ||x - y + s[(I - T - kI)x - (I - T - kI)y]|| \tag{8.17}$$

holds for all $x, y \in E$ and for all $s > 0$. Furthermore, the recursion formula $x_{n+1} = A_\varepsilon x_n$ becomes

$$x_{n+1} = (1 - \varepsilon)x_n + \varepsilon T x_n, n \geq 0. \tag{8.18}$$

Observe that $x^* = (1 + \varepsilon)x^* + \varepsilon(I - T - kI)x^* - (1 - k)\varepsilon x^*$, and from the recursion formula (8.18) that

$$\begin{aligned} x_n = (1 + \varepsilon)x_{n+1} + \varepsilon(I - T - kI)x_{n+1} - (1 - k)\varepsilon x_n \\ + (2 - k)\varepsilon^2(x_n - Tx_n) + \varepsilon(Tx_{n+1} - Tx_n), \end{aligned}$$

so that

$$\begin{aligned} x_n - x^* = (1 + \varepsilon)(x_{n+1} - x^*) \\ + \varepsilon[(I - T - kI)x_{n+1} - (I - T - kI)x^*] \\ - (1 - k)\varepsilon(x_n - x^*) \\ + (2 - k)\varepsilon^2(x_n - Tx_n) + \varepsilon(Tx_{n+1} - Tx_n). \end{aligned}$$

This implies, using inequality (8.17) with $s = \frac{\varepsilon}{1+\varepsilon}, y = x^*$ that

$$
\begin{aligned}
||x_n - x^*|| \geq\ & (1+\varepsilon)[||(x_{n+1} - x^*) \\
& +\frac{\varepsilon}{1+\varepsilon}\{(I - T - kI)x_{n+1} - (I - T - kI)x^*\}||] \\
& -(1-k)\varepsilon||x_n - x^*|| - (2-k)\varepsilon^2||x_n - Tx_n|| \\
& -\varepsilon||Tx_{n+1} - Tx_n|| \\
\geq\ & (1+\varepsilon)||x_{n+1} - x^*|| - (1-k)\varepsilon||x_n - x^*|| \\
& -(2-k)\varepsilon^2||x_n - Tx_n|| - \varepsilon||Tx_{n+1} - Tx_n||.
\end{aligned}
$$

Hence, we obtain, since $(1+\varepsilon)^{-1} < 1$, the following estimate:

$$
\begin{aligned}
||x_{n+1} - x^*|| \leq\ & \left[\frac{1 + (1-k)\varepsilon}{1+\varepsilon}\right] ||x_n - x^*|| \\
& +(2-k)\varepsilon^2||x_n - Tx_n|| + \varepsilon||Tx_{n+1} - Tx_n||.
\end{aligned}
$$

Observe that, since L denotes the Lipschitz constant of A, $||x_n - Tx_n|| \leq L||x_n - x^*||$; $||Tx_{n+1} - Tx_n|| \leq \varepsilon L(1+L)||x_n - x^*||$, so that

$$
\begin{aligned}
||x_{n+1} - x^*|| \leq\ & \left[\frac{1 + (1-k)\varepsilon}{1+\varepsilon} + (2-k)\varepsilon^2 L + \varepsilon^2 L(1+L)\right] ||x_n - x^*|| \\
\leq\ & [1 - k\varepsilon + \varepsilon^2 + (2-k)\varepsilon^2 L + \varepsilon^2 L(1+L)]||x_n - x^*|| \\
=\ & \delta||x_n - x^*|| \leq \ldots \leq \delta^n||x_0 - x^*|| \to 0 \text{ as } n \to \infty.
\end{aligned}
$$

Hence $x_n \to x^*$ as $n \to \infty$. Uniqueness follows from the strong accretivity property of A. $\qquad\square$

The following is an immediate corollary of Theorem 8.14.

Corollary 8.15 *Let E be an arbitrary real Banach space, $K \subseteq E$ be nonempty and convex. Let $T : K \to K$ be a Lipschitz and strongly pseudo-contractive map. Assume that T has a fixed point $x^* \in K$. Let the Lipschitz constant of T be denoted by $L > 0$ and assume that T satisfies inequality (8.17). Set $\varepsilon_0 := \frac{1}{2}\left[\frac{k}{1+L(3+L-k)}\right]$ and define $T_{\varepsilon_0} : K \to K$ by $T_{\varepsilon_0}x = (1 - \varepsilon_0)x + \varepsilon_0 Tx$ for each $x \in K$. For arbitrary $x_0 \in K$, define the sequence $\{x_n\}_{n=0}^{\infty}$ in K by*

$$
x_{n+1} = T_{\varepsilon_0}x_n, n \geq 0. \tag{8.19}
$$

Then, $\{x_n\}_{n=0}^{\infty}$ converges strongly to x^* with $||x_{n+1} - x^*|| \leq \rho^n||x_0 - x^*||$, where $\rho := (1 - \frac{1}{2}k\varepsilon_0) \in (0, 1)$. Moreover, x^* is unique.

Proof. Observe that x^* is a fixed point of T if and only if it is a fixed point of T_{ε_0}. Furthermore, the recursion formula (8.19) simplifies to the formula $x_{n+1} = (1 - \varepsilon_0)x_n + \varepsilon_0 Tx_n$ which is similar to equation (8.18). Following the method of computations as in the proof of the Theorem 8.14, we obtain

$$||x_{n+1} - x^*|| \leq \{1 - k\varepsilon_0 + \varepsilon_0^2[1 + L(3 + L - k)]\}||x_n - x^*||$$
$$= \rho||x_n - x^*|| \leq ... \leq \rho^n||x_0 - x^*|| \to 0 \text{ as } n \to \infty.$$

This completes the proof.   □

8.4 Generalized Phi-Accretive Self-Maps

We first prove the following technical lemmas which will be used in the sequel.

Lemma 8.16. *Let* $\{\lambda_n\}$ *be a sequence of nonnegative numbers and* $\{\alpha_n\} \subseteq$ $(0,1)$ *a sequence such that* $\sum\limits_{n=1}^{\infty} \alpha_n = \infty$. *Let the recursive inequality*

$$\lambda_{n+1}^2 \leq \lambda_n^2 - 2\alpha_n \psi(\lambda_{n+1}) + \sigma_n, n = 1, 2, ..., \qquad (8.20)$$

be given where $\psi : [0, \infty) \to [0, \infty)$ *is a strictly increasing function such that it is positive on* $(0, \infty)$, $\psi(0) = 0$ *and* $\sum\limits_{n=1}^{\infty} \sigma_n < \infty$. *Then* $\lambda_n \to 0$, *as* $n \to \infty$.

Proof. Observe that $\lambda_{n+1}^2 \leq \lambda_n^2 - 2\alpha_n \psi(\lambda_{n+1}) + \sigma_n, n = 1, 2, ...,$ implies $\lambda_{n+1}^2 \leq \lambda_n^2 + \sigma_n, n = 1, 2, ...,$ so that by Lemma 6.32, limit of $\{\lambda_n\}$ exists. Let $\lim \lambda_n = a \geq 0$.

Claim: $a = 0$. Suppose this is not the case. Then, there exists $N_0 \in \mathbb{N}$ such that $\lambda_n \geq \frac{a}{2} \; \forall \; n \geq N_0$. This implies that $\psi(\lambda_{n+1}) \geq \psi(\frac{a}{2}) > 0$ so that

$$2\psi\left(\frac{a}{2}\right)\sum_{n=1}^{\infty}\alpha_n \leq \sum_{n=1}^{\infty}\left(\lambda_n^2 - \lambda_{n+1}^2\right) + \sum_{n=1}^{\infty}\sigma_n < \infty,$$

a contradiction. Hence $a = 0$, completing the proof.   □

Lemma 8.17. *Let* $\{\lambda_n\}$ *and* $\{\gamma_n\}$ *be sequences of nonnegative numbers and* $\{\alpha_n\}$ *be a sequence of positive numbers satisfying the conditions* $\sum\limits_{n=1}^{\infty} \alpha_n = \infty$ *and* $\frac{\gamma_n}{\alpha_n} \to 0$, *as* $n \to \infty$. *Let the recursive inequality*

$$\lambda_{n+1} \leq \lambda_n - \alpha_n \psi(\lambda_n) + \gamma_n, n = 1, 2, ...,$$

be given where $\psi : [0, \infty) \to [0, \infty)$ *is a strictly increasing continuous function such that it is positive on* $(0, \infty)$ *and* $\psi(0) = 0$. *Then* $\lambda_n \to 0$, *as* $n \to \infty$.

Proof. Since $\{\lambda_n\}$ is bounded below by zero, let $\liminf \lambda_n = a \geq 0$, say.

Claim. $a = 0$. Suppose this is not the case, then $a > 0$. Hence, there exists $N_0 > 0$ such that $\lambda_n \geq \frac{a}{2} \; \forall \; n \geq N_0$. This implies, using $\frac{\gamma_n}{\alpha_n} \to 0$, that

$\lambda_{n+1} \leq \lambda_n - \alpha_n \psi(\frac{a}{2}) + \frac{1}{2}\psi(\frac{a}{2})\alpha_n$. This yields that $\sum \alpha_n < \infty$, a contradiction. Hence $a = 0$. Now, complete the proof (see Exercises 8.1, Problem 9). $\square$

We now prove the following theorem.

Theorem 8.18. *Let E be a real normed linear space. Suppose $A : E \to E$ is a generalized Φ- quasi-accretive map. For arbitrary $x_1 \in E$, define the sequence $\{x_n\}$ iteratively by*

$$x_{n+1} := x_n - \alpha_n A x_n, \ n \geq 1, \tag{8.21}$$

where $\lim\limits_{n \to \infty} \alpha_n = 0$ and $\sum \alpha_n = \infty$. For $x^* \in N(A)$, suppose that $\sigma := \inf\limits_{n \in N_0} \frac{\Phi(||x_{n+1} - x^*||)}{||x_{n+1} - x^*||} > 0$ and that $||A x_{n+1} - A x_n|| \to 0$, where $N_0 := \{n \in \mathbb{N} : x_{n+1} \neq x^*\}$ and $\Phi : [0, \infty) \to [0, \infty)$ is a strictly increasing function with $\Phi(0) = 0$. Then, $\{x_n\}$ converges strongly to the unique solution of the equation $Ax = 0$.

Proof. By the recursion formula (8.21) and inequality (4.4) we have that

$$||x_{n+1} - x^*||^2 \leq ||x_n - x^*||^2 - 2\alpha_n \langle A x_{n+1}, j(x_{n+1} - x^*) \rangle$$
$$-2\alpha_n \langle A x_n - A x_{n+1}, j(x_{n+1} - x^*) \rangle$$
$$\leq ||x_n - x^*||^2 - 2\alpha_n \Phi(||x_{n+1} - x^*||)$$
$$+2\alpha_n ||A x_{n+1} - A x_n|| ||x_{n+1} - x^*||. \tag{8.22}$$

If there exists an integer $n > 0$ such that $x_{n+1} = x^*$, then we are done. Suppose $x_{n+1} \neq x^* \ \forall n \geq 1$, then from (8.22) and our hypothesis we have that

$$||x_{n+1} - x^*||^2 \leq ||x_n - x^*||^2 - 2\alpha_n \Big(\frac{\Phi(||x_{n+1} - x^*||)}{||x_{n+1} - x^*||} - ||A x_{n+1} - A x_n|| \Big)$$
$$\times ||x_{n+1} - x^*||$$
$$\leq ||x_n - x^*||^2$$
$$- 2\alpha_n \Big(\sigma - ||A x_{n+1} - A x_n|| \Big) ||x_{n+1} - x^*||. \tag{8.23}$$

Let $N' \in N_0$ be such that $||A x_{n+1} - A x_n|| \leq \sigma \ \forall n \geq N'$, then from (8.23), for all $n \geq N'$, we get that $||x_{n+1} - x^*||^2 \leq ||x_n - x^*||^2$. Thus, the sequence $\{||x_n - x^*||\}$ is bounded. Let $\lambda_n := ||x_n - x^*||$, $b_n := 2||A x_{n+1} - A x_n||$. Then, (8.22) gives that $\lambda_{n+1}^2 \leq \lambda_n^2 - 2\alpha_n \Phi(\lambda_{n+1}) + \alpha_n b_n \lambda_{n+1}$. The conclusion now follows from Lemma 8.17. $\square$

Corollary 8.19 *Suppose K is a closed convex subset of a real normed linear space E. Suppose $T : K \to K$ is a generalized Φ-hemi-contractive map. For arbitrary $x_1 \in K$, define the sequence $\{x_n\}$ iteratively by*

$$x_{n+1} := x_n - \alpha_n(I - T)x_n, \ n \geq 1,$$

where $\{\alpha_n\} \subseteq (0,1)$, $\lim \alpha_n = 0$ and $\sum \alpha_n = \infty$. For $x^* \in F(T)$, suppose that $\sigma := \inf_{n \in N_0} \frac{\Phi(||x_{n+1} - x^*||)}{||x_{n+1} - x^*||} > 0$ and that $||Ax_{n+1} - Ax_n|| \to 0$, where $A := I - T$, and N_0, Φ are as in Theorem 8.18. Then, $\{x_n\}$ converges strongly to the unique fixed point of T.

Proof. Observe that the fixed point of T is the solution of the equation $Ax = 0$ and hence the proof follows from Theorem 8.18. □

Theorem 8.20. *Let E be a real normed linear space. Suppose $A : E \to E$ is a strongly accretive map such that the solution x^* of the equation $Ax = 0$ exists. For arbitrary $x_1 \in E$, define the sequence $\{x_n\}$ iteratively by*

$$x_{n+1} := x_n - \alpha_n A x_n, \ n \geq 1,$$

where $\{\alpha_n\}$ is a positive real sequence such that $\lim \alpha_n = 0$ and $\sum \alpha_n = \infty$. Suppose that $||Ax_n - Ax_{n+1}|| \to 0$ as $n \to \infty$. Then, $\{x_n\}$ converges strongly to the unique solution of the equation $Ax = 0$.

Proof. As in the proof of Theorem 8.18 and using the definition of strongly accretive map, we get that

$$||x_{n+1} - x^*||^2 \leq ||x_n - x^*||^2 - 2\alpha_n k ||x_{n+1} - x^*||^2$$
$$+ 2\alpha_n ||Ax_{n+1} - Ax_n|| \, ||x_{n+1} - x^*||, \qquad (8.24)$$

where k is such that $\langle Ax - Ay, j(x - y)\rangle \geq k||x - y||^2$ for all $x, y \in E$. Now, we show that $\{||x_n - x^*||\}$ is bounded. Since $||Ax_{n+1} - Ax_n|| \to 0$, there exists $M > 0$ such that $||Ax_{n+1} - Ax_n|| \leq M$ for all $n \geq 1$. If there exists n_0 such that $x_{n_0+1} = x^*$, then $x_n = x^*$ for all integers $n \geq n_0$ and so $\{||x_n - x^*||\}$ is bounded. Assume now that $||x_{n+1} - x^*|| \neq 0$ for all integers $n \geq 1$. Consider the following sets $N_1 := \{n \in \mathbb{N} : ||x_{n+1} - x^*|| \leq ||x_n - x^*||\}$ and $N_2 := \{n \in \mathbb{N} : ||x_{n+1} - x^*|| > ||x_n - x^*||\}$. Observe that $\mathbb{N} = N_1 \cup N_2$.

Case 1. If $n \in N_1$ for all positive integers $n \geq 1$, then $||x_{n+1} - x^*|| \leq ||x_1 - x^*||$. Otherwise, if n_0 is the first positive integer in N_1 then $||x_{n+1} - x^*|| \leq ||x_{n_0} - x^*|| \ \forall n \in N_1$. Hence for all $n \in N_1$, we have $||x_{n+1} - x^*|| \leq \max\{||x_{n_0} - x^*||, ||x_1 - x^*||\}$.

Case 2. If $n \in N_2$, using inequality (8.24), we obtain that $||x_{n+1} - x^*|| \leq \frac{M}{k}$ for all integers $n \in N_2$. Hence, for all positive integers n, $||x_{n+1} - x^*|| \leq \max\{||x_{n_0} - x^*||, ||x_1 - x^*||, \frac{M}{k}\}$ and so $\{||x_n - x^*||\}$ is bounded. Therefore, the conclusion follows from inequality (8.24) and Lemma 8.17. □

Using the method of proof of Theorem 8.20, the following generalization of the theorem is easily proved.

Theorem 8.21. *Let E be a real normed linear space. Suppose $A : E \to E$ is a strongly ϕ-accretive map such that the solution x^* of the equation $Ax = 0$ exists. For arbitrary $x_1 \in E$, define the sequence $\{x_n\}$ iteratively by*

$$x_{n+1} := x_n - \alpha_n A x_n, \ n \geq 1,$$

where $\{\alpha_n\}$ is a positive real sequence such that $\lim \alpha_n = 0$ and $\sum \alpha_n = \infty$. Suppose that $\|Ax_n - Ax_{n+1}\| \to 0$ as $n \to \infty$. Then, $\{x_n\}$ converges strongly to the unique solution of the equation $Ax = 0$.

We also prove the following theorem.

Theorem 8.22. *Let E be a real normed linear space. Suppose $A : E \to E$ is a generalized Φ-quasi-accretive, uniformly continuous and bounded map. For arbitrary $x_1 \in E$, define the sequence $\{x_n\}$ iteratively by*

$$x_{n+1} := x_n - \alpha_n A x_n, \ n \geq 1, \tag{8.25}$$

where $\lim \alpha_n = 0$ and $\sum \alpha_n = \infty$. Then, there exists a constant $d_0 > 0$ such that if $0 < \alpha_n \leq d_0$, $\{x_n\}$ converges strongly to the unique solution of the equation $Ax = 0$.

Proof. Let $x^* \in N(A)$ and let $r > 0$ be sufficiently large such that $x_1 \in \overline{B_r(x^*)} =: B$. Then, $A(B)$ is bounded. Let $M = \sup\{\|Ax\| : x \in B\}$ and $\Phi : [0, \infty) \to [0, \infty)$ with $\Phi(0) = 0$ be a strictly increasing function which corresponds to A. As A is uniformly continuous on E, for $\epsilon := \frac{\Phi(r)}{4r}$ there exists a $\delta > 0$ such that $x, y \in D(A)$, $\|x - y\| < \delta$ implies $\|A(x) - A(y)\| < \epsilon$. Let $d_0 := \min\{\frac{r}{M}, \frac{\delta}{2M}\}$.

Claim : $x_n \in B \ \forall n \geq 1$. We show this by induction. By our choice, $x_1 \in B$. Suppose $x_n \in B$. We show that $x_{n+1} \in B$. Suppose not, then $\|x_{n+1} - x^*\| > r$ and from (8.25) and the above estimates we have $\|x_{n+1} - x_n\| \leq \alpha_n \|Ax_n\| \leq \alpha_n M < \delta$ and hence $\|Ax_{n+1} - Ax_n\| < \epsilon$. Moreover,

$$\|x_{n+1} - x^*\| \leq \|x_n - x^*\| + \alpha_n \|Ax_n\| \leq r + \alpha_n M \leq 2r.$$

Now, as in the proof of Theorem 8.18, by Lemma 8.16 and the above estimates we have that

$$\begin{aligned}
\|x_{n+1} - x^*\|^2 &\leq \|x_n - x^*\|^2 - 2\alpha_n \Phi(\|x_{n+1} - x^*\|) \\
&\quad + 2\alpha_n \|Ax_{n+1} - Ax_n\|\|x_{n+1} - x^*\| \\
&\leq \|x_n - x^*\|^2 - \alpha_n \Phi(r) < r^2,
\end{aligned} \tag{8.26}$$

and hence $\|x_{n+1} - x^*\| < r$, a contradiction. Therefore, the claim holds. Now we show that $x_n \to x^*$. Since $\|x_{n+1} - x_n\| \to 0$, by the uniform continuity of A we have $\|Ax_{n+1} - Ax_n\| \to 0$ as $n \to \infty$. The conclusion now follows from (8.26) with the use of Lemma 8.17. $\qquad\square$

Corollary 8.23 *Suppose K is a closed convex subset of a real normed linear space E. Suppose $T : K \to K$ is a generalized Φ-hemi-contractive uniformly* *continuous and bounded map. For arbitrary $x_1 \in K$, define the sequence $\{x_n\}$ iteratively by*

$$x_{n+1} := x_n - \alpha_n(I - T)x_n, \ n \geq 1, \tag{8.27}$$

where $\{\alpha_n\} \subseteq (0,1)$, $\lim \alpha_n = 0$ and $\sum \alpha_n = \infty$. Then, there exists a constant $d_0 > 0$ such that if $0 < \alpha_n \leq d_0$, $\{x_n\}$ converges strongly to the unique fixed point of T.

Proof. The proof is similar to the proof of Corollary 8.19 with the use of Theorem 8.22 instead of Theorem 8.18. It is therefore omitted. □

8.5 Generalized Phi-Accretive Non-self Maps

We now prove the following theorem.

Theorem 8.24. *Let E be a real uniformly smooth Banach space. Suppose K is a closed convex subset of E which is a nonexpansive retract of E with P as the nonexpansive retraction. Suppose $A : K \to E$ is a bounded generalized Φ-accretive map with strictly increasing continuous function $\Phi : [0, \infty) \to [0, \infty)$ such that $\Phi(0) = 0$ and the solution x^* of the equation $Ax = 0$ exists. For arbitrary $x_1 \in K$, define the sequence $\{x_n\}$ iteratively by*

$$x_{n+1} := P\Big(x_n - \alpha_n A x_n\Big), \ n \geq 1, \tag{8.28}$$

where $\lim \alpha_n = 0$ and $\sum \alpha_n = \infty$. Then, there exists a constant $d_0 > 0$ such that if $0 < \alpha_n \leq d_0$, $\{x_n\}$ converges strongly to the unique solution of $Ax = 0$.

Proof. As in the proof of Theorem 8.22, let r be sufficiently large such that $x_1 \in B_r(x^*)$. Define $G := \overline{B_r(x^*)} \cap K$. Then, since A is bounded we have that $A(G)$ is bounded. Let $M = \sup\{||Ax|| : x \in G\}$. As j is uniformly continuous on bounded subsets of E, for $\epsilon := \frac{\psi(\frac{r}{2})}{2M}$, there exists a $\delta > 0$ such that $x, y \in D(T)$, $||x - y|| < \delta$ implies $||j(x) - j(y)|| < \epsilon$. Set $d_0 = \min\{\frac{r}{2M}, \frac{\delta}{2M}, \}$.

Claim: $\{x_n\}$ is bounded.
Suffices to show that x_n is in G for all $n \geq 1$. The proof is by induction. By our assumption, $x_1 \in G$. Suppose $x_n \in G$. We prove that $x_{n+1} \in G$. Assume for contradiction that $x_{n+1} \notin G$. Then, since $x_{n+1} \in K \ \forall n \geq 1$, we have that $||x_{n+1} - x^*|| > r$. Thus we have the following estimates:

$$||x_{n+1} - x^*|| = ||P(x_n - \alpha_n A x_n) - P x^*|| \leq ||x_n - x^* - \alpha_n A x_n||$$

and hence

$$||x_n - x^*|| \geq ||x_{n+1} - x^*|| - \alpha_n ||A x_n|| > r - \alpha_n M \geq r - \frac{r}{2} = \frac{r}{2}.$$

Set $r_n := x_n - \alpha_n A x_n$. Then from (8.28), and the above estimates we have that

$$\|x_{n+1} - x^*\|^2 \leq \|x_n - x^*\|^2 - 2\alpha_n \langle A x_n, j(x_n - x^*) \rangle$$
$$-2\alpha_n \langle A x_n, j(r_n - x^*) - j(x_n - x^*) \rangle$$
$$\leq \|x_n - x^*\|^2 - 2\alpha_n \Phi(\|x_n - x^*\|)$$
$$+2\alpha_n \|A x_n\| \|j(r_n - x^*) - j(x_n - x^*)\|. \qquad (8.29)$$

Since $\|r_n - x_n\| \leq \alpha_n \|A x_n\| \leq \alpha_n M < \delta$ we have that

$$\|j(r_n - x^*) - j(x_n - x^*)\| \leq \frac{\Phi(\frac{r}{2})}{2M}.$$

Thus, (8.29) gives that

$$\|x_{n+1} - x^*\|^2 \leq \|x_n - x^*\|^2 - 2\alpha_n \Phi(\frac{r}{2}) + \alpha_n \Phi(\frac{r}{2}) < r^2,$$

i.e., $\|x_{n+1} - x^*\| < r$, a contradiction. Therefore $x_{n+1} \in G$. Thus by induction $\{x_n\}$ is bounded. Now we show that $x_n \to x^*$. Note that $r_n - x_n \to 0$ as $n \to \infty$ and hence by the uniform continuity of j on bounded subsets of E we have that $\overline{\gamma}_n := \|j(r_n - x^*) - j(x_n - x^*)\| \to 0$ as $n \to \infty$. Let $\lambda_n := \|x_n - x^*\|^2$ and $\gamma_n := 2\alpha_n M \overline{\gamma}_n$, then from inequality (8.29) we obtain that $\lambda_{n+1} \leq \lambda_n - 2\alpha_n \overline{\Phi}(\lambda_n) + \gamma_n$, where $\overline{\Phi}(t) := \Phi(\sqrt{t})$ and $\frac{\gamma_n}{\alpha_n} \to 0$ as $n \to \infty$. Therefore, the conclusion of the theorem follows from Lemma 8.17. $\qquad \square$

Corollary 8.25 *Let E be a real uniformly smooth Banach space. Suppose K is a closed convex subset of E which is a nonexpansive retract of E with P as the nonexpansive retraction. Suppose $T : K \to E$ is a bounded generalized Φ-hemi-contractive map with strictly increasing continuous function $\Phi : [0, \infty) \to [0, \infty)$ such that $\Phi(0) = 0$. For arbitrary $x_1 \in K$, define the sequence $\{x_n\}$ iteratively by*

$$x_{n+1} := P\Big(x_n - \alpha_n(I - T)x_n\Big), \ n \geq 1,$$

where $\lim \alpha_n = 0$ and $\sum \alpha_n = \infty$. Then, there exists a constant $d_0 > 0$ such that if $0 < \alpha_n \leq d_0$, $\{x_n\}$ converges strongly to the unique fixed point of T.

Proof. The proof follows from Theorem 8.24. $\qquad \square$

If, in Theorem 8.24 and Corollary 8.25, $0 \leq \alpha_n < 1$ and the operator is a self-map, the use of the projection operator P will not be necessary. In fact, the following corollaries follow trivially.

Corollary 8.26 *Suppose E is a real uniformly smooth Banach space and $A : E \to E$ is a bounded generalized Φ-quasi-accretive map with strictly* *increasing continuous function* $\Phi : [0, \infty) \to [0, \infty)$ *such that* $\Phi(0) = 0$. *For arbitrary* $x_1 \in E$, *define the sequence* $\{x_n\}$ *iteratively by*

$$x_{n+1} := x_n - \alpha_n A x_n, \ n \geq 1,$$

where $\{\alpha_n\}$ *is a real positive sequence such that* $\lim \alpha_n = 0$ *and* $\sum \alpha_n = \infty$. *Then, there exists a constant* $d_0 > 0$ *such that if* $0 < \alpha_n \leq d_0$, $\{x_n\}$ *converges strongly to the unique solution of the equation* $Ax = 0$.

Corollary 8.27 *Suppose* K *is a closed convex subset of a real uniformly smooth Banach space* E. *Suppose* $T : K \to K$ *is a bounded generalized* Φ*-hemi-contractive map with strictly increasing continuous function* $\Phi : [0, \infty) \to [0, \infty)$ *such that* $\Phi(0) = 0$. *For arbitrary* $x_1 \in K$, *define the sequence* $\{x_n\}$ *iteratively by*

$$x_{n+1} := x_n - \alpha_n(I - T)x_n, \ n \geq 1,$$

where $\alpha_n \subseteq (0, 1), \lim \alpha_n = 0$ *and* $\sum \alpha_n = \infty$. *Then, there exists a constant* $d_0 > 0$ *such that if* $0 < \alpha_n \leq d_0$, $\{x_n\}$ *converges strongly to the unique fixed point of* T.

EXERCISE 8.1

1. Apply Proposition 3.11 to inequality (8.1) to obtain inequality (8.2); to (8.2) to obtain (8.3); and to (8.2) to obtain (8.4).
2. Show that every nonexpansive map is pseudo-contractive.
3. Verify that the map $T : [0, 1] \to \mathbb{R}$ defined by

$$Tx = \begin{cases} x - \frac{1}{2}, & \text{if } x \in [0, 1) \\ x - 1, & \text{if } x \in (\frac{1}{2}, 1], \end{cases}$$

   is pseudo-contractive and is not nonexpansive.
4. Let $T : [0, 1] \to \mathbb{R}$ be defined by $Tx = 1 - x^{\frac{2}{3}}$. Verify that T is continuous pseudo-contractive and is not nonexpansive.
5. (a) Prove that every quasi-nonexpansive map is hemi-contractive.
   (b) Verify that in a Hilbert space, inequalities (8.7) and (8.6) are equivalent.
6. By means of inequality (8.5), or otherwise, show that in a real Hilbert space, *strictly pseudo-contractive* mappings are Lipschitz.
7. Give the details to show that the map defined in Example 8.6 is a Lipschitz strongly hemi-contractive map which is not strongly pseudo-contractive.
8. Show explicitly that $-\triangle$, where $\triangle$ denotes the Laplacian in $\mathbb{R}^2$, is an m-accretive operator.
9. Complete the proof of Lemma 8.17.

8.6 Historical Remarks

Let K be a nonempty closed convex and bounded subset of a real Banach space with dual E^*. Let $T : K \to K$ be a Lipschitz strongly pseudo-contractive map with $F(T) := \{x \in K : Tx = x\} \neq \emptyset$. In [95], Chidume proved that if $E = L_p, p \geq 2$, then the Mann iteration process converges strongly to $x^* \in F(T)$. Moreover, x^* is unique. The main tool in the proof is an inequality of Bynum [61]. This result signalled the return to extensive research on inequalities in Banach spaces and their applications to nonlinear operator theory. Consequently, this theorem of Chidume has been generalized in various directions by numerous authors (see e.g., Chidume [91, 94], [104]-[114], Chidume and Moore [145, 146], Chidume and Nnoli [149], Chidume and Osilike [154]-[162], Deng [201]-[203], Zhou [554], Zhou and Jia [557]-[558], Liu [314]-[315], Qihou [390]-[395], Weng [500, 501], Xiao [507], Xu [509], Xu [521, 522], Xu [524, 523], Xu and Roach [525, 527], Xu *et al.* [529], Zhu [561]) and a host of other authors). Lemma 8.16 is due to Moore and Nnoli [336]. The proof given here is simpler than that of Moore and Nnoli. Theorem 8.14 and Corollary 8.15 are due to Chidume [118]. All the theorems of sections 8.4 and 8.5 are due to Chidume and Zegeye [169] and they are generalizations of the results in Chidume and Moore [145], [146], Chang [67], Chidume and Osilike [154]-[162], Deng [201]-[203], Ding [207]-[208], Zhou [554], Zhou and Jia [557]- [558], Liu [314], Osilike [368]-[372], Schu [437, 438], Tan and Xu [486, 487], Xu [509], Xu [521], and Xu and Roach [525, 527], Aoyama *et al.* [15], Chang *et al.* ([74], [76], [77], [80]), Chang [72], Chidume ([90], [92]), Chidume and Udomene [165], Chidume and Zegeye ([166], [167], [176]), Chidume *et al.* ([179], [182]), Cho *et al.* [188], Isac[258], Kim and Kim [278], Kohsaka and Takahashi [290], Matsuhita and Kuroiwa [326], Matsushita and Takahashi [327], Ofoedu [363], Osilike *et al.* [376], Reich [403], Suzuki [472], Udomene [493], Xu ([519], [515], [517]), Yao *et al.* [534], Zegeye and Shahzad [542], Zegeye and Prempeh [540], Zhou and Chen [556], Zhou [555], Zhou *et al.* [560], Gao *et al.* [226], Al'ber and Reich [10], Bruck[53], Schöneberg [432], Tan and Xu [486], Wei and Zhou ([502],[503]), Xu [515], Xue *et al.* [530] and a host of other authors.

Chapter 9
Iteration Processes for Zeros
of Generalized $\Phi-$Accretive Mappings

9.1 Introduction

Recall that *generalized Φ-accretive and generalized Φ-pseudo-contractive operators* have been defined in the last chapter (Definition 8.11). In this chapter, we investigate iterative methods for approximating solutions of nonlinear equations involving slight generalizations of these operators. In 1995, Liu [314] introduced what he called the *Ishikawa and Mann iteration processes with errors* as follows:

(a) For K, a nonempty subset of a normed linear space E and $T : K \to E$ any mapping, the process defined by $x_0, v_0, u_0 \in K$,

$$y_n = (1 - \beta_n)x_n + \beta_n T x_n + u_n, n \geq 0, \tag{9.1}$$

$$x_{n+1} = (1 - \alpha_n)x_n + \alpha_n T y_n + v_n, n \geq 0, \tag{9.2}$$

where $\{\alpha_n\}, \{\beta_n\} \subset [0, 1)$ are real sequences satisfying appropriate conditions and u_n, v_n are vectors in K such that $\sum ||u_n|| < \infty, \sum ||v_n|| < \infty$, he called the *Ishikawa iteration process with errors*.

(b) With the notations and definitions in (a) above, if $\beta_n \equiv 0$ then the process defined by $x_0, u_0 \in K$,

$$x_{n+1} = (1 - \alpha_n)x_n + \alpha_n T x_n + u_n, n \geq 0, \tag{9.3}$$

he called the *Mann iteration process with errors*. We observe that the recursion formulas $(9.1) - (9.3)$ may not be well defined if $K \neq E$. In [522], Xu objected to the definition given by Liu [314] on the grounds that the conditions $\sum ||u_n|| < \infty$ and $\sum ||v_n|| < \infty$ are not compatible with the randomness of the occurrence of errors (since they imply, in particular, that the errors tend to zero as n tends to infinity). He then introduced the following alternative definitions:

C. Chidume, *Geometric Properties of Banach Spaces and Nonlinear Iterations,*
Lecture Notes in Mathematics 1965,
© Springer-Verlag London Limited 2009

(A) Let K be a nonempty convex subset of E and $T : K \to K$ be any map. For any given $x_0, u_0, v_0 \in K$, the process defined by

$$y_n = a'_n x_n + b'_n T x_n + c'_n u_n, n \geq 0, \tag{9.4}$$

$$x_{n+1} = a_n x_n + b_n T y_n + c_n v_n, n \geq 0, \tag{9.5}$$

where $\{u_n\}, \{v_n\}$ are *bounded* sequences in K and the real sequences $\{a_n\}, \{b_n\}, \{c_n\}, \{a'_n\}, \{b'_n\}, \{c'_n\} \subset [0, 1]$ satisfy the conditions $a_n + b_n + c_n = 1 = a'_n + b'_n + c'_n, \forall n \geq 0$, he called the *Ishikawa iteration process with errors*.
    (B) With the same notations and definitions as in (A), if $b'_n \equiv 0 \equiv c'_n$, then the process now defined by $x_0, u_0 \in K$ and

$$x_{n+1} = a_n x_n + b_n T x_n + c_n u_n, n \geq 0, \tag{9.6}$$

he called the *Mann iteration process with errors*.
    It is easy to observe that a reasonable error term is that introduced by Xu [522]. In particular, if T is a self-mapping of a convex bounded set, then the boundedness requirement for the error term is trivially satisfied. However, under the condition that the error term is bounded, it is generally the case that whenever a theorem is proved using the Mann algorithm (without error terms), the method of proof generally carries over easily (only a slight modification in the computations will be required) to the case of *Mann process with bounded error term*. We illustrate this in what follows.

9.2 Uniformly Continuous Generalized Φ-hemi-contractive Maps

We consider the Mann iteration method with *bounded* error term and prove the following theorem.

Theorem 9.1. *Let E be a real Banach space and $T : E \to E$ be uniformly continuous and bounded. Let $\{x_n\}$ be a sequence in E defined iteratively from an arbitrary $x_0, u_0 \in E$ by :*

$$x_{n+1} = a_n x_n + b_n T x_n + c_n u_n, n \geq 0, \tag{9.7}$$

where $\{a_n\}, \{b_n\}, \{c_n\}$ are sequences in $[0, 1]$ satisfying the following conditions: $(i) a_n + b_n + c_n = 1 \ \forall \ n \geq 0; (ii) \sum_{n=0}^{\infty} b_n = \infty; (iii) \sum_{n=0}^{\infty} b_n^2 < \infty; c_n = o(b_n); (iv) \sum_{n=0}^{\infty} c_n < \infty$ and such that

$$\langle T x_n - x^*, j(x_n - x^*) \rangle \leq ||x_n - x^*||^2 - \Phi(||x_n - x^*||) \ \forall \ n \geq 0, \tag{9.8}$$

where $\Phi : [0,\infty) \to [0,\infty)$ is a strictly increasing function with $\Phi(0) = 0$, and $\{u_n\}$ is a bounded sequence in E. Then, there exists $\gamma_0 > 0$ such that if $b_n + c_n \le \gamma_0$, $\frac{c_n}{b_n + c_n} \le \gamma_0$, then $\{x_n\}$ is bounded.

Proof. If $x_0 = Tx_n \ \forall \ n \ge 0$, and $c_n \equiv 0$, then we are done. Suppose this is not the case, i.e., suppose there exists $n_0 \in \mathbb{N}$, the smallest positive integer such that $x_0 \ne Tx_{n_0}$ and let $b_n + c_n < \frac{1}{2} \ \forall \ n \ge n_0$. Without loss of generality, define $x_{n_0} := x_0$ and $a_0 := \|x_0 - Tx_0\| \cdot \|x_0 - x^*\|$. Then, from inequality (9.8), we obtain that $\|x_0 - x^*\| \le \Phi^{-1}(a_0)$. Let $N^* := \sup_n \|u_n - x^*\|$. Define

$$M_0 := \sup_n \{\|x - Tx\| : \|x - x^*\| \le 8\Phi^{-1}(a_0)\} + N^*.$$

Set $\epsilon_0 := \frac{\Phi(2\Phi^{-1}(a_0))}{14\Phi^{-1}(a_0)} > 0$. Since T is uniformly continuous, there exists a $\delta > 0$ such that $\|Tx - Ty\| < \epsilon_0$ whenever $\|x - y\| < \delta$. Now define $\alpha_n := b_n + c_n$;

$$\gamma_0 := \frac{1}{2}\min_n\left\{1, \frac{\Phi(2\Phi^{-1}(a_0))}{1 + 28\Phi^{-1}(a_0)[3M_0 + 2\Phi^{-1}(a_0)]}, \frac{\delta}{3M_0 + 2\Phi^{-1}(a_0)}, \frac{\Phi^{-1}(a_0)}{M_0 + 1}\right\}.$$

Claim. $\|x_n - x^*\| \le 2\Phi^{-1}(a_0) \ \forall \ n \ge 0$. The proof of this claim is by induction. Clearly, the claim holds for $n = 0$. Assume now it holds for some n i.e., assume that $\|x_n - x^*\| \le 2\Phi^{-1}(a_0)$. We prove that $\|x_{n+1} - x^*\| \le 2\Phi^{-1}(a_0)$. Suppose this is not the case. Then $\|x_{n+1} - x^*\| > 2\Phi^{-1}(a_0)$. This implies that $\Phi(\|x_{n+1} - x^*\|) > \Phi(2\Phi^{-1}(a_0))$. With $\alpha_n := b_n + c_n$, equation (9.7) becomes

$$x_{n+1} = (1 - \alpha_n)x_n + \alpha_n Tx_n + c_n U_n, \tag{9.9}$$

where $U_n := (u_n - Tx_n)$. Set $M_1 := 14\Phi^{-1}(a_0)[3M_0 + 2\Phi^{-1}(a_0)]$. Observe that $\|U_n\| \le N^* + \|x_n - x^*\| + \|x_n - Tx_n\| \le 2M_0 + 2\Phi^{-1}(a_0)$. Furthermore,

$$\|x_{n+1} - x^*\| \le \|x_n - x^*\| + \alpha_n\|x_n - Tx_n\| + c_n\|U_n\|$$
$$\le 2\Phi^{-1}(a_0) + \gamma_0 M_0 + \gamma_0(2M_0 + 2\Phi^{-1}(a_0)) \le 7\Phi^{-1}(a_0).$$

Set $\rho := 14[2M_0 + 2\Phi^{-1}(a_0)]\Phi^{-1}(a_0)$. Also,

$$\|x_{n+1} - x_n\| \le \alpha_n\{\|x_n - Tx_n\| + \|U_n\|\}, \ c_n < \alpha_n$$
$$\le \alpha_n[3M_0 + 2\Phi^{-1}(a_0)]$$
$$\le \gamma_0[3M_0 + 2\Phi^{-1}(a_0)] < \delta,$$

so that $\|Tx_{n+1} - Tx_n\| < \epsilon_0$. Using these estimates and the recursion formula (9.9), we now obtain the following estimates:

$$\|x_{n+1} - x^*\|^2 \le \|x_n - x^*\|^2 - 2\alpha_n\langle x_n - Tx_n, j(x_{n+1} - x^*)\rangle$$
$$+ 2c_n\|U_n\| \cdot \|x_{n+1} - x^*\|$$
$$\le \|x_n - x^*\|^2 - 2\alpha_n\Phi(\|x_{n+1} - x^*\|)$$

$$+2\alpha_n||x_{n+1} - x_n||.||x_{n+1} - x^*||$$
$$+2\alpha_n||Tx_{n+1} - Tx_n||.||x_{n+1} - x^*|| + c_n\rho$$
$$\leq ||x_n - x^*||^2 - 2\alpha_n\Phi(2\Phi^{-1}(a_0)) + \alpha_n^2 M_1$$
$$+14\Phi^{-1}(a_0)\alpha_n||Tx_{n+1} - Tx_n|| + c_n\rho.$$

Hence, we obtain that

$$||x_{n+1} - x^*||^2 \leq ||x_n - x^*||^2 - \alpha_n\Phi(2\Phi^{-1}(a_0)) + M_1\alpha_n^2 + c_n\rho. \quad (9.10)$$

Observe that $M_1\alpha_n^2 < \frac{1}{4}\Phi(2\Phi^{-1}(a_0))\alpha_n$; $c_n\rho < \frac{1}{4}\Phi(2\Phi^{-1}(a_0))\alpha_n$. Hence, we obtain from (9.10) that $||x_{n+1} - x^*|| < ||x_n - x^*|| \leq \Phi(2\Phi^{-1}(a_0))$, a contradiction. Hence, $\{x_n\}$ is bounded. $\qquad\square$

Theorem 9.2. *Let E be a real normed linear space, K be a nonempty subset of E and $T : K \to E$ be a bounded uniformly continuous generalized Φ−hemi-contractive mapping, i.e., there exist $x^* \in F(T)$ and a strictly increasing function $\Phi : [0,\infty) \to [0,\infty), \Phi(0) = 0$ such that for all $x \in K$, there exists $j(x - x^*) \in J(x - x^*)$ such that*

$$\langle Tx - x^*, j(x - x^*)\rangle \leq ||x - x^*||^2 - \Phi(||x - x^*||). \quad (9.11)$$

(a) If $y^* \in K$ is a fixed point of T, then $y^* = x^*$ and so T has at most one fixed point in K; (b) Suppose there exist $x_0, u_0 \in K$ such that the sequence $\{x_n\}$ defined by

$$x_{n+1} = a_nx_n + b_nTx_n + c_nu_n \ \forall \ n \geq 0 \quad (9.12)$$

is contained in K, where $\{a_n\}, \{b_n\}$ and $\{c_n\}$ are real sequences satisfying the following conditions, (i) $a_n + b_n + c_n = 1$; (ii) $\sum_{n=0}^{\infty} b_n = \infty$; (iii) $\sum_{n=0}^{\infty} b_n^2 < \infty$; (iv) $\sum_{n=0}^{\infty} c_n < \infty$; $c_n = o(b_n)$ and $\{u_n\}$ is a bounded sequence in E. Then, there exists $\gamma_0 > 0$ such that if $b_n + c_n \leq \gamma_0$, $\frac{c_n}{b_n+c_n} \leq \gamma_0$, then $\{x_n\}$ converges strongly to x^*. In particular, if y^* is a fixed point of T in K, then $\{x_n\}$ converges strongly to y^*.

Proof. (a) If $y^* \in K$ is a fixed point of T then by (9.11), there exists $j(y^* - x^*) \in J(y^* - x^*)$ such that

$$||y^* - x^*||^2 = \langle y^* - x^*, j(y^* - x^*)\rangle = \langle Ty^* - x^*, j(y^* - x^*)\rangle$$
$$\leq ||y^* - x^*||^2 - \Phi(||y^* - x^*||).$$

This implies, $y^* = x^*$. Hence T has at most one fixed point. (b) By Theorem 9.1, $\{x_n\}$ is a bounded sequence in K. Hence, $\{Tx_n\}$ is bounded. As in the proof of Theorem 9.1, set $\alpha_n := b_n + c_n$. Then, equation (9.12) becomes

$$x_{n+1} = (1 - \alpha_n)x_n + \alpha_nTx_n + c_nU_n, n \geq 0,$$

where $U_n := (u_n - Tx_n)$. Set

$$M_1 := 2\sup_n ||U_n||.||x_{n+1} - x^*||, \; M_2 := 2\sup_n\{||x_n - Tx_n|| + ||U_n||\} \times ||x_{n+1} - x^*||$$

and $M_3 := 2\sup_n ||x_{n+1} - x^*||$. Then,

$$
\begin{aligned}
||x_{n+1} - x^*||^2 &\leq ||x_n - x^*||^2 - 2\alpha_n\langle x_n - Tx_n, j(x_{n+1} - x^*)\rangle \\
&\quad + 2c_n\langle U_n, j(x_{n+1} - x^*)\rangle \\
&\leq ||x_n - x^*||^2 - 2\alpha_n\Phi(||x_{n+1} - x^*||) + c_n M_1 \\
&\quad -2\alpha_n\langle -(x_{n+1} - x_n) + (Tx_{n+1} - Tx_n), j(x_{n+1} - x^*)\rangle \\
&\leq ||x_n - x^*||^2 - 2\alpha_n\Phi(||x_{n+1} - x^*||) \\
&\quad + \alpha_n M_3||Tx_{n+1} - Tx_n|| + c_n M_1 + \alpha_n^2 M_2 \\
&= ||x_n - x^*||^2 - 2\alpha_n\Phi(||x_{n+1} - x^*||) + \sigma_n \quad\quad (9.13)
\end{aligned}
$$

where $\sigma_n := \alpha_n M_3||Tx_{n+1} - Tx_n|| + c_n M_1 + \alpha_n^2 M_2$. The result now follows from Lemma 8.17. $\qquad\square$

Corollary 9.3 *Let E be a real normed linear space and let $A : E \to E$ be a bounded uniformly continuous generalized $\Phi-$quasi-accretive mapping, i.e., there exists $x^* \in N(A)$ such that for all $x \in E$, there exist $j(x - x^*) \in J(x - x^*)$ and a strictly increasing function $\Phi : [0, \infty) \to [0, \infty), \Phi(0) = 0$ such that $\langle Ax - Ax^*, j(x - x^*)\rangle \geq \Phi(||x - x^*||)$. For arbitrary $x_0, u_0 \in E$, define the sequence $\{x_n\}$ iteratively by*

$$x_{n+1} = a_n x_n + b_n S x_n + c_n u_n, n \geq 0,$$

where $S : E \to E$ is defined by $Sx := x - Ax$ for all $x \in E$; and $\{a_n\}, \{b_n\}, \{c_n\}$ are real sequences in $[0, 1]$ satisfying the following conditions:

(i) $a_n + b_n + c_n = 1$;

$$(ii)\sum_{n=0}^{\infty} b_n = \infty; \; (iii) \sum_{n=0}^{\infty} b_n^2 < \infty; \; c_n = o(b_n); (iv)\sum_{n=0}^{\infty} c_n < \infty;$$

and $\{u_n\}$ is a bounded sequence in E. Then, $\{x_n\}$ converges strongly to x^*.

Proof. We simply observe that S is a uniformly continuous and generalized $\Phi-$hemi-contractive mapping of E into E. The result follows from Theorem 9.2. $\qquad\square$

In the recursion formulas (9.7) and (9.12), it must be noted that $\{u_n\}$ is *not* known and is included in the recursion formula only as a *noise*. The actual iteration process is $x_0 \in K$, $x_{n+1} = a_n x_n + b_n T x_n, n \geq 0$, where $c_n \equiv 0 \; \forall n$.

9.3 Generalized Lipschitz, Generalized Φ-quasi-accretive Mappings

Let E be a real linear space. A mapping $T : D(T) \subset E \to E$ is said to be *generalized Lipschitz* if there exists $L > 0$ such that $\|Tx - Ty\| \leq L(1 + \|x - y\|) \ \forall \ x, \ y \in D(T)$. Clearly, every Lipschitz map is generalized Lipschitz. Furthermore, any map with bounded range is a generalized Lipschitz map. The following example (see e.g., [76]) shows that the class of generalized Lipschitz maps properly includes the class of Lipschitz maps and that of mappings with bounded range.

Example 9.4. Let $E = (-\infty, +\infty)$ and $T : E \to E$ be defined by

$$
Tx = \begin{cases} x - 1, & \text{if } x \in (-\infty, -1), \\ x - \sqrt{1 - (x + 1)^2}, & \text{if } x \in [-1, 0), \\ x + \sqrt{1 - (x - 1)^2}, & \text{if } x \in [0, 1], \\ x + 1, & \text{if } x \in (1, +\infty). \end{cases}
$$

Then, T is a generalized Lipschitz map which is not Lipschitz and whose range is not bounded.

We now prove the following theorems.

Theorem 9.5. *Let E be a real uniformly smooth Banach space and $A : E \to E$ be a mapping with $N(A) \neq \emptyset$. Suppose A is a generalized Lipschitz, generalized Φ–quasi-accretive mapping. Let $\{a_n\}, \{b_n\},$ and $\{c_n\}$ be real sequences in [0,1] satisfying the following conditions, (i) $a_n + b_n + c_n = 1; (ii) \sum b_n = \infty; (iii) \sum c_n < \infty; (iv) c_n = o(b_n)$. Let $\{x_n\}$ be generated iteratively from arbitrary $x_0, u_0 \in E$ by,*

$$
x_{n+1} = a_n x_n + b_n S x_n + c_n u_n, n \geq 0, \tag{9.14}
$$

where $S : E \to E$ is defined by $Sx := x - Ax \ \forall x \in E$ and $\{u_n\}$ is an arbitrary bounded sequence in E. Then, there exists a constant $\gamma_0 \in \mathbb{R}^+$ such that if $b_n + c_n \leq \gamma_0 \ \forall \ n \geq 0$, the sequence $\{x_n\}$ converges strongly to the unique solution of the equation $Au = 0$.

Proof. Let $x^* \in N(A)$ be arbitrary. Observe that $Ax^* = 0$ if and only if $Sx^* = x^*$. Since A is a generalized Lipschitz, generalized Φ–quasi-accretive mapping, there exist a strictly increasing function $\Phi : [0, \infty) \to [0, \infty), \Phi(0) = 0$ and $j(x - x^*) \in J(x - x^*)$ such that the following inequalities hold:

$$
\|Sx_n - Sx^*\| \leq L(1 + \|x_n - x^*\|), \tag{9.15}
$$

and,

$$
\langle Sx_n - Sx^*, j(x_n - x^*) \rangle \leq \|x_n - x^*\|^2 - \Phi(\|x_n - x^*\|), \tag{9.16}
$$

where $L := L^* + 1$ and L^* is the generalized Lipschitz constant of A. Uniqueness of the fixed point follows from inequality (9.16), as in the proof of Theorem 9.2. Set $\alpha_n := b_n + c_n$. Then, the recursion formula (9.14) reduces to the formula

$$x_{n+1} = (1 - \alpha_n)x_n + \alpha_n Sx_n + U_n, n \geq 0, \qquad (9.17)$$

where $U_n := c_n(u_n - Sx_n)$. We first prove that the sequence $\{x_n\}$ is bounded. If $x_0 = Sx_n \ \forall \ n \geq 0$ and $c_n \equiv 0$, then we are done. Suppose this is not the case, i.e., suppose there exists $n_0 \in \mathbb{N}$, the smallest positive integer such that $x_{n_0} \neq Sx_{n_0}$. Without loss of generality define $x_{n_0} = x_0$ and $a_0 := ||x_0 - Sx_0||.||x_0 - x^*||$. From

$$\langle Sx_0 - Sx^*, j(x_0 - x^*) \rangle \leq ||x_0 - x^*||^2 - \Phi(||x_0 - x^*||),$$

we obtain that $||x_0 - x^*|| \leq \Phi^{-1}(a_0)$. Since j is uniformly continuous on bounded subsets of E, given

$$\epsilon_0 := \frac{a_0}{2[2(1 + L)\Phi^{-1}(a_0) + L]} > 0, \ \exists \delta > 0$$

such that $||x - y|| < \delta \Rightarrow ||j(x) - j(y)|| < \epsilon \ \forall \ x, y \in B_R(0)$, for some $R > 0$. Set $N^* := \sup_n ||u_n - x^*||$. Define,

$$M_1 := [N^* + L(1 + 2\Phi^{-1}(a_0))]2\Phi^{-1}(a_0); \quad M := M_1 + 2[N^* + L\{1 + 2\Phi^{-1}(a_0)\}]\epsilon_0.$$

Without loss of generality, we may assume (since $c_n = o(b_n)$) that there exists $n^* \geq n_0$ such that $c_n \leq \frac{\alpha_n a_0}{2M + \alpha_n a_0} \ \forall \ n \geq n^*$. Now, define

$$\gamma_0 := \min_n \frac{1}{2}\left\{1, \frac{\Phi^{-1}(a_0)}{3[2(1 + 2L)\Phi^{-1}(a_0) + 2L]}, \frac{\delta}{2(1 + L)\Phi^{-1}(a_0) + L}, \frac{\Phi^{-1}(a_0)}{3N^*}\right\}.$$

Claim. $||x_n - x^*|| \leq 2\Phi^{-1}(a_0) \ \forall \ n \geq n^*$.
The claim clearly holds for $n = n^*$. Assume $||x_n - x^*|| \leq 2\Phi^{-1}(a_0)$ holds for some $n \geq n^*$. We prove $||x_{n+1} - x^*|| \leq 2\Phi^{-1}(a_0)$. Suppose this is not the case, i.e., suppose that $||x_{n+1} - x^*|| > 2\Phi^{-1}(a_0)$. Then, we obtain the following inequalities:

(a) $||x_n - Sx_n|| \leq [2(1 + L)\Phi^{-1}(a_0) + L]$,
(b) $||x_{n+1} - x^*|| \leq 3\Phi^{-1}(a_0)$,
(c) $||x_n - x^*|| \geq \Phi^{-1}(a_0)$.

We also obtain the following inequality: (d) $||U_n||.||x_n - x^*|| \leq c_n M_1$. Using the recursion formula (9.17), we compute as follows:

$$\|x_{n+1} - x^*\|^2$$
$$\leq \|x_n - x^*\|^2 - 2\alpha_n \langle x_n - Sx_n, j(x_{n+1} - x^*) - j(x_n - x^*) \rangle$$
$$+ 2\|U_n\|.\|j(x_{n+1} - x^*) - j(x_n - x^*)\|$$
$$- 2\alpha_n \langle x_n - Sx_n, j(x_n - x^*) \rangle + 2\|U_n\|.\|x_n - x^*\|$$
$$\leq \|x_n - x^*\|^2 - 2\alpha_n \Phi(\|x_n - x^*\|) + 2\alpha_n \|x_n - Sx_n\| \times$$
$$\|j(x_{n+1} - x^*) - j(x_n - x^*)\|$$
$$+ 2\|U_n\|.\|j(x_{n+1} - x^*) - j(x_n - x^*)\| + 2\|U_n\|.\|x_n - x^*\|$$
$$\leq \|x_n - x^*\|^2 - 2\alpha_n a_0 + 2\alpha_n [2(1+L)\Phi^{-1}(a_0) + L] \times$$
$$\|j(x_{n+1} - x^*) - j(x_n - x^*)\|$$
$$+ M_1 c_n + c_n 2[N^* + L\{1 + 2\Phi^{-1}(a_0)\}].\|j(x_{n+1} - x^*) - j(x_n - x^*)\|.$$

Observe that $(x_{n+1} - x^*), (x_n - x^*) \in B_R(0)$, where $R := 3\Phi^{-1}(a_0) > 0$. Moreover,

$$\|x_{n+1} - x_n\| = \alpha_n \|x_n - Sx_n\| + \|U_n\| \leq 2\gamma_0 [2(1+L)\Phi^{-1}(a_0) + L] < \delta,$$

so that $\|j(x_{n+1} - x^*) - j(x_n - x^*)\| < \epsilon_0$. Hence, we obtain that

$$\|x_{n+1} - x^*\|^2 \leq \|x_n - x^*\|^2 - \alpha_n a_0 + c_n M$$
$$\leq \|x_n - x^*\|^2 - \alpha_n a_0 + \frac{\alpha_n a_0}{2M + \alpha_n a_0} M$$
$$< \|x_n - x^*\|^2,$$

a contradiction. Hence, claim 1 holds so that $\{x_n\}$ is bounded and so, $\{Sx_n\}$ is also bounded. Set $\rho := \sup_n \|x_n - Sx_n\|$; $\beta := \sup_n \|u_n - Sx_n\|$.

$$\sigma := \sup_n \{2\|u_n - Sx_n\|.\|j(x_{n+1} - x^*) - j(x_n - x^*)\| + 2\|u_n - Sx_n\|.\|x_n - x^*\|\}.$$

Then,

$$\|x_{n+1} - x^*\|^2 \leq \|x_n - x^*\|^2 - 2\alpha_n \Phi(\|x_n - x^*\|) + 2\alpha_n \|x_n - Sx_n\| \times$$
$$\|j(x_{n+1} - x^*) - j(x_n - x^*)\|$$
$$+ 2\|U_n\|.\|j(x_{n+1} - x^*) - j(x_n - x^*)\| + 2\|U_n\|.\|x_n - x^*\|$$
$$\leq \|x_n - x^*\|^2 - 2\alpha_n \Phi(\|x_n - x^*\|)$$
$$+ 2\alpha_n \rho.\|j(x_{n+1} - x^*) - j(x_n - x^*)\| + c_n \sigma_n.$$
$$\leq \|x_n - x^*\|^2 - 2\alpha_n \Phi(\|x_n - x^*\|) + \sigma_n$$

where $\sigma_n := 2\alpha_n \rho.\|j(x_{n+1} - x^*) - j(x_n - x^*)\| + c_n \sigma$. The result now follows from Lemma 8.17. $\qquad \square$

Theorem 9.6. *Let E be a uniformly smooth real Banach space. Let $F(T) := \{x \in E : Tx = x\} \neq \emptyset$. Suppose $T : E \to E$ is a generalized Lipschitz and generalized $\Phi-$hemi-contractive mapping. Let $\{a_n\}, \{b_n\}$, and $\{c_n\}$ be real sequences in $[0,1)$ satisfying the following conditions,*
(i) $a_n + b_n + c_n = 1$;

$$(ii) \ \sum b_n = \infty; \ c_n = o(b_n); (iii) \ \sum c_n < \infty; (iv) \ \lim b_n = 0.$$

Let $\{x_n\}$ be generated iteratively from arbitrary $x_0, u_0 \in E$ by,

$$x_{n+1} = a_n x_n + b_n T x_n + c_n u_n, n \geq 0,$$

where $\{u_n\}$ is an arbitrary bounded sequence in E. Then, there exists a constant $\gamma_0 \in \mathbb{R}^+$ such that if $b_n + c_n \leq \gamma_0 \ \forall \ n \geq 0$, the sequence $\{x_n\}$ converges strongly to the unique fixed point of T.

Proof. Since T is a generalized $\Phi-$hemi-contractive mapping, it follows that $\langle Tx - Tx^*, j(x - x^*) \rangle \leq ||x - x^*||^2 - \Phi(||x - x^*||) \ \forall \ x \in E, x^* \in F(T)$. In particular, $\langle Tx_n - Tx^*, j(x_n - x^*) \rangle \leq ||x_n - x^*||^2 - \Phi(||x_n - x^*||)$. Since T is generalized Lipschitz, we have $||Tx_n - Tx^*|| \leq L(1 + ||x_n - x^*||)$. The rest now follows exactly as in the proof of Theorem 9.5, with S replaced by T. $\square$

EXERCISES 9.1

1. Verify that every Lipschitz map is generalized Lipschitz and every map with bounded range is generalized Lipschitz.
2. Let $E = (-\infty, +\infty)$ and $T : E \to E$ be defined by

$$Tx = \begin{cases} x - 1, & \text{if } x \in (-\infty, -1), \\ x - \sqrt{1 - (x+1)^2}, & \text{if } x \in [-1, 0), \\ x + \sqrt{1 - (x-1)^2}, & \text{if } x \in [0, 1], \\ x + 1, & \text{if } x \in (1, +\infty). \end{cases}$$

Show that T is a generalized Lipschitz map which is not Lipschitz and whose range is not bounded.
3. Let K be a nonempty closed convex subset of a real Banach space E, $T : K \to K$ be an asymptotically quasi-nonexpansive self-mapping. For arbitrary $x_0 \in K$, let $\{x_n\}$ be iteratively defined as follows: $x_{n+1} := a_n x_n + b_n T^n x_n + c_n u_n$, where $\{a_n\}, \{b_n\}$ and $\{c_n\}$ are sequences of real numbers satisfying the following conditions: $(i) a_n + b_n + c_n = 1$, $(ii) \ 0 < b_n + c_n < 1$, $(iii) \sum b_n = \infty$, $(iv) \sum c_n < \infty$ and $\{u_n\}$ is a bounded sequence in K. Prove that $\{x_n\}$ converges strongly to a fixed point of T if and only if $\liminf_{n \to \infty} d(x_n, F(T)) = 0$, where $d(y, K)$ denotes the distance of y from the set K, i.e. $d(y, K) := \inf_{x \in K} ||y - x||$.

For the next exercise, the following notion will be needed.

Let $T : K \to K$ be a uniformly continuous map, then we have that the set $\{\|Tx - Ty\| : \|x - y\| < \delta\}$ is bounded and the function $\omega(\delta) := \sup_n \{\|Tx - Ty\| : \|x - y\| < \delta\}$ is called the *modulus of continuity* of T. Also, the uniform continuity of T implies that $\omega(\delta) \to 0$ as $\delta \to 0$. Now solve the following exercise.

4. Let K be a nonempty closed convex subset of a real reflexive Banach space E possessing a uniformly Gâteaux differentiable norm for which every closed convex and bounded subset has the fixed point property (f.p.p.) for nonexpansive self-mappings. Let $T : K \to K$ be a uniformly continuous pseudo-contraction with modulus of continuity ω. Let $\{\lambda_n\}$, $\{\theta_n\}$ and $\{c_n\}$ be real sequences in $(0, 1]$ satisfying the following conditions: *(i)* $\{\theta_n\}$ is non-increasing and $\lim_{n\to\infty} \theta_n = 0$; *(ii)* $\lambda_n(1 + \theta_n) + c_n \le 1$, $\sum \lambda_n \theta_n = \infty$, $\lim_{n\to\infty} \frac{\lambda_n}{\theta_n} = 0$; *(iii)* $\lim_{n\to\infty} \frac{\left(\frac{\theta_{n-1}}{\theta_n} - 1\right)}{\lambda_n \theta_n} = 0$. *(iv)* $c_n = o(\lambda_n \theta_n)$. Let $F(T) \ne \emptyset$, and let a sequence $\{x_n\}$ be generated from arbitrary $x_1 \in K$ by $x_{n+1} := (1 - \lambda_n)x_n + \lambda_n T x_n - \lambda_n \theta_n (x_n - x_1) - c_n(x_n - u_n)$, $n \in \mathbb{N}$, where $\{u_n\}$ is a bounded sequence in K. Suppose that $\{x_n\}$ and $\{\frac{\omega(M\lambda_n)}{\lambda_n}\}$ (for any $M > 0$) are bounded. Prove that $\{x_n\}$ converges strongly to a fixed point of T. Give examples of λ_n and θ_n which satisfy conditions $(i) - (iv)$.

9.4 Historical Remarks

Numerous convergence results have been proved on iterative methods for approximating zeros of *Lipschitz* Φ−strongly accretive-type (or fixed points of Φ− strongly pseudo-contractive-type) nonlinear mappings and their stability (see e.g., Chang *et al.* [75], Chang and Tan [83], Chidume ([88, 89], [97]- [102], [115, 116]), Chidume and Zegeye [174]; Deng [202, 203]; Deng and Ding [204]; Lim [305]; Reich [406]; Rhoades [422]; Schu [434]; Shahzad and Zegeye [446], Takahashi [474], and the references contained therein). Also, many authors have proved convergence theorems under the assumption that these operators have *bounded range* (see e.g., Browder and Petryshyn [52], Hirano and Huang [252] and the references contained therein).

Some of these results have been extended to *uniformly continuous* mappings. The most general results for uniformly continuous ϕ−hemi-contractive mappings seem to be the following theorems (stated here as have been published).

Theorem G1 ([244], Theorem 2.1) *Let E be a real normed linear space, K be a nonempty subset of E and $T : K \to E$ be a uniformly continuous Φ−pseudo-contractive-type operator, i.e., there exist $x^* \in K$ and a strictly increasing function $\Phi : [0,\infty) \to [0,\infty)$, $\Phi(0) = 0$ such* *that for all* $x \in K$, *there exists* $j(x - x^*) \in J(x - x^*)$ *satisfying* $\langle Tx - x^*, j(x - x^*) \rangle \leq ||x - x^*||^2 - \Phi(||x - x^*||)$. *(a) If* $y^* \in K$ *is a fixed point of* T, *then* $y^* = x^*$, *and so* T *has at most one fixed point in* K; *(b) Suppose there exists* $x_0 \in K$ *such that both the Ishikawa iterative sequence* $\{x_n\}$ *with error and the auxiliary* $y_n = (1 - \beta_n)x_n + \beta_n Tx_n + v_n$, $n \geq 0$, $x_{n+1} = (1 - \alpha_n)x_n + \alpha_n Ty_n + u_n$, *are contained in* K, *where* $\{u_n\}$, $\{v_n\}$ *are two sequences in* E *and* $\{\alpha_n\}, \{\beta_n\}$ *are two sequences in* $[0, 1]$ *satisfying the following conditions:* $(i)\alpha_n, \beta_n \to 0$ $(n \to \infty)$ *and* $\sum \alpha_n = \infty$; (ii) $||u_n|| = o(\alpha_n)$ *and* $||v_n|| \to 0$ $(n \to \infty)$. *If* $\{x_n\}$ *is a bounded sequence in* K, *then* $\{x_n\}$ *converges strongly to* x^*. *In particular, if* y^* *is a fixed point of* T *in* K, *then* $\{x_n\}$ *converges strongly to* y^*.

Theorem G2 ([244], Theorem 2.2) *Let* E *be a real normed linear space,* K *be a nonempty subset of* E *such that* $K + K \subset K$. *Let* $T : K \to K$ *be a uniformly continuous* $\Phi-$*pseudo-contractive-type operator. Let* $\{u_n\}, \{v_n\}, \{\alpha_n\}, \{\beta_n\}$ *be as in Theorem G1. For any given* $x_0 \in K$, *the Ishikawa iterative sequence* $\{x_n\}$ *with errors is defined as in Theorem G1. (a) If* $y^* \in K$ *is a fixed point of* T, *then* $y^* = x^*$, *and so* T *has at most one fixed point in* K; *(b) If* $\{x_n\}$ *is a bounded sequence, then* $\{x_n\}$ *converges strongly to* x^*. *In particular, if* y^* *is a fixed point of* T *in* K, *then* $\{x_n\}$ *converges strongly to* y^*.

Theorem CCZ1 ([76], Theorem 7.2.1, p.248) *Let* E *be a real normed linear space,* K *be a nonempty convex subset of* E *such that* $K + K \subset K$, *and* $T : K \to K$ *be a uniformly continuous and* $\phi-$*hemi-contractive mapping. Let* $\{\alpha_n\}$, $\{\beta_n\}$ *be two real sequences in* $(0, 1)$ *satisfying the following conditions:* $(i)\alpha_n, \beta_n \to 0$, $(n \to \infty)$; $(ii) \sum \alpha_n = \infty$. *Assume that* $\{u_n\}, \{v_n\}$ *are two sequences in* K *satisfying the following conditions:* $u_n = u_n' + u_n''$ *for any sequences* $\{u_n'\}, \{u_n''\}$ *in* K *with* $\sum ||u_n'|| < \infty$; $||u_n''|| = o(\alpha_n)$ *and* $||v_n|| \to 0$ *as* $n \to \infty$. *Define the Ishikawa iterative sequence with mixed errors in* K *by* $x_0 \in K$, $y_n = (1-\beta_n)x_n + \beta_n Tx_n + v_n$, $n \geq 0$, $x_{n+1} = (1-\alpha_n)x_n + \alpha_n Ty_n + u_n$. *If* $\{Ty_n\}$ *is bounded, then the sequence* $\{x_n\}$ *converges strongly to the unique fixed point of* T.

Theorem CCZ2 ([76], Theorem 7.2.2, p.251) *Let* E *be a real normed linear space, and* $T : E \to E$ *be a uniformly continuous and strongly* $\phi-$*quasi-accretive mapping. Let* $\{\alpha_n\}$, $\{\beta_n\}$ *be two real sequences in* $(0, 1)$ *satisfying the following conditions:* $(i)\alpha_n, \beta_n \to 0, (n \to \infty)$; $(ii) \sum \alpha_n = \infty$. *Let* $\{u_n\}, \{v_n\}$ *be as in Theorem CCZ1. Define a mapping* $S : E \to E$ *by* $Sx := x - Tx$ *for each* $x \in E$. *For an arbitrary* $x_0 \in E$, *define the Ishikawa iterative sequence* $\{x_n\}$ *with mixed errors by* $x_0 \in E$, $y_n = (1 - \beta_n)x_n + \beta_n Tx_n + v_n$, $n \geq 0$, $x_{n+1} = (1 - \alpha_n)x_n + \alpha_n Ty_n + u_n$. *If* $\{Sy_n\}$ *is bounded, then the sequence* $\{x_n\}$ *converges strongly to the unique fixed point of* T.

Theorems G1, G2, CCZ1, and CCZ2 are important generalizations of several results. We observe that the class of mappings considered in Theorems CCZ1 and CCZ2 is a *proper subclass* of the class of mappings studied in Theorems G1 and G2 in which $\Phi(s) = s\phi(s)$. However, the requirement that $\{x_n\}$ be bounded imposed in Theorems G1 and G2 is stronger than the requirement that $\{Ty_n\}$ or $\{Sy_n\}$ be bounded imposed in Theorems CCZ1 and CCZ2, respectively.

For operators with bounded range, the following interesting results have been proved.

Theorem HH1 ([252], Corollary 1) *Let E be a uniformly smooth Banach space and $T : E \rightarrow 2^E$ be a multi-valued $\phi-$hemi-contractive operator with bounded range. Suppose $\{a_n\}, \{b_n\}, \{c_n\}$ are real sequences in $[0, 1]$ satisfying the following conditions: (i) $a_n+b_n+c_n=1 \,\forall\, n\geq 0$; (ii) $\lim b_n=0$; (iii) $\sum b_n = \infty$; (iv) $c_n = o(b_n)$. For arbitrary $x_1, u_1 \in E$, define the sequence $\{x_n\}$ by $x_{n+1} = a_n x_n + b_n \nu_n + c_n u_n, \exists \nu_n \in Tx_n, n \geq 1$; where $\{u_n\}$ is an arbitrary bounded sequences in E. Then, $\{x_n\}$ converges strongly to the unique fixed point of T.*

Theorem HH2 ([252]) *Let E be a uniformly smooth Banach space and $T : E \rightarrow 2^E$ be a multi-valued $\phi-$strongly accretive operator with bounded range. For a given $f \in E$, let x^* denote a solution of the inclusion $f \in Tx$. Define the operator $S : E \rightarrow E$ by $Sx = f + x - Tx$. Suppose $\{a_n\}, \{b_n\}, \{c_n\}$ are real sequences in $[0, 1]$ satisfying conditions (i)-(iv) of Theorem HH1. For arbitrary $x_1, u_1 \in E$, define the sequence $\{x_n\}$ by $x_{n+1} = a_n x_n + b_n \nu_n + c_n u_n, \exists \nu_n \in Sx_n, n \geq 1$; where $\{u_n\}$ is an arbitrary bounded sequences in E. Then, $\{x_n\}$ converges strongly to the unique solution of the inclusion $f \in Tx$.*

Theorem HH2 is an immediate corollary of Theorem 2 of [252]. Theorems HH1 and HH2 include several recent important results as special cases.

All the theorems of this chapter are due to Chidume and Chidume [129, 131].

Chapter 10
An Example; Mann Iteration for Strictly Pseudo-contractive Mappings

10.1 Introduction and a Convergence Theorem

We have seen (Chapter 6) that the Mann iteration method has been successfully employed in approximating fixed points (when they exist) of nonexpansive mappings. This success has not carried over to the more general class of pseudo-contractions. If K is a compact convex subset of a Hilbert space and $T : K \rightarrow K$ is Lipschitz, then, by Schauder fixed point theorem, T has a fixed point in K. All efforts to approximate such a fixed point by means of the Mann sequence when T is also assumed to be pseudo-contractive proved abortive. In 1974, Ishikawa introduced a new iteration scheme and proved the following theorem.

Theorem 10.1. *If K is a compact convex subset of a Hilbert space H, $T : K \mapsto K$ is a Lipschitzian pseudo-contractive map and x_0 is any point of K, then the sequence $\{x_n\}_{n\geq0}$ converges strongly to a fixed point of T, where x_n is defined iteratively for each positive integer $n \geq 0$ by*

$$x_{n+1} = (1 - \alpha_n)x_n + \alpha_n Ty_n; \quad y_n = (1 - \beta_n)x_n + \beta_n Tx_n, \qquad (10.1)$$

where $\{\alpha_n\}$, $\{\beta_n\}$ are sequences of positive numbers satisfying the conditions (i) $0 \leq \alpha_n \leq \beta_n < 1$; (ii) $\lim\limits_{n\to\infty} \beta_n = 0$; (iii) $\sum\limits_{n\geq0} \alpha_n\beta_n = \infty$.

The recursion formula (10.1) with conditions $(i), (ii)$ and (iii) is generally referred to as the *Ishikawa iteration process*.

10.2 An Example

Since its publication in 1974, it had remained an open question (see e.g., Borwein and Borwein [33], Chidume and Moore [145], Hicks and Kubicek [251]) whether or not the Mann recursion formula defined by (6.3), which is certainly simpler than the Ishikawa recursion formula (10.1), converges under the setting of theorem 10.1 to a fixed point of T if the operator T is pseudo-contractive and Lipschitz. Hicks and Kubicek [251], gave an example of a *discontinuous* pseudo-contraction with a unique fixed point for which the Mann iteration does not always converge. Borwein and Borwein [33], (Proposition 8), gave an example of a Lipschitz map (which is not pseudo-contractive) with a unique fixed point for which the Mann sequence fails to converge. The problem for *Lipschitz* pseudo-contraction still remained open. This was eventually resolved in the negative by Chidume and Mutangadura [148] in the following example.

Example 10.2. Let X be the real Hilbert space $\mathbb{R}^2$ under the usual Euclidean inner product. If $x = (a, b) \in X$ we define $x^\perp \in X$ to be $(b, -a)$. Trivially, we have $\langle x, x^\perp \rangle = 0$, $||x^\perp|| = ||x||$, $\langle x^\perp, y^\perp \rangle = \langle x, y \rangle$, $||x^\perp - y^\perp|| = ||x - y||$ and $\langle x^\perp, y \rangle + \langle x, y^\perp \rangle = 0$ for all $x, y \in X$. We take our closed and bounded convex set K to be the closed unit ball in X and put $K_1 = \{x \in X : ||x|| \leq \frac{1}{2}\}$, $K_2 = \{x \in X : \frac{1}{2} \leq ||x|| \leq 1\}$. We define the map $T : K \longrightarrow K$ as follows:

$$Tx = \begin{cases} x + x^\perp, & \text{if } x \in K_1. \\ \frac{x}{||x||} - x + x^\perp, & \text{if } x \in K_2. \end{cases}$$

Then, T is a Lipschitz pseudo-contractive map of a compact convex set into itself with a unique fixed point for which no Mann sequence converges.

We notice that, for $x \in K_1 \cap K_2$, the two possible expressions for Tx coincide and that T is continuous on both of K_1 and K_2. Hence T is continuous on all of K. We now show that T is, in fact, Lipschitz. One easily shows that $||Tx - Ty|| = \sqrt{2}||x - y||$ for $x, y \in K_1$. For $x, y \in K_2$, we have

$$\left|\left| \frac{x}{||x||} - \frac{y}{||y||} \right|\right|^2 = \frac{2}{||x||||y||}(||x||||y|| - \langle x, y \rangle)$$
$$= \frac{1}{||x||||y||}\{||x - y||^2 - (||x|| - ||y||)^2\}$$
$$\leq \frac{1}{||x||||y||} 2||x - y||^2 \leq 8||x - y||^2.$$

Hence, for $x, y \in K_2$, we have

$$||Tx - Ty|| \leq \left|\left| \frac{x}{||x||} - \frac{y}{||y||} \right|\right| + ||x - y|| + ||x^\perp - y^\perp|| \leq 5||x - y||,$$

so that T is Lipschitz on K_2. Now let x and y be in the interiors of K_1 and K_2 respectively. Then there exist $\lambda \in (0, 1)$ and $z \in K_1 \cap K_2$ for which $z = \lambda x + (1 - \lambda)y$. Hence

$$||Tx - Ty|| \leq ||Tx - Tz|| + ||Tz - Ty||$$
$$\leq \sqrt{2}||x - z|| + 5||z - y||$$
$$\leq 5||x - z|| + 5||z - y|| = 5||x - y||.$$

Thus $||Tx - Ty|| \leq 5||x - y||$ for all x, $y \in K$, as required. The origin is clearly a fixed point of T. For $x \in K_1$, $||Tx||^2 = 2||x||^2$, and for $x \in K_2$, $||Tx||^2 = 1 + 2||x||^2 - 2||x||$. From these expressions and from the fact that $Tx = x^\perp \neq x$ if $||x|| = 1$, it is easy to show that the origin is the only fixed point of T. We now show that no Mann iteration sequence for T is convergent for any nonzero starting point.

First, we show that no such Mann sequence converges to the fixed point. Let $x \in K$ be such that $x \neq 0$. Then, in case $x \in K_1$, any Mann iterate of x is actually further away from the fixed point of T than x is. This is because $||(1 - \lambda)x + \lambda Tx||^2 = (1 + \lambda^2)||x||^2 > ||x||^2$ for $\lambda \in (0, 1)$. If $x \in K_2$ then, for any $\lambda \in (0, 1)$,

$$||(1 - \lambda)x + \lambda Tx||^2 = ||(\tfrac{\lambda}{||x||} + 1 - 2\lambda)x + \lambda x^\perp||^2$$
$$= [(\tfrac{\lambda}{||x||} + 1 - 2\lambda)^2 + \lambda^2]||x||^2 > 0.$$

More generally, it is easy to see that for the recursion formula

$$x_0 \in K, \ x_{n+1} = (1 - c_n)x_n + c_n Tx_n, \ n \geq 0, \tag{10.2}$$

if $x_0 \in K_1$ then $||x_{n+1}|| > ||x_n||$ for all integers $n \geq 0$, and if $x_0 \in K_2$, then $||x_{n+1}|| \geq \tfrac{\sqrt{2}}{2}||x_n||$ for all integers $n \geq 0$. We therefore conclude that, in addition, any Mann iterate of any non zero vector in K is itself non zero. Thus any Mann sequence $\{x_n\}$, starting from a nonzero vector, must be infinite. For such a sequence to converge to the origin, x_n would have to lie in the neighborhood K_1 of the origin for all $n > N_0$, for some real N_0. This is not possible because, as already established for K_1, $||x_n|| < ||x_{n+1}||$ for all $n > N_0$.

We now show that no Mann sequence converges to $x \neq 0$. We do this in the form of a general lemma.

Lemma 10.3. *Let M be a nonempty, closed and convex subset of a real Banach space E and let $S : M \to M$ be any continuous function. If a Mann sequence for S is norm convergent, then the corresponding limit is a fixed point for S.*

Proof. Let $\{x_n\}$ be a Mann sequence in M for S, as defined in the recursion formula (10.2). Assume, for proof by contradiction, that the sequence converges, in norm, to x in M, where $Sx \neq x$. For each $n \in \mathbb{N}$, put $\epsilon_n = x_n - Sx_n - x + Sx$. Since S is continuous, the sequence ϵ_n converges to 0. Pick $p \in \mathbb{N}$ such that, if $m \geq p$ and $n \geq p$, then $||\epsilon_n|| < \tfrac{1}{3}||x - Sx||$ and $||x_n - x_m|| < \tfrac{1}{3}||x - Sx||$. Pick any positive integer q such that $\sum\limits_{n=p}^{p+q} c_n \geq 1$. We have that

$$||x_p - x_{p+q+1}|| = \left|\left|\sum_{n=p}^{p+q}(x_n - x_{n+1})\right|\right| = \left|\left|\sum_{n=p}^{p+q} c_n(x - Sx + \epsilon_n)\right|\right|$$

$$\geq \left|\left|\sum_{n=p}^{p+q} c_n(x - Sx)\right|\right| - \left|\left|\sum_{n=p}^{p+q} c_n \epsilon_n\right|\right|$$

$$\geq \sum_{n=p}^{p+q} c_n\left(||x - Sx|| - \tfrac{1}{3}||x - Sx||\right) \geq \tfrac{2}{3}||x - Sx||.$$

The contradiction proves the result. □

We now show that T is a pseudo-contraction. First, we note that we may put $j(x) = x$, since X is Hilbert. For $x, y \in K$, put $\Gamma(x;y) = ||x - y||^2 - \langle Tx - Ty, x - y\rangle$ and, if x and y are both non zero, put $\lambda(x;y) = \dfrac{\langle x, y\rangle}{||x||||y||}$. Hence, to show that T is a pseudo-contraction, we need to prove that $\Gamma(x;y) \geq 0$ for all $x, y \in K$. We only need examine the following three cases:

1. $x, y \in K_1$:   An easy computation shows that $\langle Tx - Ty, x - y\rangle = ||x - y||^2$ so that $\Gamma(x;y) = 0$; thus we are home and dry for this case.
2. $x, y \in K_2$:   Again, a straightforward calculation shows that

$$\langle Tx - Ty, x - y\rangle = ||x|| - ||x||^2 + ||y|| - ||y||^2$$
$$+\langle x, y\rangle\left(2 - \frac{1}{||x||} - \frac{1}{||y||}\right)$$
$$= ||x|| - ||x||^2 + ||y|| - ||y||^2$$
$$+\lambda(x;y)(2||x||||y|| - ||x|| - ||y||).$$

Hence $\Gamma(x;y) = 2||x||^2 + 2||y||^2 - ||x|| - ||y|| - \lambda(x;y)(4||x||||y|| - ||x|| - ||y||)$. It is not hard to establish that $(4||x||||y|| - ||x|| - ||y||) \geq 0$ for all $x, y \in K_2$. Hence, for fixed $||x||$ and $||y||$, $\Gamma(x;y)$ has a minimum when $\lambda(x;y) = 1$. This minimum is therefore $2||x||^2 + 2||y||^2 - 4||x||||y|| = 2(||x|| - ||y||)^2$. Again, we have that $\Gamma(x;y) \geq 0$ for all $x, y \in K_2$ as required.
3. $x \in K_1, y \in K_2$:   We have
$\langle Tx - Ty, x - y\rangle = ||x||^2 + ||y|| - ||y||^2 - \lambda(x;y)||x||$.
Hence $\Gamma(x;y) = 2||y||^2 - ||y|| + (||x|| - 2||x||||y||)\lambda(x;y)$. Since $||x|| - 2||x||||y|| \leq 0$ for $x \in K_1$ and $y \in K_2$, $\Gamma(x;y)$ has its minimum, for fixed $||x||$ and $||y||$ when $\lambda(x;y) = 1$. We conclude that

$$\Gamma(x;y) \geq 2||y||^2 - ||y|| + ||x|| - 2||x||||y|||$$
$$= (||y|| - ||x||)(2||y|| - 1)$$
$$\geq 0 \text{ for all } x \in K_1, y \in K_2.$$

This completes the proof. □

10.3 Mann Iteration for a Class of Lipschitz Pseudo-contractive Maps

A mapping $T : K \to E$ is said to be *strictly pseudo-contractive in the sense of Browder and Petryshyn* [52] if

$$\langle Tx - Ty, j(x - y) \rangle \leq \|x - y\|^2 - \lambda \|x - y - (Tx - Ty)\|^2 \qquad (10.3)$$

holds for $x, y \in K$ and for some $\lambda > 0$.

It is easy to see that such mappings are Lipschitz with Lipschitzian constant $L = \dfrac{1 + \lambda}{\lambda}$.

This class of mappings was introduced in 1967 by Browder and Petryshyn [52] who actually defined it in a Hilbert space as follows: Let K be a nonempty subset of real Hilbert space. A mapping $T : K \to K$ is said to be strictly pseudo-contractive if

$$\|Tx - Ty\|^2 \leq \|x - y\|^2 + k\|x - y - (Tx - Ty)\|^2 \qquad (10.4)$$

holds for all $x, y \in K$ and for some $0 < k < 1$. Clearly, nonexpansive mappings satisfy (10.4) and it is also easy to see that in real Hilbert spaces, inequalities (10.3) and (10.4) are equivalent. The class of strictly pseudo-contractive mappings is a subclass of the class of Lipshitz pesudo-contractive ones.

From now on, we shall denote the class of mappings which are *strictly pseudo-contractive* in the sense of Browder and Petryshyn by $\mathcal{M}$. For $\mathcal{M}$, Browder and Petryshyn proved the following theorem.

Theorem 10.4. *(Browder and Petryshyn, [52]) Let K be a bounded closed convex subset of a real Hilbert space H and let $T : K \to K$ be a mapping belonging to $\mathcal{M}$. Then, for any $x_0 \in K$ and any fixed γ such that $1 - k < \gamma < 1$, the sequence $\{x_n\}$ defined by*

$$x_{n+1} = \gamma T x_n + (1 - \gamma)x_n, \ n = 1, 2, ...,$$

converges weakly to some fixed point of T in K. If, in addition, T is demi-compact, then $\{x_n\}$ converges strongly to some fixed point of T in K.

Maruster [325] proved, for $T \in \mathcal{M}$, $F(T) \neq \emptyset$, in a real Hilbert space, that a Halpern-type iteration process converges to a fixed point of T. Chidume [101] extended this result of Maruster to L_p spaces, $p \geq 2$. Interest in iterative methods for approximating fixed points of mappings belonging to $\mathcal{M}$ has continued to grow (see, e.g., [255],[256],[120],[374] and the references contained therein).

While the example of Chidume and Mutangadura shows that the Mann sequence will not always converge to a fixed point of a Lipschitz pseudo-contractive mapping defined even on a compact convex subset of a Hilbert space, we show in this section that it can be used to approximate a fixed point of the *important subclass* of the class of Lipschitz pseudo-contractive maps consisting of mappings which are strictly pseudo-contractive in the sense of Browder and Petryshyn.

We begin with the following lemma.

Lemma 10.5. *Let $\{\sigma_n\}$ and $\{\beta_n\}$ be sequences of non-negative real numbers satisfying the following inequality $\beta_{n+1} \leq (1 + \sigma_n)\beta_n$, $n \geq 0$. If $\sum\limits_{n \geq 0} \sigma_n < \infty$ then $\lim\limits_{n \to \infty} \beta_n$ exists and if there exists a subsequence of $\{\beta_n\}$ converging to 0, then $\lim\limits_{n \to \infty} \beta_n = 0$.*

We now prove the following results.

In what follows L will denote the Lipschitz constant of T, and λ is the constant appearing in inequality (10.3).

Lemma 10.6. *Let E be a real Banach space. Let K be a nonempty closed and convex subset of E. Let $T : K \to K$ be a strictly pseudo-contractive map in the sense of Browder and Petryshyn with $F(T) := \{x \in K : Tx = x\} \neq \emptyset$. Let $x^* \in F(T)$. For a fixed $x_0 \in K$, define a sequence $\{x_n\}$ by*

$$x_{n+1} = (1 - \alpha_n)x_n + \alpha_n T x_n \tag{10.5}$$

where $\{\alpha_n\}$ is a real sequence in $[0, 1]$ satisfying the following conditions: (i) $\sum\limits_{n=1}^{\infty} \alpha_n = \infty$, (ii) $\sum\limits_{n=1}^{\infty} \alpha_n^2 < \infty$. Then, (a) $\{x_n\}$ is bounded; (b) $\lim \|x_n - x^*\|$ exists for some $x^* \in F(T)$; (c) $\liminf\limits_{n \to \infty} \|x_n - Tx_n\| = 0$.

Proof. Since T is strictly pseudo-contractive in the sense of Browder and Petryshyn, we have $\langle Tx_n - Tx^*, j(x_n - x^*) \rangle \leq \|x_n - x^*\|^2 - \lambda\|x_n - x^* - (Tx_n - Tx^*)\|^2$. Using the recursion formula (10.5) and inequality (4.4), we obtain that

$$\begin{aligned}
\|x_{n+1} - x^*\|^2 &\leq \|x_n - x^*\|^2 - 2\alpha_n\langle x_{n+1} - x^* + x^* - Tx_{n+1}, j(x_{n+1} - x^*) \rangle \\
&\quad - 2\alpha_n\langle x_n - x_{n+1} + Tx_{n+1} - Tx_n, j(x_{n+1} - x^*) \rangle \\
&\leq \|x_n - x^*\|^2 - 2\alpha_n\|x_{n+1} - x^*\|^2 \\
&\quad + 2\alpha_n\langle Tx_{n+1} - x^*, j(x_{n+1} - x^*) \rangle \\
&\quad + 2\alpha_n^2(1 + L)\|x_n - Tx_n\|\,\|x_{n+1} - x^*\| \\
&\leq \|x_n - x^*\|^2 - 2\alpha_n\lambda\|x_{n+1} - Tx_{n+1}\|^2 \\
&\quad + 2\alpha_n^2(1 + L)\|x_n - Tx_n\|\,\|x_{n+1} - x^*\| \\
&\leq \|x_n - x^*\|^2 + 2\alpha_n^2(1 + L) \times \\
&\quad \left(\|x_n - x^*\| + L\|x_n - x^*\| \right)\|x_{n+1} - x^*\|
\end{aligned}$$

$$\leq \|x_n - x^*\|^2 + 2\alpha_n^2(1+L)^2\|x_n - x^*\| \times$$
$$\left[(1+\alpha_n(1+L))\|x_n - x^*\|\right]$$
$$= \left[1 + 2\alpha_n^2(1+L)^2(1+\alpha_n(1+L))\right]\|x_n - x^*\|^2$$
$$\leq \left(1 + C\alpha_n^2)\right)\|x_n - x^*\|^2 \leq \|x_1 - x^*\|^2 e^{C\sum_{n=1}^{\infty}\alpha_n^2} < \infty, \quad (*)$$

for some constant $C > 0$. Thus, $\{x_n\}$ is bounded and from $(*)$, Lemma 10.5, and condition (ii), we obtain that $\lim_{n \to \infty}\|x_n - x^*\|$ exists. Furthermore,

$$\|x_{n+1} - x^*\|^2 \leq \|x_n - x^*\|^2 - 2\alpha_n\lambda\|x_{n+1} - Tx_{n+1}\|^2$$
$$+ 2\alpha_n^2(1+L)\|x_n - Tx_n\| \, \|x_{n+1} - x^*\|$$
$$\leq \|x_n - x^*\|^2 - 2\alpha_n\lambda\|x_{n+1} - Tx_{n+1}\|^2 + \alpha_n^2 M,$$

where $M := 2(1+L)\sup(\|x_n - Tx_n\| \, \|x_{n+1} - x^*\|)$. The last inequality now implies that

$$\sum \alpha_n\lambda\|x_{n+1} - Tx_{n+1}\|^2 \leq \sum \left(\|x_n - x^*\|^2 - \|x_{n+1} - x^*\|^2\right) + \sum \alpha_n^2 M < \infty.$$

Condition (i) now implies that $\liminf\|x_{n+1} - Tx_{n+1}\| = 0$, completing the proof. $\qquad\square$

Theorem 10.7. *Let E be a real Banach space. Let K be a nonempty closed and convex subset of E. Let $T : K \to K$ be a strictly pseudo-contractive map in the sense of Browder and Petryshyn with $F(T) := \{x \in K : Tx = x\} \neq \emptyset$. For a fixed $x_0 \in K$, define a sequence $\{x_n\}$ by*

$$x_{n+1} = (1 - \alpha_n)x_n + \alpha_n Tx_n \qquad (10.6)$$

where $\{\alpha_n\}$ is a real sequence satisfying the following conditions: $(i) \sum \alpha_n = \infty$ and $(ii) \sum \alpha_n^2 < \infty$. If T is demicompact, then $\{x_n\}$ converges strongly to some fixed point of T in K.

Proof. By lemma 10.6, $\liminf_{n \to \infty}\|x_n - Tx_n\| = 0$. Hence, there exists a subsequence of $\{x_n\}$, say $\{x_{n_k}\}$, such that $\lim_{k \to \infty}\|x_{n_k} - Tx_{n_k}\| = 0$. From the fact that T is demicompact, the continuity of T and passing to a subsequence which we still denote by $\{x_{n_k}\}$ we obtain that $\{x_{n_k}\}$ converges to some fixed point, say q, of T. Since $\lim\|x_n - q\|$ exists, we have that $\{x_n\}$ converges strongly to q. The proof is complete. $\qquad\square$

Corollary 10.8 *Let E be a real Banach space. Let K be a nonempty compact and convex subset of E. Let $T : K \to K$ be a strictly pseudo-contractive map in the sense of Browder and Petryshyn. For a fixed $x_0 \in K$, define a sequence $\{x_n\}$ by*

$$x_{n+1} = (1 - \alpha_n)x_n + \alpha_n T x_n \qquad (10.7)$$

where $\{\alpha_n\}$ is a real sequence in $(0,1)$ satisfying the following conditions: (i) $\sum \alpha_n = \infty$ and (ii) $\sum \alpha_n^2 < \infty$. Then, $\{x_n\}$ converges strongly to some fixed point of T in K.

EXERCISES 10.1

1. Let H be the complex plane and $K := \{z \in H : |z| \leq 1\}$. Define $T : K \to K$ by

$$T(re^{i\theta}) = \begin{cases} 2re^{i(\theta + \frac{\pi}{3})}, & \text{for } 0 \leq r \leq \frac{1}{2}, \\ e^{i(\theta + \frac{2\pi}{3})}, & \text{for } \frac{1}{2} < r \leq 1. \end{cases}$$

   Prove that

   (*i*) Zero is the only fixed point of T.
   (*ii*) T is pseudo-contractive.
   Let $c_n := \frac{1}{n+1}$ for all integers $n \geq 0$. Define the sequence $\{z_n\}$ by $z_{n+1} := (1 - c_n)z_n + c_n T z_n, z_0 \in K, n \geq 0$.
   (*iii*) Prove $\{z_n\}$ does not converge to zero.
   (*iv*) Prove T is not continuous.

2. Let K be a compact convex subset of a real Hilbert space, H; $T : K \to K$ a continuous hemi-contractive map (i.e., $F(T) := \{x \in K : Tx = x\} \neq \emptyset$ and $\|Tx - x^*\|^2 \leq \|x - x^*\|^2 + \|x - Tx\|^2 \ \forall \ x \in H, x^* \in F(T)$). Let $\{a_n\}, \{b_n\}, \{c_n\}, \{a_n'\}, \{b_n'\}$ and $\{c_n'\}$ be real sequences in $[0,1]$ satisfying the following conditions:

   (*i*) $a_n + b_n + c_n = a_n' + b_n' + c_n' = 1 \forall n \geq 1$;
   (*ii*) $\lim b_n = \lim b_n' = 0$;
   (*iii*) $\sum c_n < \infty; \sum c_n' < \infty$;
   (*iv*) $0 \leq \alpha_n \leq \beta_n < 1, \forall n \geq 1$, where $\alpha_n := b_n + c_n; \beta_n := b_n' + c_n'$.
   (*v*) $\sum \alpha_n \beta_n = \infty; \sum \alpha_n \beta_n \delta_n < \infty$, where $\delta_n := \|Tx_n - Ty_n\|^2$.

   For arbitrary $x_1, u_1, v_1 \in K$, define the sequence $\{x_n\}$ iteratively by $x_{n+1} = a_n x_n + b_n T y_n + c_n u_n; \ y_n := a_n' x_n + b_n' T x_n + c_n' v_n, n \geq 1$, where $\{u_n\}$ and $\{v_n\}$ are arbitrary bounded sequences in K. Prove that $\{x_n\}$ converges strongly to a fixed point of T.
   (Hint. Chidume and Moore [145], Theorem 1).

3. Prove Lemma 10.5.

4. Let $T : K \to E$ be strictly pseudocontractive in the sense of Browder and Petryshyn. Prove the statement made at the begining of section 10.3 that T is Lipschitz with Lipschitz constant $L := \frac{1}{\lambda}(1 + \lambda)$.

5. Prove that in a real Hilbert space, inequalities (10.3) and (10.4) are equivalent.

10.4 Historical Remarks

Qihou [389] extended Theorem 10.1 to the slightly more general class of Lipschitz hemi-contractive maps. In [390] he proved that if K is a compact convex subset of *a Hilbert space* and $T : K \to K$ is a *continuous* pseudo-contractive map *with a finite number of fixed points* then the Ishikawa iteration sequence defined by (10.1) converges strongly to a fixed point of T. Consequently, while the Mann sequence does not converge to the fixed point of T in Example 10.2, the Ishikawa sequence does. Chidume and Moore [145] also extended Theorem 10.1 to continuous maps under additional assumption that $\sum \alpha_n \beta_n \delta_n < \infty$, where $\delta_n := ||Tx_n - Ty_n||^2$. Conditions similar to this had been imposed in the literature. Reich [408] imposed the condition $\sum_{n=0}^{\infty} c_n^2 ||Tx_n||^2 < \infty$ (where $\{c_n\}$ is a real sequence in (0,1) satisfying appropriate conditions). Furthermore, if T is Lipschitz, then $\sum \alpha_n \beta_n \delta_n < \infty$, for suitable choices of α_n, β_n (see Chidume and Moore, [145]).

While the example of Chidume and Mutangadura [148] shows that the Mann iteration method cannot always be used to approximate a fixed point of Lipschitz pseudo-contractive maps even in a compact convex domain, it has been shown in this section that for the important subclass $\mathcal{M}$ of the class of Lipschitz pseudo-contractions, Mann iteration method can always be used. Marino and Xu [321] recently proved *weak* convergence of the Mann process to a fixed point of $T \in \mathcal{M}$ in *a real Hilbert space*. Theorem 10.7 and Corollary 10.8 yield *strong* convergence in *arbitrary real Banach spaces*. All the results of section 10.3 are due to Chidume, Abbas and Ali [120].

Chapter 11
Approximation of Fixed Points of Lipschitz Pseudo-contractive Mappings

11.1 Iteration Methods for Lipschitz Pseudo-contractions

In Chapter 10, we stated the Ishikawa iteration theorem (Theorem 10.1) which converges strongly to a fixed point of a Lipschitz pseudo-contractive map T defined on a compact convex subset of a Hilbert space.

It is still an open question whether or not Theorem 10.1 can be extended to Banach spaces more general than Hilbert spaces. However, two other iteration methods have been introduced and have successfully been employed to approximate fixed points of Lipschitz pseudo-contractive mappings in certain Banach spaces *more general than Hilbert spaces*.

One of these iteration processes turns out to be simpler than the Ishikawa process and has been proved to converge to a fixed point of a Lipschitz pseudo-contractive map *in arbitrary real Banach spaces*. We explore these iteration processes in this chapter. We begin with the following definition.

Definition 11.1. (see e.g., Schu, [438]) Let $\alpha_n \in (0, \infty)$, $\mu_n \in (0, 1)$ for all nonnegative integers n. Then, $(\{\alpha_n\}, \{\mu_n\})$ is said to have *property (A)* if and only if the following conditions hold: (i) $\{\alpha_n\}$ is decreasing and $\{\mu_n\}$ is strictly increasing; (ii) there is a sequence $\{\beta_n\} \subseteq I\!N$, strictly increasing such that (a) $\lim_{n\to\infty} \beta_n (1 - \mu_n) = \infty$; (b) $\lim_{n\to\infty} \frac{1 - \mu_{(n+\beta_n)}}{1 - \mu_n} = 1$; (c) $\lim_{n\to\infty} \frac{\alpha_n - \alpha_{(n+\beta_n)}}{1 - \mu_n} = 0$.

Then we have the following theorem.

Theorem 11.2. *(Schu, [438]). Let K be a nonempty closed bounded and convex subset of a Hilbert space H. Suppose that (i) $T : K \to K$ is pseudo-contractive and Lipschitzian with constant $L \geq 0$; (ii) $\{\lambda_n\}_{n\in I\!N} \subset (0,1)$ with $\lim_{n\to\infty} \lambda_n = 1$, $\{\alpha_n\}_{n\in I\!N} \subset (0,1)$ with $\lim_{n\to\infty} \alpha_n = 0$, such that $(\{\alpha_n\}, \{\mu_n\})$ has property (A), $((1 - \mu_n)(1 - \lambda_n)^{-1})$ is bounded and $\lim_{n\to\infty} \alpha_n^{-1}(1 - \mu_n) = 0$, where $k_n = (1 + \alpha_n^2(1 + L)^2)^{\frac{1}{2}}$ and $\mu_n := \lambda_n k_n^{-1}$ for all $n \in I\!N$. For arbitrary vectors $z_0, w \in K$, define, for all $n \in I\!N$,*

$$z_{n+1} = (1 - \mu_{n+1})w + \mu_{n+1}y_n; \quad y_n = (1 - \alpha_n)z_n + \alpha_n T z_n. \quad (11.1)$$

Then $\{z_n\}$ converges strongly to the unique fixed point of T closest to w.

Unlike the Ishikawa method, Theorem 11.2 has been extended to real Banach spaces more general than Hilbert spaces. Chidume [116] extended it to *real Banach spaces possessing weakly sequentially continuous duality maps* (e.g., l_p spaces, $1 < p < \infty$). In this extension the iteration parameters $\alpha_n \in (0, \infty)$ and $\mu_n \in (0, 1)$ are such that the pair $(\{\alpha_n\}, \{\mu_n\})$ has the so-called *property (A)*. However, it is known that L_p spaces, $1 < p < \infty, p \neq 2$ do not possess weakly sequentially continuous duality maps. This leads to the following question.

Question* Is it possible to construct an iterative algorithm which converges to a fixed point of a Lipschitz pseudo-contractive map in Banach spaces that include the L_p spaces, $1 < p < \infty$?

A second iteration scheme for approximating fixed points of Lipschitz pseudo-contractive mappings was implicitly introduced by Bruck [56] who actually applied the scheme, still in Hilbert spaces, to approximate a solution of the inclusion $0 \in Au$, where A is an *m-accretive* operator, i.e., A is accretive and $R(I + \lambda A) = H$ for all $\lambda > 0$.

Definition 11.3. Two real sequences $\{\lambda_n\}$ and $\{\theta_n\}$ are called *acceptably paired* if they satisfy the following conditions: $\{\theta_n\}$ is decreasing, $\lim\limits_{n \to \infty} \theta_n = 0$ and there exists a strictly increasing sequence $\{n(i)\}_{i=1}^{\infty}$ of positive integers such that (i) $\liminf\limits_{i} \theta_{n(i)} \sum\limits_{j=n(i)}^{n(i+1)} \lambda_j > 0$, (ii) $\lim\limits_{i} [\theta_{n(i)} - \theta_{n(i+1)}] \sum\limits_{j=n(i)}^{n(i+1)} \lambda_j = 0$,

(iii) $\lim\limits_{i} \sum\limits_{j=n(i)}^{n(i+1)} \lambda_j^2 = 0$.

An example of acceptably paired sequences is $\lambda_n = n^{-1}$, $\theta_n = (log\ log\ n)^{-1}$, $n(i) = i^i$ (see e.g., Bruck, [56]). Nevanlinna [351] gave a technique for constructing acceptably paired sequences.

Let H be a Hilbert space, $A : H \to H$ be an *m*-accretive operator with $0 \in R(A)$, the range of A. For arbitrary $z \in H$, Bruck considered the sequence $\{x_n\}$ in H defined by $x_0 \in H$,

$$x_{n+1} = x_n - \lambda_n(Ax_n + \theta_n(x_n - x_1)), \quad (11.2)$$

and proved that if $\{x_n\}$ and $\{Ax_n\}$ are bounded then $\{x_n\}$ converges strongly to some x^*, solution of $0 \in Au$, provided λ_n and θ_n are *acceptably paired sequences*.

The ideas of sequences with property (A) and acceptably paired sequences are due to Halpern [245]. Reich [405, 409] also studied the recursion formula (11.2) for Lipschitz accretive operators on real uniformly convex Banach spaces *with a duality mapping that is weakly sequentially continuous at zero* but with λ_n and θ_n not necessarily being acceptably paired.

Motivated by the papers of Reich [405, 410], Chidume and Zegeye [168] studied the following perturbation of the Mann recurrence relation to approximate fixed points of Lipschitz pseudo-contractive mappings in Banach spaces much more general than Hilbert spaces.

Let E be a real normed space, K be a nonempty convex subset of E, $T : K \to K$ be a Lipschitz pseudo-contractive map. For arbitrary $x_1 \in K$, let the sequence $\{x_n\}$ be defined iteratively by

$$x_{n+1} := (1 - \lambda_n)x_n + \lambda_n T x_n - \lambda_n \theta_n (x_n - x_1), \tag{11.3}$$

where λ_n and θ_n are real sequences in $(0,1)$ satisfying the following conditions: *(i)* $\lim_{n \to \infty} \theta_n = 0$; *(ii)* $\lambda_n(1 + \theta_n) < 1$, $\sum \lambda_n \theta_n = \infty$, $\lim_{n \to \infty} \frac{\lambda_n}{\theta_n} = 0$; *(iii)* $\lim_{n \to \infty} \frac{\left(\frac{\theta_{n-1}}{\theta_n} - 1\right)}{\lambda_n \theta_n} = 0$. Examples of real sequences which satisfy these conditions are $\lambda_n = \frac{1}{(n+1)^a}$ and $\theta_n = \frac{1}{(n+1)^b}$, where $0 < b < a$ and $a + b < 1$ (see e.g., Reich, [405]).

For the rest of this section λ_n and θ_n will be assumed to satisfy conditions $(i) - (iii)$.

In what follows, we shall make use of the following lemma.

Lemma 11.4. *(Reich [409], Morales and Jung [341], Takahashi and Ueda [478]) Let K be closed convex subset of a reflexive Banach space E with a uniformly Gâteaux differentiable norm. Let $T : K \to K$ be continuous pseudo-contractive mapping with $F(T) \neq \emptyset$. Suppose that every closed convex and bounded subset of K has the fixed point property for nonexpansive self-mappings. Then for $u \in K$, the path $t \to y_t \in K, t \in [0,1)$, satisfying, $y_t = (1 - t)Ty_t + tu$, converges strongly to a fixed point Qu of T as $t \to 0$, where Q is the unique sunny nonexpansive retraction from K onto $F(T)$.*

If, in Lemma 11.4, we have that K is bounded then the conclusion of the lemma holds without the requirement that $F(T) \neq \emptyset$ (see, eg., Morales and Jung, [341]).

We note also that in Lemma 11.4 if, in addition, $F(T) \neq \emptyset$, then $\{y_t\}$ is bounded. Furthermore, if E is assumed to have a uniformly Gâteaux differentiable norm and is such that every closed convex and bounded subset of K has the fixed point property for nonexpansive self-mappings, then as $t \to 1$, the path converges strongly to a fixed point of T. In what follows, $\{y_n\}$ denotes the sequence defined by $y_n := y_{t_n} = t_n T y_{t_n} + (1 - t_n)x_1$, where $t_n = \frac{1}{1+\theta_n}$ $\forall n \geq 1$, guaranteed by Lemma 11.4.

We now prove the following theorems.

Theorem 11.5. *Let K be a nonempty closed convex subset of a real Banach space E. Let $T : K \to K$ be a Lipschitz pseudo-contractive map with Lipschitz constant $L > 0$ and $F(T) \neq \emptyset$. Let a sequence $\{x_n\}$ be generated from arbitrary $x_1 \in K$ by*

$$x_{n+1} := (1 - \lambda_n)x_n + \lambda_n T x_n - \lambda_n \theta_n (x_n - x_1), \qquad (11.4)$$

for all positive integers n. Then, $||x_n - Tx_n|| \to 0$ as $n \to \infty$.

Proof. Since $\frac{\lambda_n}{\theta_n} \to 0$, there exists $N_0 > 0$ such that $\frac{\lambda_n}{\theta_n} \leq d := \frac{1}{2(\frac{5}{2}+L)(2+L)}$ $\forall\, n \geq N_0$. Let $x^* \in F(T)$ and $r > 0$ be sufficiently large such that $x_{N_0} \in B_r(x^*)$ and $x_1 \in B_{\frac{r}{2}}(x^*)$. We split the proof into three parts.

(i) *We prove $\{x_n\}$ is bounded.* It suffices to show by induction that $\{x_n\}$ belongs to $B := \overline{B_r(x^*)}$ for all integers $n \geq N_0$. Now, $x_{N_0} \in B$ by construction. Hence we may assume $x_n \in B$ for any $n > N_0$ and prove that $x_{n+1} \in B$. Suppose x_{n+1} is not in B. Then $||x_{n+1} - x^*|| > r$ and thus from the recursion formula (11.4) and inequality (4.4) we get that

$$
\begin{aligned}
||x_{n+1} - x^*||^2 &\leq ||x_n - x^*||^2 - 2\lambda_n \langle (x_n - Tx_n) + \theta_n(x_n - x_1), \\
&\qquad j(x_{n+1} - x^*) \rangle \\
&= ||x_n - x^*||^2 - 2\lambda_n \theta_n ||x_{n+1} - x^*||^2 \\
&\qquad + 2\lambda_n \langle \theta_n(x_{n+1} - x_n) - (x_n - Tx_n) + \theta_n(x_1 - x^*) \\
&\qquad + (x_{n+1} - Tx_{n+1}) \\
&\qquad - (x_{n+1} - Tx_{n+1}), j(x_{n+1} - x^*) \rangle. \qquad (11.5)
\end{aligned}
$$

But since T is pseudo-contractive, we have $\langle x_{n+1} - Tx_{n+1}, j(x_{n+1} - x^*) \rangle \geq 0$. Thus, (11.5) gives

$$
\begin{aligned}
||x_{n+1} - x^*||^2 &\leq ||x_n - x^*||^2 - 2\lambda_n \theta_n ||x_{n+1} - x^*||^2 \\
&\quad + 2\lambda_n \langle \theta_n(x_{n+1} - x_n) + \theta_n(x_1 - x^*) \\
&\quad + (x_{n+1} - Tx_{n+1}) - (x_n - Tx_n), j(x_{n+1} - x^*) \rangle \\
&\leq ||x_n - x^*||^2 - 2\lambda_n \theta_n ||x_{n+1} - x^*||^2 \\
&\quad + 2\lambda_n \Big[(2+L)||x_{n+1} - x_n|| + \theta_n ||x_1 - x^*|| \Big] ||x_{n+1} - x^*|| \\
&\leq ||x_n - x^*||^2 - 2\lambda_n \theta_n ||x_{n+1} - x^*||^2 \\
&\quad + 2\lambda_n \Big[(2+L)\lambda_n \big((2+L)||x_n - x^*|| + \theta_n ||x_1 - x^*|| \big) \\
&\quad + \theta_n ||x_1 - x^*|| \Big] ||x_{n+1} - x^*|| \\
&\leq ||x_n - x^*||^2 - 2\lambda_n \theta_n ||x_{n+1} - x^*||^2 \\
&\quad + 2\lambda_n \Big[\lambda_n(2+L)\Big(\frac{5}{2} + L\Big)r + \frac{\theta_n}{2}r \Big] ||x_{n+1} - x^*||, \qquad (11.6)
\end{aligned}
$$

since $x_n \in B$ and $x_1 \in B_{\frac{r}{2}}(x^*)$. But $||x_{n+1} - x^*|| > ||x_n - x^*||$, so we have from (11.6) that $\theta_n ||x_{n+1} - x^*|| \leq \left[\lambda_n (2 + L)(\frac{5}{2} + L)r + \frac{\theta_n}{2} r \right]$, and hence $||x_{n+1} - x^*|| \leq r$, since $\frac{\lambda_n}{\theta_n} \leq \frac{1}{2(\frac{5}{2}+L)(2+L)}$ $\forall n \geq N_0$. Thus we get a contradiction. Therefore, $x_n \in B$ for all positive integers $n \geq N_0$ and hence the sequence $\{x_n\}$ is bounded.

    (ii) We prove $||x_n - y_n|| \to 0$ as $n \to \infty$. From the recursion formula (11.4) and inequality (4.4) we have that

$$
\begin{aligned}
||x_{n+1} - y_n||^2 &\leq ||x_n - y_n||^2 - 2\lambda_n \theta_n \langle (x_{n+1} - y_n), j(x_{n+1} - y_n) \rangle \\
&\quad + 2\lambda_n \langle \theta_n (x_{n+1} - y_n) - (x_n - Tx_n) - \theta_n (x_n - x_1), \\
&\quad\ \ j(x_{n+1} - y_n) \rangle \\
&= ||x_n - y_n||^2 - 2\lambda_n \theta_n ||x_{n+1} - y_n||^2 + 2\lambda_n \langle \theta_n (x_{n+1} - x_n) \\
&\quad + [\theta_n (x_1 - y_n) - (y_n - Ty_n)] - [(x_{n+1} - Tx_{n+1}) \\
&\quad - (y_n - Ty_n)] + [(x_{n+1} - Tx_{n+1}) - (x_n - Tx_n)], \\
&\quad\ \ j(x_{n+1} - y_n) \rangle .
\end{aligned}
\tag{11.7}
$$

Observe that by the property of y_n and pseudo-contractivity of T we have $\theta_n (x_1 - y_n) - (y_n - Ty_n) = 0$ and $\langle (x_{n+1} - Tx_{n+1}) - (y_n - Ty_n), j(x_{n+1} - y_n) \rangle \geq 0$ for all $n \geq 1$. Thus, we have from (11.7) that

$$
\begin{aligned}
||x_{n+1} - y_n||^2 &\leq ||x_n - y_n||^2 - 2\lambda_n \theta_n ||x_{n+1} - y_n||^2 + 2\lambda_n \langle \theta_n (x_{n+1} - x_n) \\
&\quad + [(x_{n+1} - Tx_{n+1}) - (x_n - Tx_n)], j(x_{n+1} - y_n) \rangle \\
&\leq ||x_n - y_n||^2 - 2\lambda_n \theta_n ||x_{n+1} - y_n||^2 \\
&\quad + 2\lambda_n (2 + L) ||x_{n+1} - x_n|| . ||x_{n+1} - y_n|| \\
&\leq ||x_n - y_n||^2 - 2\lambda_n \theta_n ||x_{n+1} - y_n||^2 \\
&\quad + 2\lambda_n^2 (2 + L) ||x_n - Tx_n \\
&\quad + \theta_n (x_n - x_1)|| . ||x_{n+1} - y_n|| .
\end{aligned}
\tag{11.8}
$$

But since $F(T) \neq \emptyset$, by proposition 2 of [341] we have that $\{y_n\}$ is bounded. Therefore, there exists $M_1 > 0$ such that $\sup_n \{||x_{n+1} - y_n|| . ||x_n - Tx_n + \theta_n (x_n - x_1)||\} \leq M_1$. From (11.8) we get that

$$
\begin{aligned}
||x_{n+1} - y_n||^2 &\leq ||x_n - y_n||^2 - 2\lambda_n \theta_n ||x_{n+1} - y_n||^2 \\
&\quad + 2\lambda_n^2 (2 + L) M_1 .
\end{aligned}
\tag{11.9}
$$

Moreover, by the pseudo-contractivity of T we have that

$$
\begin{aligned}
||y_{n-1} - y_n|| &\leq ||y_{n-1} - y_n + \frac{1}{\theta_n} ((y_{n-1} - Ty_{n-1}) - (y_n - Ty_n))|| \\
&\leq \left(\frac{\theta_{n-1}}{\theta_n} - 1 \right) (||y_{n-1}|| + ||x_1||) .
\end{aligned}
\tag{11.10}
$$

Thus, from (11.9) and (11.10) we get that

$$||x_{n+1} - y_n||^2 \leq ||x_n - y_{n-1}||^2 - 2\lambda_n\theta_n||x_{n+1} - y_n||^2$$

$$+ M\left(\frac{\theta_{n-1}}{\theta_n} - 1\right) + 2\lambda_n^2(2+L)M, \qquad (11.11)$$

for some constant $M > 0$. By Lemma 8.17 and the conditions on $\{\lambda_n\}$ and $\{\theta_n\}$ we get $||x_{n+1} - y_n|| \to 0$. Consequently, $||x_n - y_n|| \to 0$ as $n \to \infty$.
(iii) We prove $||x_n - Tx_n|| \to 0$ as $n \to \infty$. Since $\{y_n\}$ (and hence $\{Ty_n\}$) is bounded, we have $||y_n - Ty_n|| \leq (1 - t_n)|||Ty_n|| + (1 - t_n)||x_1|| \to 0$ as $n \to \infty$. Hence,

$$||x_n - Tx_n|| \leq ||x_n - y_n|| + ||y_n - Ty_n|| + ||Ty_n - Tx_n||$$

$$\leq (1 + L)||x_n - y_n|| + ||y_n - Ty_n|| \to 0, \qquad (11.12)$$

as $n \to \infty$. This completes the proof of the theorem.   □

Theorem 11.6. *Let K be a nonempty closed convex subset of a real reflexive Banach space E with a uniformly Gâteaux differentiable norm. Let $T : K \to K$ be a Lipschitz pseudo-contractive map with Lipschitz constant $L > 0$ and $F(T) \neq \emptyset$. Suppose every closed convex and bounded subset of K has the fixed point property for nonexpansive self-mappings. Let a sequence $\{x_n\}$ be generated from arbitrary $x_1 \in K$ by*

$$x_{n+1} := (1 - \lambda_n)x_n + \lambda_n Tx_n - \lambda_n\theta_n(x_n - x_1), \ \forall \ n \geq 1. \qquad (11.13)$$

Then, $\{x_n\}$ converges strongly to a fixed point of T.

Proof. As in the proof of Theorem 11.5, and using the notations of that theorem, we get that $||x_n - y_n|| \to 0$ as $n \to \infty$, and by Theorem 1 of [341] (see also remarks following Lemma 11.4), we have that $y_n \to x^* \in F(T)$. The conclusion follows.   □

Corollary 11.7 *Let E be a real reflexive Banach space with a uniformly Gâteaux differentiable norm. Let $A : E \to E$ be a Lipschitzian accretive operator with Lipschitz constant $L' \geq 0$ and $N(A) \neq \emptyset$, where $N(A) := \{x \in E : Ax = 0\}$. Suppose every closed convex and bounded subset of E has the fixed point property for nonexpansive self-mappings. Let a sequence $\{x_n\}$ be generated from arbitrary $x_1 \in E$ by*

$$x_{n+1} = x_n - \lambda_n Ax_n - \lambda_n\theta_n(x_n - x_1),$$

for all positive integers n. Then, $\{x_n\}$ converges strongly to a solution of the equation $Ax = 0$.

Proof. Since $T := (I - A)$ is a Lipschitz pseudo-contractive map with Lipschitz constant $L := (L' + 1)$ and the fixed point of T is the solution of the equation $Ax = 0$, the conclusion follows from Theorem 11.6.   □

Corollary 11.8 *Let K be a nonempty closed convex subset of a real uniformly smooth Banach space E. Let $T : K \to K$ be a Lipschitz pseudocontractive map with Lipschitz constant $L \geq 0$ and $F(T) \neq \emptyset$. Let a sequence $\{x_n\}$ be generated from arbitrary $x_0 \in K$ by*

$$x_{n+1} = (1 - \lambda_n)x_n + \lambda_n T x_n - \lambda_n \theta_n (x_n - x_1),$$

for all positive integers n. Then, $\{x_n\}$ converges strongly to a fixed point of T.

Proof. Since every uniformly smooth Banach space is reflexive and has uniformly Gâteaux differentiable norm and every closed bounded convex subset of K has the fixed point property for nonexpansive self-mappings (see e.g., Turett [492]), the conclusion follows from Corollary 11.7. $\square$

Remark 11.9. In Theorem 11.6, if in addition, K is bounded, by Proposition 2 of [341], the sequence $\{y_n\}$ is bounded. Therefore, the condition that $F(T) \neq \emptyset$ will not be required in the proof. Hence, we have the conclusions of Corollary 11.8 and Corollary 11.7 without the assumption that $F(T) \neq \emptyset$.

Remark 11.10. Theorem 11.6, Corollaries 11.7 and 11.8 provide convergence results for fixed points of Lipschitz pseudo-contractive maps in real Banach spaces much more general than Hilbert spaces. In particular, Theorem 11.6 and Corollary 11.8 independently provide affirmative answer to Question*.

EXERCISES 11.1

1. Verify that the sequence $\{x_n\}$ given by the recursion (11.4) is well defined.
2. Verify that the real sequences given by $\lambda_n = \frac{1}{(n+1)^a}, \theta_n = \frac{1}{(n+1)^b}$, where $0 < b < a$ and $a + b < 1$ satisfy the conditions of Theorem 11.6.

11.2 Historical Remarks

1. Example 10.2 shows that for the iterative approximation of fixed points of Lipschitz pseudo-contractions, the Mann iteration method does not always converge and so the Ishikawa iteration process was certainly desirable. However, it is clear that whenever a Mann sequence converges, it is preferred to the Ishikawa process because of its simpler form. But, because the Ishikawa method was successfully applied to approximate fixed points of Lipschitz pseudo-contractive maps in Hilbert spaces, the following modification of it (which we shall refer to as *Ishikawa-type*) was introduced: If K is a nonempty convex subset of a real normed linear space and $T : K \to K$ is a map, then for arbitrary $x_0 \in K$, the sequence $\{x_n\}$ in K is defined by

$$x_{n+1} = (1 - \alpha_n)x_n + \alpha_n T y_n, \quad y_n = (1 - \beta_n)x_n + \beta_n T x_n, \quad (11.14)$$

$n \geq 0$, where, (i) $0 \leq \alpha_n, \beta_n < 1$ for all integers $n \geq 0$; (ii) $\lim \alpha_n = \lim \beta_n = 0$; (iii) $\sum \alpha_n = \infty$.

We emphasize the following observations:

a) The condition $0 \leq \alpha_n \leq \beta_n < 1$ in the Ishikawa scheme is replaced with (i) where α_n and β_n are independent.
b) The condition $\sum \alpha_n \beta_n = \infty$ is replaced with condition (iii) and the condition $\lim \alpha_n = 0$ is introduced.

It turns out, however, that if the Ishikawa-type sequence *defined by (11.14) with conditions* $(i) - (iii)$ converges, then the Mann sequence $\{x_n\}$ defined by $x_0 \in K$, $x_{n+1} = (1 - \alpha_n)x_n + \alpha_n T x_n, n \geq 0$, where, (a) $0 \leq \alpha_n < 1$ for all integers $n \geq 0$; (b) $\lim \alpha_n = 0, (iii)$ $\sum \alpha_n = \infty$, converges. One simply sets $\beta_n \equiv 0$ for all integers $n \geq 0$. In view of this remark and Example 10.2, it is clear that the sequence $\{x_n\}$ defined by the recursion formula (11.14), with α_n and β_n being independent, will not always converge to a fixed point of Theorem 10.1. This re-asserts that the recursion formula (11.14) with conditions $(i) - (iv)$ where α_n and β_n are independent *is not* the Ishikawa iteration scheme. But since the recursion *formula* (11.14) is that introduced by Ishikawa, we decided to call any iteration scheme involving the formula (11.14)with α_n and β_n independent, *Ishikawa-type*. Since the mid 1970's, both the Mann iteration process and the Ishikawa-type method have been studied extensively by numerous authors in order to compare their *rates of convergence* (see., e.g., Rhoades [419] and the references contained therein). By the late 1980's and early 1990's, it became evident that there was no significant difference in the rates of convergence of the two schemes. In fact, the canonical choice of α_n is $\alpha_n = \frac{1}{n+1}, n \geq 0$ for the Mann process, and for the Ishikawa process, it is $\alpha_n = \beta_n = \frac{1}{\sqrt{(n+1)}}$. Furthermore, several authors have recently proved that the Mann and *Ishikawa-type* (many have referred to this as Ishikawa) iteration processes are equivalent for various classes of nonlinear mappings (see e.g., [425], [426], [427], [428], [424]). In the light of the above remarks, the conclusions obtained by these authors confirm the equivalence of the Mann and the Ishikawa-type process for the classes of operators they considered. However, since the Mann sequence always converges whenever the Ishikawa-type sequence converges, the Mann iteration process, because of its simpler form, is superior to this Ishikawa-type method.

2. We have observed that the Ishikawa-type iteration method does not, in general, converge to a fixed point of a Lipschitz pseudo-contractive map. We have seen that the following *three explicit iteration formulas* have been introduced and studied for approximating fixed points of Lipschitz pseudo-contractive mappings.

(I) The iteration formula (10.1) of Theorem 10.1 which is defined by

$$x_{n+1} = (1 - \alpha_n)x_n + \alpha_n T y_n, \quad y_n = (1 - \beta_n)x_n + \beta_n T x_n, \ (11.15)$$

$n \geq 0$. Here, the parameters, α_n and β_n can be chosen easily at the beginning of the process as $\alpha_n = \beta_n = (n+1)^{-\frac{1}{2}}$ for all $n \geq 0$. Furthermore, the sequence $\{x_n\}$ defined by this formula has been proved to converge to a fixed point of T only under the following conditions: $(i)E = H$ is a Hilbert space; $(ii)K \subset H$ is compact and convex; $(iii)T : K \to K$ is Lipschitz and pseudo-contractive.

(II) The iteration formula (11.1) of Theorem 11.2 defined by

$$z_{n+1} = (1 - \mu_{n+1})\omega + \mu_{n+1}y_n;$$
$$y_n = (1 - \alpha_n)z_n + \alpha_n T z_n, \tag{11.16}$$

$n \geq 0$, where the parameters $(\{\alpha_n\}, \{\mu_n\})$ are required to have the so-called *Property (A)*. The choices of the parameters α_n and μ_n with this property are nontrivial. For example, if $T : K \to K$ is Lipschitz and pseudo-contractive, where K is a nonempty closed convex and bounded subset of a Hilbert space, one can choose $\alpha_n := (n + 1)^{-\frac{1}{4}}(L + 1)^{-1}$, $\mu_n := 1 - (n + 1)^{-\frac{1}{2}}$.

(III) The iteration formula (11.13) of Theorem 11.6 defined by

$$x_{n+1} = (1 - \lambda_n)x_n + \lambda_n T x_n - \lambda_n \theta_n(x_n - x_1), n \geq 1. \tag{11.17}$$

Here the parameters λ_n and θ_n can easily be chosen as $\lambda_n = \frac{1}{(n+1)^a}, \theta_n = \frac{1}{(n+1)^b}$, where $0 < b < a$ and $a + b < 1$. The sequence $\{x_n\}$ generated by this formula has been proved to converge to a fixed point of T under the following setting (Theorem 11.6). $(i)E$ is real Banach space with a uniformly Gâteaux differentiable norm; $(ii)K \subset E$ is nonempty, closed and convex; $(iii)T : K \to K$ is Lipschitz pseudo-contractive with $F(T) \neq \emptyset$. (iv) Every nonempty closed convex and bounded subset of K has the fixed point property for nonexpansive self-mappings.

Observations

- It is easy to see that the recursion formula of Method (III) is much simpler than the formula given in (I) or (II) and consequently, is more efficient, requiring less computer time than the other two formulas.
- The iteration parameters λ_n and θ_n in method (III) can be chosen at the beginning of the iteration process as easily as those in (I).
- The iteration method (III), while being simpler and superior to those of methods (I) and (II), is also applicable in the much more general Banach spaces considered in Theorem 11.6.

Conclusion From these observations, it is easy to draw the following conclusions.

1. The study of the *Ishikawa-type* iteration process is no longer of interest. The simpler Mann process which always converges whenever the Ishikawa-type process converges is preferred.
2. The iteration process of Theorem 11.6 is superior to the Ishikawa iteration process of Theorem 10.1 for the approximation of fixed points of Lipschitz pseudo-contractive mappings.

Theorem 11.5 and Theorem 11.6 are due to Chidume and Zegeye [174].

Chapter 12
Generalized Lipschitz Accretive and Pseudo-contractive Mappings

12.1 Introduction

We have seen in Chapter 11 that several convergence results have been proved on iterative methods for approximating zeros of *Lipschitz* accretive-type (or, equivalently, fixed points of Lipschitz pseudo-contractive-type) nonlinear mappings. We have also seen (Chapter 9) that a natural generalization of the class of Lipschitz mappings and the class of mappings with bounded range is that of *generalized Lipschitz mappings*. In this chapter, by means of an iteration process introduced by Chidume and Ofoedu [152], we prove convergence theorems for fixed points of *generalized* Lipschitz pseudo-contractive mappings in real Banach spaces.

12.2 Convergence Theorems

We first prove the following theorem.

Theorem 12.1. *Let K be a nonempty closed convex subset of a real Banach space E. Let $T : K \to K$ be a generalized Lipschitz pseudo-contractive mapping such that $F(T) := \{x \in K : Tx = x\} \neq \emptyset$. Let $\{\alpha_n\}_{n\geq 1}$, $\{\lambda_n\}_{n\geq 1}$ and $\{\theta_n\}_{n\geq 1}$ be real sequences in (0, 1) such that $\alpha_n = o(\theta_n)$, $\lim\limits_{n\to\infty} \lambda_n = 0$ and $\lambda_n(\alpha_n + \theta_n) < 1$. From arbitrary $x_1 \in K$, let the sequence $\{x_n\}_{n\geq 1}$ be iteratively generated by*

$$x_{n+1} = (1 - \lambda_n\alpha_n)x_n + \lambda_n\alpha_n Tx_n - \lambda_n\theta_n(x_n - x_1), \quad n \geq 1. \quad (12.1)$$

Then, $\{x_n\}_{n\geq 1}$ is bounded.

Proof. Let $x^* \in F(T)$. If $x_n = x^*$ for all integers $n \geq 1$, then we are done. So, let N_* be the first integer such that $x_{N_*} \neq x^*$. Clearly, there exist $N_0 \geq N_*$

C. Chidume, *Geometric Properties of Banach Spaces and Nonlinear Iterations*, Lecture Notes in Mathematics 1965,
© Springer-Verlag London Limited 2009

and $r > 0$ such that $x_{N_0} \in \overline{B_r(x^*)} := \{x \in E : ||x - x^*|| \leq r\}$, $x_1 \in \overline{B_{\frac{r}{2}}(x^*)} := \{x \in E : ||x - x^*|| \leq \frac{r}{2}\}$, and for all $n \geq N_0$, $\alpha_n < \theta_n$, $\lambda_n \leq \frac{r}{8[L(1+r)+\frac{5}{2}r]}$ and $\frac{\alpha_n}{\theta_n} \leq \frac{r}{8L[2+L(1+r)+\frac{9}{2}r]}$. It suffices to show by induction that x_n belongs to $B := \overline{B_r(x^*)}$ for all integers $n \geq N_0$. By construction $x_{N_0} \in B$. Assume $x_n \in B$ for $n > N_0$. We prove $x_{n+1} \in B$. Suppose that x_{n+1} is not in B. Then $||x_{n+1} - x^*|| > r$. From the recursion formula (12.1) and inequality (4.4) we have

$$\begin{aligned}
||x_{n+1} - x^*||^2 &= \left|\left|x_n - x^* - \lambda_n\Big(\alpha_n x_n - \alpha_n T x_n + \theta_n(x_n - x_1)\Big)\right|\right|^2 \\
&\leq ||x_n - x^*||^2 \\
&\quad -2\lambda_n\langle\alpha_n(x_n - Tx_n) + \theta_n(x_n - x_1), j(x_{n+1} - x^*)\rangle \\
&= ||x_n - x^*||^2 - 2\lambda_n\theta_n||x_{n+1} - x^*||^2 \\
&\quad +2\lambda_n\langle\theta_n(x_{n+1} - x_n) + \theta_n(x_1 - x^*) + \alpha_n(x_{n+1} - x_n) \\
&\quad +\alpha_n(x^* - Tx_{n+1}) - \alpha_n(x^* - Tx_n), j(x_{n+1} - x^*)\rangle \\
&\quad -2\lambda_n\alpha_n\langle x_{n+1} - Tx_{n+1}, j(x_{n+1} - x^*)\rangle.
\end{aligned}$$

Since T is pseudo-contractive, it follows that $\langle x_{n+1} - Tx_{n+1}, j(x_{n+1} - x^*)\rangle \geq 0$, so that

$$\begin{aligned}
||x_{n+1} - x^*||^2 &\leq ||x_n - x^*||^2 - 2\lambda_n\theta_n||x_{n+1} - x^*||^2 \\
&\quad +2\lambda_n(\theta_n + \alpha_n)||x_{n+1} - x_n||\,||x_{n+1} - x^*|| \\
&\quad +2\lambda_n\theta_n||x_1 - x^*||\,||x_{n+1} - x^*|| \\
&\quad +2\lambda_n\alpha_n L(1 + ||x_{n+1} - x^*||)||x_{n+1} - x^*|| \\
&\quad +2\lambda_n\alpha_n L(1 + ||x_n - x^*||)||x_{n+1} - x^*|| \\
&\leq ||x_n - x^*||^2 - 2\lambda_n\theta_n||x_{n+1} - x^*||^2 \\
&\quad +2\lambda_n^2(\theta_n + \alpha_n)\Big(\alpha_n L(1 + ||x_n - x^*||) + \alpha_n||x_n - x^*|| \\
&\quad +\theta_n||x_n - x^*|| + \theta_n||x_1 - x^*||\Big)||x_{n+1} - x^*|| \\
&\quad +2\lambda_n\theta_n||x_1 - x^*||\,||x_{n+1} - x^*|| \\
&\quad +2\lambda_n\alpha_n L\Big(1 + \alpha_n L(1 + ||x_n - x^*||) \\
&\quad +\alpha_n||x_n - x^*|| + \theta_n||x_n - x^*|| \\
&\quad +\theta_n||x_1 - x^*|| + ||x_n - x^*||\Big)||x_{n+1} - x^*|| \\
&\quad +2\lambda_n\alpha_n L(1 + ||x_n - x^*||)||x_{n+1} - x^*||.
\end{aligned}$$

Now, since $\|x_n - x^*\| \leq \|x_{n+1} - x^*\|$, we have

$$2\lambda_n\theta_n\|x_{n+1} - x^*\| \leq 4\lambda_n^2\theta_n\Big(L(1+r) + \frac{5}{2}r\Big)$$
$$+ 2\lambda_n\theta_n\frac{r}{2} + 2\lambda_n\alpha_nL\Big(2 + L(1+r) + \frac{9}{2}r\Big).$$

This implies

$$\|x_{n+1} - x^*\| \leq 2\lambda_n\Big(L(1+r) + \frac{5}{2}r\Big) + \frac{r}{2} + \frac{\alpha_n}{\theta_n}L\Big(2 + L(1+r) + \frac{9}{2}r\Big).$$

Hence, $\|x_{n+1} - x^*\| \leq \frac{r}{4} + \frac{r}{2} + \frac{r}{8} < r$, a contradiction. Therefore $x_n \in B \; \forall \; n \geq N_0$ and so $\{x_n\}_{n\geq 1}$ is bounded. This completes the proof. $\qquad\square$

Remark 12.2. Since $\{x_n\}_{n\geq 1}$ is bounded, there exists $R > 0$ sufficiently large such that $x_n \in B^* := \overline{B_R(x^*)} \; \forall \; n \in \mathbb{N}$. Furthermore, the set $K \cap B^*$ is a bounded closed and convex nonempty subset of E. If we define a map $\varphi : E \to \mathbb{R}$ by $\varphi(y) = \mu_n\|x_{n+1} - y\|^2$, where μ_n denotes a Banach limit, then, φ is continuous, convex and $\varphi(y) \to +\infty$ as $\|y\| \to +\infty$. Thus, if E is a reflexive Banach space, then there exists $x_0 \in K \cap B^*$ such that $\varphi(x_0) = \min_{y \in K \cap B^*} \varphi(y)$. So, the set $K_{min} := \{x \in K \cap B^* : \varphi(x) = \min_{y \in K \cap B^*} \varphi(y)\} \neq \emptyset$.

For the remainder of this chapter, $\{\alpha_n\}_{n\geq 1}$, $\{\lambda_n\}_{n\geq 1}$ and $\{\theta_n\}_{n\geq 1}$ will be as in Remark 12.2 and additionally, $\sum_{n=1}^{\infty} \lambda_n\theta_n = \infty$ will be assumed. We now prove the following theorems.

Theorem 12.3. *Let E be a real reflexive Banach space with uniformly Gâteaux differentiable norm. Let K be a nonempty closed and convex subset of E and $T : K \to K$ be a generalized Lipschitz pseudo-contractive mapping such that $F(T) \neq \emptyset$. Let $\{x_n\}_{n\geq 1}$ be iteratively generated by $x_1 \in K$,*

$$x_{n+1} = (1 - \lambda_n\alpha_n)x_n + \lambda_n\alpha_nTx_n - \lambda_n\theta_n(x_n - x_1), \quad n \geq 1. \quad (12.2)$$

Suppose that $K_{min} \cap F(T) \neq \emptyset$. Then, $\{x_n\}_{n\geq 1}$ converges strongly to some $x^* \in F(T)$.

Proof. Let $x^* \in K_{min} \cap F(T)$ and $t \in (0, 1)$. Then by the convexity of $K \cap B^*$ we have that $(1-t)x^* + tx_1 \in K \cap B^*$. It then follows from Remark 12.2 that $\varphi(x^*) \leq \varphi((1-t)x^* + tx_1)$. Using inequality (4.4), we have that

$$\|x_n - x^* - t(x_1 - x^*)\|^2 \leq \|x_n - x^*\|^2 - 2t\langle x_1 - x^*, j(x_n - x^* - t(x_1 - x^*))\rangle.$$

Thus, taking Banach limits over $n \geq 1$ gives

$$\mu_n\|x_n - x^* - t(x_1 - x^*)\|^2 \leq \mu_n\|x_n - x^*\|^2$$
$$- 2t\mu_n\langle x_1 - x^*, j(x_n - x^* - t(x_1 - x^*))\rangle.$$

This implies,

$$2t\mu_n\langle x_1 - x^*, j(x_n - x^* - t(x_1 - x^*)))\rangle \leq \varphi(x^*) - \varphi((1-t)x^* + tx_1) \leq 0.$$

This therefore implies that $\mu_n\langle x_1 - x^*, j(x_n - x^* - t(x_1 - x^*)))\rangle \leq 0 \ \forall \ n \geq 1$. Since the normalized duality mapping is norm-to-weak* uniformly continuous on bounded subsets of E, we obtain, as $t \to 0$, that

$$\langle x_1 - x^*, j(x_n - x^*)\rangle - \langle x_1 - x^*, j(x_n - x^* - t(x_1 - x^*))\rangle \to 0.$$

Hence, for all $\varepsilon > 0$, there exists $\delta > 0$ such that $\forall t \in (0, \delta)$ and for all $n \geq 1$,

$$\langle x_1 - x^*, j(x_n - x^*)\rangle < \langle x_1 - x^*, j(x_n - x^* - t(x_1 - x^*)))\rangle + \varepsilon.$$

Consequently,

$$\mu_n\langle x_1 - x^*, j(x_n - x^*)\rangle \leq \mu_n\langle x_1 - x^*, j(x_n - x^* - t(x_1 - x^*)) + \varepsilon \leq \varepsilon.$$

Since ε is arbitrary, we have $\mu_n\langle x_1 - x^*, j(x_n - x^*)\rangle \leq 0$. On the other hand, as $\{x_n\}$ and $\{Tx_n\}$ are bounded, using $\lim \lambda_n = 0$, we have that as $n \to \infty$,

$$||x_{n+1} - x_n|| \leq \lambda_n(\alpha_n||x_n|| + \alpha_n||Tx_n|| + \theta_n||x_n - x_1||) \to 0.$$

Therefore, from the norm-to-weak* uniform continuity of j on bounded sets, we obtain that

$$\lim_{n \longrightarrow \infty} (\langle x_1 - x^*, j(x_{n+1} - x^*)\rangle - \langle x_1 - x^*, j(x_n - x^*)\rangle) = 0.$$

Thus, the sequence $\{\langle x_1 - x^*, j(x_n - x^*)\rangle\}$ satisfies the conditions of Lemma 7.7. Hence, we obtain that

$$\limsup_{n\to\infty}\langle x_1 - x^*, j(x_n - x^*)\rangle \leq 0.$$

Define

$$\varepsilon_n := \max \{\langle x_1 - x^*, j(x_{n+1} - x^*)\rangle, 0\}.$$

Then, $\lim \varepsilon_n = 0$, and $\langle x_1 - x^*, j(x_{n+1} - x^*)\rangle \leq \varepsilon_n$. Now, using inequality (4.4), the recursion formula (12.2) and the last inequality, we obtain that

$$\begin{aligned}
||x_{n+1} - x^*||^2 &= ||x_n - x^* - \lambda_n[\alpha_n x_n - \alpha_n Tx_n + \theta_n(x_n - x_1)]||^2 \\
&\leq ||x_n - x^*||^2 - 2\lambda_n\langle \alpha_n(x_n - Tx_n) \\
&\quad + \theta_n(x_n - x_1), j(x_{n+1} - x^*)\rangle \\
&\leq ||x_n - x^*||^2 - 2\lambda_n\theta_n||x_{n+1} - x^*||^2 \\
&\quad + 2\lambda_n\theta_n||x_{n+1} - x_n||\,||x_{n+1} - x^*||
\end{aligned}$$

$$+2\lambda_n\alpha_n\|x_n - Tx_n\|\|x_{n+1} - x^*\|$$
$$+2\lambda_n\theta_n\langle x_1 - x^*, j(x_{n+1} - x^*)\rangle$$
$$\leq \|x_n - x^*\|^2 - 2\lambda_n\theta_n\|x_{n+1} - x^*\|^2$$
$$+(\lambda_n^2\theta_n + \lambda_n\alpha_n)M + 2\lambda_n\theta_n\varepsilon_n.$$

Thus, $\|x_{n+1} - x^*\|^2 \leq \|x_n - x^*\|^2 - 2\lambda_n\theta_n\|x_{n+1} - x^*\|^2 + \sigma_n$, where $\sigma_n :=$ $(\lambda_n^2\theta_n + \lambda_n\alpha_n)M + 2\lambda_n\theta_n\varepsilon_n = o(\lambda_n\theta_n)$ for some constant $M > 0$. Hence, by Lemma 8.17, we have that the sequence $\{x_n\}_{n\geq 1}$ converges strongly to $x^* \in F(T)$. This completes the proof. □

Theorem 12.4. *Let E be a real reflexive Banach space with uniformly Gâteaux differentiable norm, $A : E \to E$ be a generalized Lipschitz accretive operator such that $N(A) := \{x \in E : Ax = 0\} \neq \emptyset$. Let $x_1 \in E$ be fixed. Let $\{x_n\}_{n\geq 1}$ be iteratively generated by*

$$x_{n+1} = x_n - \lambda_n\alpha_n Ax_n - \lambda_n\theta_n(x_n - x_1), \ n \geq 1. \quad (12.3)$$

Suppose that $K_{min} \cap N(A) \neq \emptyset$. Then, $\{x_n\}_{n\geq 1}$ converges strongly to some $x^* \in N(T)$.

Proof. Recall that A is accretive if and only if $T := I - A$ is pseudo-contractive. Again, $x^* \in K_{min} \cap N(A)$ implies $Ax^* = 0$. This implies $Tx^* = x^*$. Thus, $K_{min} \cap F(T) \neq \emptyset$. Moreover, it is easy to see that T is also generalized Lipschitz pseudo-contractive mapping. Hence, if we replace A by $(I - T)$ in (12.3), then boundedness of $\{x_n\}_{n\geq 1}$ follows as in Theorem 12.1. The rest follows as in the proof of Theorem 12.3. This completes the proof. □

We now give an example in which condition $K_{min} \cap F(T) \neq \emptyset$ is easily satisfied.

Corollary 12.5 *Let E be a real Banach space with uniformly Gâteaux differentiable norm possessing uniform normal structure. Let K be a nonempty closed and convex subset of E and $T : K \to K$ be a nonexpansive mapping such that $F(T) \neq \emptyset$. Let $\{x_n\}_{n\geq 1}$ be iteratively generated by $x_1 \in K$, $x_{n+1} = (1 - \lambda_n\alpha_n)x_n + \lambda_n\alpha_n Tx_n - \lambda_n\theta_n(x_n - x_1)$, $n \geq 1$. Then, $\{x_n\}_{n\geq 1}$ converges strongly to $x^* \in F(T)$.*

Proof. Since E has uniform normal structure, it is reflexive. $K_{min} = \{x \in K \cap B^* : \varphi(x) = \min_{y \in K \cap B^*} \varphi(y)\}$ is a nonempty, closed and convex subset of K that has *property (P)*. This follows as in [307] (since every nonexpansive mapping is asymptotically nonexpansive with $k_n \equiv 1 < N(E)^{\frac{1}{2}} \ \forall \ n \in N$). Hence, $K_{min} \cap F(T) \neq \emptyset$. The result therefore follows. This completes the proof. □

12.3 Some Applications

Following the methods of proofs used in this chapter, the following theorems
are obtained.

Theorem 12.6. *Let E be a real reflexive Banach space with uniformly
Gâteaux differentiable norm. Let K be a nonempty closed and convex subset
of E and $T : K \to K$ be a generalized Lipschitz hemi-contractive mapping.
Let $\{x_n\}_{n\geq 1}$ be iteratively generated by $x_1 \in K$,*

$$x_{n+1} = (1 - \lambda_n \alpha_n)x_n + \lambda_n \alpha_n T x_n - \lambda_n \theta_n(x_n - x_1), \quad n \geq 1. \quad (12.4)$$

*Suppose that $K_{min} \cap F(T) \neq \emptyset$. Then, $\{x_n\}_{n\geq 1}$ converges strongly to some
$x^* \in F(T)$.*

Proof. Boundedness of $\{x_n\}_{n\geq 1}$ follows as in the proof of Theorem 12.1. The
proof of Theorem 12.6 then follows as in the proof of Theorem 12.3.   □

Theorem 12.7. *Let E be a real reflexive Banach space with uniformly
Gâteaux differentiable norm. Let $A : E \to E$ be a generalized Lipschitz quasi-
accretive operator. Let $\{x_n\}_{n\geq 1}$ be iteratively generated by $x_1 \in E$,*

$$x_{n+1} = x_n - \lambda_n \alpha_n A x_n - \lambda_n \theta_n(x_n - x_1), \quad n \geq 1.$$

*Suppose that $K_{min} \cap N(A) \neq \emptyset$. Then, $\{x_n\}_{n\geq 1}$ converges strongly to some
$x^* \in N(A)$.*

Proof. The mapping A is quasi-accretive if and only if $T := (I - A)$ is hemi-
contractive. Thus, replacing A by $I - T$ in (12.3), the result follows as in
Theorem 12.4.   □

Remark 12.8. Prototypes for the iteration parameters of the theorems of this
chapter are, for example, $\lambda_n = \frac{1}{(n+1)^b}$, $\alpha_n = \frac{1}{(n+1)^b}$ and $\theta_n = \frac{1}{(n+1)^a}, 0 <
a < b; \ a + b < 1, \ b < \frac{1}{2}$.

Remark 12.9. The addition of bounded error terms in any of the recursion
formulas (12.1), (12.2), (12.3) and (12.4) leads to no further generalization.

Remark 12.10. All the theorems in this chapter carry over trivially to the
so-called viscosity process.

12.4 Historical Remarks

We saw in Chapter 11 an iteration process studied by Chidume and Zegeye
which was used to approximate a fixed point of a Lipschitz pseudo-contractive
map in real reflexive Banach spaces with uniformly Gateaux differentiable

norm (in particular, in L_p spaces, $1 < p < \infty$). Moreover, this iteration process seems simpler than the Ishikawa process.

In this chapter, we have studied an iteration process introduced by Chidume and Ofoedu [152] for approximating a fixed point of a *generalized* Lipschitz pseudo-contraction in real reflexive Banach spaces with uniformly Gateaux differentiable norms. The following question is of interest.

Question *Can the recursion formula (11.4) be used to approximate a fixed point (assuming existence) of a generalized Lipschitz pseudo-contractive mapping in the setting of Theorem 12.3?*

All the theorems of this chapter are due to Chidume and Ofoedu [152].

Chapter 13
Applications to Hammerstein Integral Equations

13.1 Introduction

In this chapter, we shall examine iterative methods for approximating solutions of important nonlinear integral equations involving accretive-type operators. In particular, we examine iteration methods for solving *nonlinear integral equations of Hammerstein type*.

13.2 Solution of Hammerstein Equations

Let X be a real normed linear space. In this section, we study iterative methods for solutions of the so-called *equations of Hammerstein type* (see e.g., Hammerstein [246], Pascali and Sburlan [379], or Zeidler [549]). A nonlinear integral equation of Hammerstein-type is one of the form:

$$u(x) + \int_{\Omega} K(x,y)f(y,u(y))dy = h(x), \qquad (13.1)$$

where dy is a σ-finite measure on the measure space Ω; the real kernel K is defined on $\Omega \times \Omega$, f is a real-valued function defined on $\Omega \times \mathbb{R}$ and is, in general, nonlinear and h is a given function on Ω. If we now define an operator K by $Kv(x) := \int_{\Omega} K(x,y)v(y)dy$; $x \in \Omega$, and the so-called *superposition* or *Nemytskii operator* by $Fu(y) := f(y,u(y))$ then, the integral equation (13.1) can be put in operator theoretic form as follows:

$$u + KFu = 0, \qquad (13.2)$$

where, without loss of generality, we have taken $h \equiv 0$. It is now obvious that the equation (13.1) is a special case of equation $Au = 0$ in which $A := I + KF$.

C. Chidume, *Geometric Properties of Banach Spaces and Nonlinear Iterations*,
Lecture Notes in Mathematics 1965,
© Springer-Verlag London Limited 2009

Interest in equation (13.2) stems mainly from the fact that several problems that arise in differential equations, for instance, elliptic boundary value problems whose linear parts possess Greens functions can, as a rule, be transformed into the form (13.2) (see e.g., Pascali and Sburlan [379], Chapter IV). Equations of Hammerstein type play a crucial role in the theory of feedback control systems (see e.g., Dolezale [210]). Several existence and uniqueness theorems have been proved for equations of the Hammerstein type (see e.g., Brezis and Browder [35, 36, 37], Browder, De Figueiredo and Gupta [49], Browder and Gupta [50], Sh. Chepanovich [86], Deimling [199]). We have seen that for the iterative approximation of solutions of the equation $Au = 0$, the *monotonicity/accretivity* of A is crucial and that the Mann iteration scheme has successfully been employed. Attempts to apply this method to equation (13.2) have not provided satisfactory results. In particular, the recursion formulas obtained involved K^{-1} and this is not convenient in applications (see e.g., Chidume and Zegeye [173], [175]). Part of the difficulty is the fact that the composition of two monotone operators need not be monotone. In the special case in which the operators are defined on subsets D of X which are compact, or more generally, *angle-bounded* (see e.g., Brezis and Browder [35] or Browder [47] for definition), Brezis and Browder [35] have proved the strong convergence of a suitably defined *Galerkin approximation* to a solution of (13.2) (see also Brezis and Browder [37]).

In this chapter, we study a method that contains an auxiliary operator, defined in an appropriate real Banach space in terms of K and F which, under certain conditions, is accretive whenever K and F are, and whose zeros are solutions of equation (13.2). Moreover, the operators K and F need not be defined on compact or angle-bounded subset of X. Furthermore, the method which does not involve K^{-1} provides an explicit algorithm for the computation of solutions of equation (13.2). The proofs of convergence theorems are obtained as applications of some of the theorems we have studied in previous chapters.

Combining Theorem 5.7 and Corollary 5.8 we obtain the following theorem.

Theorem HK(Xu, [509]) *Let $q > 1$ and X be a real Banach space. Then the following are equivalent.*

(i) X is q-uniformly smooth.

(ii) There exists a constant $d_q > 0$ such that for all $x, y \in X$,

$$||x + y||^q \le ||x||^q + q\langle y, j_q(x)\rangle + d_q||y||^q.$$

(iii) There exists a constant $c_q > 0$ such that for all $x, y \in X$ and $\lambda \in [0, 1]$,

$$||(1 - \lambda)x + \lambda y||^q \ge (1 - \lambda)||x||^q + \lambda||y||^q - w_q(\lambda)c_q||x - y||^q,$$

where $w_q(\lambda) := \lambda^q(1 - \lambda) + \lambda(1 - \lambda)^q$.

We first prove the following important results.

Lemma 13.1. *For $q > 1$, let X be a real q-uniformly smooth Banach space. Let $E := X \times X$ with norm*

$$||z||_E := \left(||u||_X^q + ||v||_X^q \right)^{\frac{1}{q}},$$

for arbitrary $z = [u, v] \in E$. Let $E^* := X^* \times X^*$ denote the dual space of E. For arbitrary $x = [x_1, x_2] \in E$ define the map $j_q^E : E \to E^*$ by

$$j_q^E(x) = j_q^E[x_1, x_2] := [j_q^X(x_1), j_q^X(x_2)],$$

so that for arbitrary $z_1 = [u_1, v_1], z_2 = [u_2, v_2]$ in E the duality pairing $\langle ., . \rangle$ is given by

$$\langle z_1, j_q^E(z_2) \rangle = \langle u_1, j_q^X(u_2) \rangle + \langle v_1, j_q^X(v_2) \rangle.$$

Then, (a) j_q^E is a duality mapping on E; (b) E is q-uniformly smooth.

Proof. (a) For arbitrary $x = [x_1, x_2] \in E$, let $j_q^E(x) = j_q^E[x_1, x_2] = \psi_q$. Then $\psi_q = [j_q^X(x_1), j_q^X(x_2)]$ in E^*. Observe that for $p > 1$ such that $\frac{1}{p} + \frac{1}{q} = 1$,

$$||\psi_q||_{E^*} = \left(||[j_q^X(x_1), j_q^X(x_2)]|| \right)^{\frac{1}{p}} = \left(||j_q(x_1)||_{X^*}^p + ||j_q(x_2)||_{X^*}^p \right)^{\frac{1}{p}}$$

$$= \left(||x_1||_X^{(q-1)p} + ||x_2||_X^{(q-1)p} \right)^{\frac{1}{p}} = \left(||x_1||_X^q + ||x_2||_X^q \right)^{\frac{q-1}{q}}$$

$$= ||x||_E^{q-1}.$$

Hence, $||\psi_q||_{E^*} = ||x||_E^{q-1}$. Furthermore,

$$\langle x, \psi_q \rangle = \langle [x_1, x_2], [j_q^X(x_1), j_q^X(x_2)] \rangle = \langle x_1, j_q^X(x_1) \rangle + \langle x_2, j_q^X(x_2) \rangle$$

$$= ||x_1||_X^q + ||x_2||_X^q = \left(||x_1||_X^q + ||x_2||_X^q \right)^{\frac{1}{q}} \left(||x_1||_X^q + ||x_2||_X^q \right)^{\frac{q-1}{q}}$$

$$= ||x||_E . ||\psi||_{E^*} = ||x||_E^q.$$

Hence, j_q^E is a single-valued normalized duality mapping on E.
For (b), let $x = [x_1, x_2]$, $y = [y_1, y_2]$ be arbitrary elements of E. It suffices to show that x and y satisfy condition (ii) of Theorem HK. We compute as follows:

$$||x + y||_E^q = ||[x_1 + y_1, x_2 + y_2]||_E^q = ||x_1 + y_1||_X^q + ||x_2 + y_2||_X^q$$

$$\leq ||x_1||_X^q + ||x_2||_X^q + d_q \left(||y_1||_X^q + ||y_2||_X^q \right)$$

$$+ q \left\{ \langle y_1, j_q^X(x_1) \rangle + \langle y_2, j_q^X(x_2) \rangle \right\}$$

for some constant $d_q > 0$ (using (ii) of Theorem HK, since X is q-uniformly smooth). It follows that $||x + y||_E^q \leq ||x||_E^q + q\langle y, j_q^E(x) \rangle + d_q ||y||_E^q$. So, the result follows from Theorem HK. $\quad \square$

In the sequel, we shall take the norm on $E := X \times X$ to be given by $||z||_E^q = ||u||_X^q + ||v||_X^q$ for $z = [u, v] \in E$.

Lemma 13.2. *Let X be a real q-uniformly smooth Banach space. Let $F, K : X \to X$ be maps with $D(K) = F(X) = X$ such that the following conditions hold: (i) For each $u_1, u_2 \in X$, there exists a strictly increasing function $\phi_1 : [0, \infty) \to [0, \infty)$, $\phi_1(0) = 0$ such that*

$$\langle Fu_1 - Fu_2, j_q(u_1 - u_2) \rangle \geq \phi_1(||u_1 - u_2||)||u_1 - u_2||^{q-1}.$$

(ii) For each $u_1, u_2 \in X$, there exists a strictly increasing function $\phi_2 : [0, \infty) \to [0, \infty)$, $\phi_2(0) = 0$ such that

$$\langle Ku_1 - Ku_2, j_q(u_1 - u_2) \rangle \geq \phi_2(||u_1 - u_2||)||u_1 - u_2||^{q-1}.$$

(iii) $(1 + d_q)(1 + c_q) \geq 2^q$, where d_q, c_q are the constants appearing in inequalities (ii) and (iii), respectively, of Theorem HK. Define a map $T : E \to E$ by

$$Tz := T[u, v] = [Fu - v, u + Kv], \text{ for } [u, v] \in E.$$

Then, for each $z_1 = [u_1, u_2], z_2 = [v_1, v_2] \in E$, the following inequality holds:

$$\langle Tz_1 - Tz_2, j_q^E(z_1 - z_2) \rangle \geq \phi_1(||u_1 - u_2||)||u_1 - u_2||^{q-1}$$
$$+ \phi_2(||v_1 - v_2||)||v_1 - v_2||^{q-1}.$$
$$- q^{-1}\left[(1 + d_q) - \frac{2^q}{(1 + c_q)}\right]||z_1 - z_2||^q$$

Proof. We compute as follows:

$$\langle Tz_1 - Tz_2, j_q^E(z_1 - z_2) \rangle$$
$$= \langle Fu_1 - Fu_2, j_q(u_1 - u_2) \rangle - \langle v_1 - v_2, j_q(u_1 - u_2) \rangle$$
$$+ \langle Kv_1 - Kv_2, j_q(v_1 - v_2) \rangle + \langle u_1 - u_2, j_q(v_1 - v_2) \rangle$$
$$\geq \phi_1(||u_1 - u_2||)||u_1 - u_2||^{q-1} + \phi_2(||v_1 - v_2||)||v_1 - v_2||^{q-1}$$
$$- \langle v_1 - v_2, j_q(u_1 - u_2) \rangle + \langle u_1 - u_2, j_q(v_1 - v_2) \rangle.$$

Since X is real q-uniformly smooth, inequality of part (iii) of Theorem HK holds for each $x, y \in X$. Setting $\lambda = \frac{1}{2}$ in this inequality yields the following estimates:

$$||x + y||^q \geq 2^{q-1}(||x||^q + ||y||^q) - c_q||x - y||^q,$$
$$||x - y||^q \geq 2^{q-1}(||x||^q + ||y||^q) - c_q||x + y||^q,$$

so that

$$||x + y||^q + ||x - y||^q \geq 2^q(||x||^q + ||y||^q) - c_q(||x + y||^q + ||x - y||^q),$$

which yields, $\forall\, x, y \in E$,

$$||x + y||^q + ||x - y||^q \geq \frac{2^q}{(1 + c_q)}(||x||^q + ||y||^q).$$

Furthermore, from inequality of part (ii) of Theorem HK, replacing y by $-y$ we obtain the following inequality:

$$-\langle y, j_q(x)\rangle \geq q^{-1}(||x - y||^q - ||x||^q - d_q||y||^q).$$

Using the above estimates, we obtain the following:

$$\langle Tz_1 - Tz_2, j_q^E(z_1 - z_2)\rangle$$
$$\geq \phi_1(||u_1 - u_2||)||u_1 - u_2||^{q-1} + \phi_2(||v_1 - v_2||)||v_1 - v_2||^{q-1}$$
$$+ q^{-1}\Big(||v_1 - v_2 - (u_1 - u_2)||^q - ||u_1 - u_2||^q - d_q||v_1 - v_2||^q\Big)$$

$$+ q^{-1}\Big(||v_1 - v_2 + (u_1 - u_2)||^q - ||v_1 - v_2||^q - d_q||u_1 - u_2||^q\Big)$$
$$= \phi_1(||u_1 - u_2||)||u_1 - u_2||^{q-1} + \phi_2(||v_1 - v_2||)||v_1 - v_2||^{q-1}$$
$$+ q^{-1}\Big[||v_1 - v_2 + (u_1 - u_2)||^q + ||v_1 - v_2 - (u_1 - u_2)||^q$$
$$- (||u_1 - u_2||^q + ||v_1 - v_2||^q) - d_q(||v_1 - v_2||^q + ||u_1 - u_2||^q)\Big]$$
$$\geq \phi_1(||u_1 - u_2||)||u_1 - u_2||^{q-1} + \phi_2(||v_1 - v_2||)||v_1 - v_2||^{q-1}$$
$$- q^{-1}\Big[(1 + d_q) - \Big(\frac{2^q}{1 + c_q}\Big)\Big]||z_1 - z_2||^q,$$

completing the proof of Lemma 13.2. $\qquad\qquad\square$

Corollary 13.3 *Let X be a real q-uniformly smooth Banach space. Let $F, K :$ $X \to X$ be maps with $D(K) = F(X) = X$ such that the following conditions hold: (i) For each $u_1, u_2 \in X$, there exists $\alpha > 0$ such that*

$$\langle Fu_1 - Fu_2, j_q(u_1 - u_2)\rangle \geq \alpha||u_1 - u_2||^q.$$

(ii) *For each $u_1, u_2 \in X$, there exists $\beta > 0$ such that*

$$\langle Ku_1 - Ku_2, j_q(u_1 - u_2)\rangle \geq \beta||u_1 - u_2||^q.$$

(iii) $(1 + d_q)(1 + c_q) \geq 2^q$, *where c_q and d_q are as in Theorem HK. Let E and T be defined as in Lemma 13.2. Then, for $z_1, z_2 \in E$, the following inequality holds:*

$$\langle Tz_1 - Tz_2, j_q^E(z_1 - z_2)\rangle \geq \Big[\gamma - q^{-1}\Big[(1 + d_q) - \Big(\frac{2^q}{1 + c_q}\Big)\Big]\Big]||z_1 - z_2||^q,$$

where we assume that $\gamma := \min\ \{\alpha,\ \beta\} > q^{-1}\left[(1+d_q) - \left(\frac{2^q}{(1+c_q)}\right)\right]$.

Proof. Define $\phi_1(t) := \alpha t$; $\phi_2(t) := \beta t$. Then, the result follows immediately from Lemma 13.2.   □

Corollary 13.4 *Let* $X = L_p, 2 \le p < \infty$. *Let* $F, K : X \to X$ *be maps with* $D(K) = F(X) = X$ *such that the following conditions hold: (i) For each* $u_1, u_2 \in X$, *there exists a strictly increasing function* $\phi_1 : [0, \infty) \to [0, \infty)$, $\phi_1(0) = 0$ *such that*

$$\langle Fu_1 - Fu_2, j_q(u_1 - u_2)\rangle \ge \phi_1(||u_1 - u_2||)||u_1 - u_2||^{q-1}.$$

(ii) For each $u_1, u_2 \in X$, *there exists a strictly increasing function* $\phi_2 : [0, \infty) \to [0, \infty)$, $\phi_2(0) = 0$ *such that*

$$\langle Ku_1 - Ku_2, j_q(u_1 - u_2)\rangle \ge \phi_2(||u_1 - u_2||)||u_1 - u_2||^{q-1}.$$

Define a map $T : E \to E$ *by*

$$Tz := T[u, v] = [Fu - v, u + Kv],\ \text{for } [u, v] \in E.$$

Then, for each $z_1 = [u_1, u_2], z_2 = [v_1, v_2] \in E$, *the following inequality holds:*

$$\langle Tz_1 - Tz_2, j_q^E(z_1 - z_2)\rangle \ge \phi_1(||u_1 - u_2||)||u_1 - u_2||^{q-1}$$

$$+ \phi_2(||u_1 - u_2||)||u_1 - u_2||^{q-1}.$$

$$- q^{-1}\left[(1 + d_q) - \frac{2^q}{(1 + c_q)}\right]||z_1 - z_2||^q$$

Proof. For L^p spaces, $2 \le p < \infty$, we note that $c_q = d_q = (p-1), q = 2$ and so condition *(iii)* of Lemma 13.2 is easily satisfied. Hence, the corollary follows.   □

Corollary 13.5 *Let* H *be a real Hilbert space. Let* $F, K : H \to H$ *be maps with* $D(K) = F(X) = H$ *such that conditions (i) and (ii) of Corollary 13.3 are satisfied. Let* $\alpha, \beta > 0$, E *and* T *be defined as in Corollary 13.3. Then, for* $z_1, z_2 \in E$ *we have that*

$$\langle Tz_1 - Tz_2, j(z_1 - z_2)\rangle \ge \gamma||z_1 - z_2||^2,$$

where $\gamma := \min\{\alpha, \beta\}$.

Proof. Since, for real Hilbert spaces, the duality mapping j_q^E is the identity map, and $q = 2$, $d_q = 1$, $c_q = 1$, the result follows from Corollary 13.3.   □

13.2.1 Convergence Theorems for Lipschitz Maps

Remark 13.6. If K and F are Lipschitzian maps with positive constants L_K and L_F, respectively, then, T is a Lipschitzian map.

Indeed, if $z_1 = [u_1, v_1])$, $z_2 = [u_2, v_2]$ in E then we have that

$$\|Tz_1 - Tz_2\|^q$$
$$\leq (L_F\|u_1 - u_2\| + \|v_1 - v_2\|)^q + (\|u_1 - u_2\| + L_K\|v_1 - v_2\|)^q$$
$$\leq d[L_F^q\|u_1 - u_2\|^q + \|v_1 - v_2\|^q + \|u_1 - u_2\|^q + L_K^q\|v_1 - v_2\|^q]$$
$$\text{for some } d > 0$$
$$\leq d \max\{L_F^q + 1, L_K^q + 1\}[\|u_1 - u_2\|^q + \|v_1 - v_2\|^q]$$
$$= d \max\{L_F^q + 1, L_K^q + 1\}\|z_1 - z_2\|^q,$$

and so T is Lipschitzian.

Consequently, we have the following theorem.

Theorem 13.7. *Let X be a real q-uniformly smooth Banach space. Let $F, K : X \to X$ be Lipschitzian maps such that $D(K) = F(X) = X$. Suppose the following conditions are satisfied: (i) there exists $\alpha > 0$ such that*

$$\langle Fu_1 - Fu_2, j_q(u_1 - u_2)\rangle \geq \alpha\|u_1 - u_2\|^q \ \forall \ u_1, u_2 \in X.$$

(ii) there exists $\beta > 0$ such that

$$\langle Ku_1 - Ku_2, j_q(u_1 - u_2)\rangle \geq \beta\|u_1 - u_2\|^q \ \forall \ u_1, u_2 \in X.$$

(iii) $(1+d_q)(1+c_q) \geq 2^q$, where d_q, c_q are the constants appearing in inequalities (ii) and (iii) of Theorem HK, respectively. Assume that $u + KFu = 0$ has solution $u^* \in E$. Define the map $T : E \to E$ by $Tz := T[u, v] = [Fu - v, Kv + u]$. Let L be Lipschitz constant of T and $\varepsilon := \frac{1}{2}\left(\frac{\gamma}{1+L(3+L-\gamma)}\right)$. Define the map $A_\varepsilon : E \to E$ by $A_\varepsilon z := z - \varepsilon Tz$ for each $z \in E$. For arbitrary $z_0 \in E$, define the Picard sequence $\{z_n\}$ in E by

$$z_{n+1} := A_\varepsilon z_n, n \geq 0.$$

Then, $\{z_n\}$ converges strongly to $z^* = [u^*, v^*]$, the unique solution of the equation $Tz = 0$, with

$$\|z_{n+1} - z^*\| \leq \delta^n\|z_1 - z^*\|, \delta := (1 - \frac{1}{2}\gamma\varepsilon) \in (0,1),$$

where $v^* = Fu^*$ and u^* is the solution of the equation $u + KFu = 0$.

Proof. Observe that u^* is a solution of $u + KFu = 0$ if and only if $z^* = [u^*, v^*]$ is a solution of $Tz = 0$ for $v^* = Fu^*$. Hence $Tz = 0$ has a solution $z^* = [u^*, v^*]$

in E. Since T is Lipschitz and by Corollary 13.3 it is strongly accretive, the conclusion follows from Theorem 8.14.   □

Remark 13.8. For L_p spaces, $2 \leq p < \infty$, we note that $c_q = d_q = (p-1), q = 2$ and so condition (iii) of Theorem 13.7 is easily satisfied. Hence, we have the following corollaries.

Corollary 13.9 *Let $X = L_p$, $2 \leq p < \infty$. Let $F, K : X \to X$ be Lipschitzian maps such that $D(K) = F(X) = X$. Suppose the following conditions are satisfied: (i) there exists $\alpha > 0$ such that*

$$\langle Fu_1 - Fu_2, j(u_1 - u_2) \rangle \geq \alpha ||u_1 - u_2||^q, \ \forall \ u_1, u_2 \in X;$$

(ii) there exists $\beta > 0$ such that

$$\langle Ku_1 - Ku_2, j(u_1 - u_2) \rangle \geq \beta ||u_1 - u_2||^q, \ \forall \ u_1, u_2 \in X.$$

Assume that $u + KFu = 0$ has solution $u^* \in E$. Define the map $T : E \to E$ by $Tz := T[u, v] = [Fu - v, Kv + u]$. Let L be Lipschitz constant of T and let $\varepsilon := \frac{1}{2}\left(\frac{\gamma}{1 + L(3 + L - \gamma)}\right)$. Define the map $A_\varepsilon : E \to E$ by $A_\varepsilon z := z - \varepsilon Tz$ for each $z \in E$. For arbitrary $z_0 \in E$, define the Picard sequence $\{z_n\}$ in E by $z_{n+1} := A_\varepsilon z_n, n \geq 0$. Then, $\{z_n\}$ converges strongly to $z^* = [u^*, v^*]$, the unique solution of the equation $Tz = 0$, with

$$||z_{n+1} - z^*|| \leq \delta^n ||z_1 - z^*||, \delta := (1 - \frac{1}{2}\gamma\varepsilon) \in (0, 1),$$

where $v^* = Fu^*$ and u^* is the solution of the equation $u + KFu = 0$.

Corollary 13.10 *Let H be a real Hilbert space. Let $F, K : X \to X$ be Lipschitzian maps such that $D(K) = F(X) = H$. Suppose the following conditions are satisfied: (i) there exists $\alpha > 0$ such that*

$$\langle Fu_1 - Fu_2, (u_1 - u_2) \rangle \geq \alpha ||u_1 - u_2||^2, \ \forall \ u_1, u_2 \in D(F);$$

(ii) there exists $\beta > 0$ such that

$$\langle Ku_1 - Ku_2, (u_1 - u_2) \rangle \geq \beta ||u_1 - u_2||^2, \ \forall \ u_1, u_2 \in D(K).$$

Assume that $u + KFu = 0$ has solution $u^* \in E$. Define the map $T : E \to E$ by $Tz := T[u, v] = [Fu - v, Kv + u]$. Let L be Lipschitz constant of T and let $\varepsilon := \frac{1}{2}\left(\frac{\gamma}{1 + L(3 + L - \gamma)}\right)$. Define the map $A_\varepsilon : E \to E$ by $A_\varepsilon z := z - \varepsilon Tz$ for each $z \in E$. For arbitrary $z_0 \in E$, define the Picard sequence $\{z_n\}$ in E by $z_{n+1} := A_\varepsilon z_n, n \geq 0$. Then, $\{z_n\}$ converges strongly to $z^* = [u^*, v^*]$, the unique solution of the equation $Tz = 0$, with

$$||z_{n+1} - z^*|| \leq \delta^n ||z_1 - z^*||, \delta := (1 - \frac{1}{2}\gamma\varepsilon) \in (0, 1),$$

where $v^* = Fu^*$ and u^* is the solution of the equation $u + KFu = 0$.

13.2.2 Convergence Theorems for Bounded Maps

Theorem 13.11. *Let X be a real q-uniformly smooth Banach space. Let $F, K : X \to X$ be maps with $D(K) = F(X) = X$. Let F, K be bounded maps such that the following conditions hold: (i) For each $u_1, u_2 \in X$, there exists a strictly increasing function $\phi_1 : [0, \infty) \to [0, \infty)$, $\phi_1(0) = 0$ such that*

$$\langle Fu_1 - Fu_2, j_q(u_1 - u_2) \rangle \geq \phi_1(||u_1 - u_2||)||u_1 - u_2||^{q-1}.$$

(ii) For each $u_1, u_2 \in X$, there exists a strictly increasing function $\phi_2 : [0, \infty) \to [0, \infty)$, $\phi_2(0) = 0$ such that

$$\langle Ku_1 - Ku_2, j_q(u_1 - u_2) \rangle \geq \phi_2(||u_1 - u_2||)||u_1 - u_2||^{q-1}.$$

(iii) $(1 + d_q)(1 + c_q) \geq 2^q$ where d_q, c_q are the constants appearing in inequalities (ii) and (iii) of Theorem HK, respectively. Assume that $0 = u + KFu$ has a solution u^* in X. Define the map $T : E \to E$ by $Tz := T[u, v] = [Fu - v, u + Kv]$. Let $\{\alpha_n\}$ be a real sequence satisfying the following conditions: (i) $\lim_{n \to \infty} \alpha_n = 0$; (ii) $\sum \alpha_n = \infty$. Then, for arbitrary $z_0 \in E$, there exists $d_0 > 0$ such that if $0 < \alpha_n \leq d_0$, the sequence $\{z_n\}$, defined by $z_{n+1} := z_n - \alpha_n T z_n$, $n \geq 0$, converges strongly to $z^* = [u^*, v^*]$, where $v^* = Fu^*$ and u^* is the unique solution of $0 = u + KFu$.

Proof. Observe that since K and F are bounded maps we have that T is a bounded map. Observe also that u^* is the solution of $0 = u + KFu$ in X, if and only if $z^* = [u^*, v^*]$ is a solution of $0 = Tz$ in E for $v^* = Fu^*$. Thus we obtain that $N(T)$ (null space of T) $\neq \emptyset$. Also by Lemma 13.2, T is $\phi-$ strongly accretive. Therefore the conclusion follows from Corollary 8.27. □

Following the method of proof of Theorem 13.11 and making use of Corollary 13.9 we obtain the following theorem.

Theorem 13.12. *Let X be a real q-uniformly smooth Banach space. Let $F, K : X \to X$ be maps with $D(K) = F(X) = X$. Let F, K be bounded maps such that the following conditions hold: (i) For each $u_1, u_2 \in X$, there exists $\alpha > 0$ such that*

$$\langle Fu_1 - Fu_2, j_q(u_1 - u_2) \rangle \geq \alpha ||u_1 - u_2||^q.$$

(ii) For each $u_1, u_2 \in X$, there exists $\beta > 0$ such that

$$\langle Ku_1 - Ku_2, j_q(u_1 - u_2) \rangle \geq \beta ||u_1 - u_2||^q.$$

(iii) $(1 + d_q)(1 + c_q) \geq 2^q$ where d_q, c_q are the constants appearing in inequalities (ii) and (iii) of Theorem HK, respectively. Assume that $0 = u + KFu$ has a solution u^*. Let E, T and $\{z_n\}$ be defined as in Theorem 13.11. Then, the conclusion of Theorem 13.11 holds.

Corollary 13.13 *Let $X = L_p$, $(1 < p < \infty)$, and $F, K : X \to X$ be maps with $D(K) = F(X) = X$. Let K, F be bounded maps such that (i) and (ii) of Theorem 13.12 hold. Assume that $0 = u + KFu$ has solution u^*. Let E, T and $\{z_n\}$ be defined as in Theorem 13.12. Then, the conclusion of Theorem 13.12 holds.*

Proof. The proof follows from Theorem 13.12 with Remark 13.8. $\square$.

Corollary 13.14 *Let H be real Hilbert space. Let F and K be as in Corollary 13.13. Assume that $0 = u + KFu$ has solution u^*. Let E, T and $\{z_n\}$ be defined as in Corollary 13.13. Then the conclusion of Corollary 13.13 holds.*

Proof. The proof follows from Corollary 13.13 with $p = 2$. $\square$

13.2.3 Explicit Algorithms

The method of our proofs provides the following explicit algorithms for computing the solution of the equation $0 = u + KFu$ in the space X.

(a) **For Lipschitz operators** (Theorem 13.7, Corollary 13.9 and Corollary 13.10) with initial values $u_0, v_0 \in X$, define the sequences $\{u_n\}$ and $\{v_n\}$ in X as follows:

$$u_{n+1} = u_n - \varepsilon\Big(F(u_n) - v_n\Big);$$

$$v_{n+1} = v_n - \varepsilon\Big(K(v_n) + u_n\Big),$$

where ε is as defined in this theorem and the corollaries. Then $u_n \to u^*$ in X, the unique solution u^* of $0 = u + KFu$ with $v^* = Fu^*$, where ε is as defined in Theorem 13.7

(b) **For bounded operators** (Theorem 13.11, Theorem 13.12, Corollary 13.13 and Corollary 13.14) with initial values $u_0, v_0 \in X$, define the sequences $\{u_n\}$ and $\{v_n\}$ in X as follows:

$$u_{n+1} = u_n - \alpha_n\Big(Fu_n - v_n\Big);$$

$$v_{n+1} = v_n - \alpha_n\Big(Kv_n + u_n\Big),$$

where α_n is as defined in these theorems and corollaries. Then $u_n \to u^*$ in X, the unique solution u^* of $0 = u + KFu$ with $v^* = Fu^*$, where α_n is as defined in Theorem 13.11.

13.3 Convergence Theorems with Explicit Algorithms

In this section, we give direct proofs that the explicit algorithms of the last sub-section defined in the space X converges to a solution of the Hammerstein equation. For the remainder of this chapter, c_q and d_q will denote the constants appearing in Theorem HK.

13.3.1 Some Useful Lemmas

Lemma 13.15. *(Chidume and Djitte, [137]) Let $q > 1$ and X be a real q-uniformly smooth Banach space. Then, for all $x, y \in X$ we have the following inequality:*

$$\|x + y\|^q + \|x - y\|^q \geq \frac{2^q}{1 + c_q}\left(\|x\|^q + \|y\|^q\right). \tag{13.3}$$

Proof. This has been proved as part of the proof of Lemma 13.2.

We observe that in L_p spaces, $2 \leq p < \infty$, $c_p = d_q = p - 1$ and $q = 2$. So, the following corollary is immediate.

Corollary 13.16 *Let $E = L_p$, $2 \leq p < \infty$. Then for all $x, y \in E$, we have:*

$$\|x + y\|^2 + \|x - y\|^2 \geq \frac{4}{p}\left(\|x\|^2 + \|y\|^2\right). \tag{13.4}$$

Lemma 13.17. *Assume that X is a real q-uniformly smooth Banach space, $q > 1$. Let x_1, x_2, y_1, y_2 be points in X. Then, we have the following estimate:*

$$A(x_1, x_2, y_1, y_2) := \langle y_1 - y_2, j_q(x_1 - x_2)\rangle - \langle x_1 - x_2, j_q(y_1 - y_2)\rangle$$
$$\leq \frac{(1 + c_q)(1 + d_q) - 2^q}{q(1 + c_q)}\left(\|x_1 - x_2\|_X^q + \|y_1 - y_2\|_X^q\right).$$

Proof. Since X is real q-uniformly smooth, inequality of part (ii) of Theorem HK holds for each $x, y \in X$. Replacing y by $-y$ in this inequality yields the following estimate:

$$\langle y, j_q(x)\rangle \leq \frac{1}{q}\left(\|x\|_X^q + d_q\|y\|_X^q - \|x - y\|_X^q\right). \tag{13.5}$$

From (13.5), it follows that

$$\langle y_1 - y_2, j_q(x_1 - x_2)\rangle \leq \frac{1}{q}\left(\|x_1 - x_2\|_X^q + d_q\|y_1 - y_2\|_X^q\right) \tag{13.6}$$
$$-\frac{1}{q}\left(\|(x_1 - x_2) - (y_1 - y_2)\|_X^q\right)$$

and, similarly,

$$\langle x_1 - x_2, j_q(y_2 - y_1) \rangle \leq \frac{1}{q}\left(\|y_1 - y_2\|_X^q + d_q\|x_1 - x_2\|_X^q \right) \qquad (13.7)$$
$$- \frac{1}{q}\left(\|(x_1 - x_2) + (y_1 - y_2)\|_X^q \right).$$

Using (13.3), (13.6) and (13.7), we obtain:

$$A(x_1, x_2, y_1, y_2) \leq \frac{1}{q}(1 + d_q)\left(\|x_1 - x_2\|_X^q + \|y_1 - y_2\|_X^q \right)$$
$$- \frac{1}{q}\left(\|(x_1 - x_2) + (y_1 - y_2)\|_X^q + \|(x_1 - x_2) - (y_1 - y_2)\|_X^q \right)$$
$$\leq \frac{(1 + c_q)(1 + d_q) - 2^q}{q(1 + c_q)}\left(\|x_1 - x_2\|_X^q + \|y_1 - y_2\|_X^q \right).$$

$\square$

13.3.2 Convergence Theorems with Coupled Schemes for the Case of Lipschitz Maps

For $q > 1$, let X be a real q- uniformly smooth Banach space and $F, K : X \to X$ be maps with $D(K) = F(K) = X$ such that the following conditions hold:
(i) F is Lipschitzian with constant L_F and there exists a positive constant $\alpha > 0$ such that

$$\langle Fu_1 - Fu_2, j_q(u_1 - u_2) \rangle \geq \alpha\|u_1 - u_2\|_X^q \quad \forall\, u_1, u_2 \in X; \qquad (13.8)$$

(ii) K is Lipschitzian with constant L_K and there exists a positive constant $\beta > 0$ such that

$$\langle Ku_1 - Ku_2, j_q(u_1 - u_2) \rangle \geq \beta\|u_1 - u_2\|_X^q \quad \forall\, u_1, u_2 \in X; \qquad (13.9)$$

(iii) $(1 + c_q)(1 + d_q) \geq 2^q$, $\gamma > \frac{\gamma_q}{q}$, where $\gamma_q := \frac{\left[(1+c_q)(1+d_q) - 2^q \right]}{1 + c_q}$, $L := \max\{L_F, L_K\}$, $\gamma := \min\{\alpha, \beta\}$, $\delta_q := d_q^2 L + d_q L + \frac{q}{q'}d_q L + d_q$ and q' is the Holder's conjugate of q.

With these notations, we prove the following theorem.

Theorem 13.18. *Let $\{u_n\}$ and $\{v_n\}$ be sequences in X defined as follows:*

$$u_0 \in X, \;\; u_{n+1} = u_n - \varepsilon(Fu_n - v_n), \;\; n \geq 1;$$
$$v_0 \in X, \;\; v_{n+1} = v_n - \varepsilon(Kv_n + u_n), \;\; n \geq 1; \qquad (13.10)$$

with

$$0 < \varepsilon < \min \left\{ \frac{1}{\Lambda_q}, \left(\frac{q\gamma}{\lambda_0 \delta_q} \right)^{\frac{1}{q-1}} \right\}, \quad \Lambda_q := \frac{1}{\frac{q\gamma(\lambda_0 - 1)}{\lambda_0} - \gamma_q},$$

where λ_0 is any number such that $\lambda_0 > q\gamma / [q\gamma - \gamma_q]$ and the symbols have their usual meanings. Assume that $u + KFu = 0$ has a solution u^*. Then, $\{u_n\}$ converges to u^* in X, $\{v_n\}$ converges to v^* in X and u^* is the unique solution of $u + KFu = 0$ with $v^* = Fu^*$.

Proof. Let $E := X \times X$ with the norm $\|z\|_E = (\|u\|_X^q + \|v\|_X^q)^{1/q}$ where $z = (u, v)$. E is clearly a Banach space. Define the sequence $\{w_n\}$ in E by $w_n = (u_n, v_n)$. It suffices to show that $\{w_n\}$ is Cauchy in E. Let $n, m \in \mathbb{N}$. We compute as follows:

$$\|w_{n+1} - w_{m+1}\|^q = \|u_{n+1} - u_{m+1}\|_X^q + \|v_{n+1} - v_{m+1}\|_X^q.$$

From Theorem HK, it follows that

$$
\begin{aligned}
\|u_{n+1} - u_{m+1}\|_X^q &= \|u_n - u_m + \varepsilon(Fu_m - F_n + v_n - v_m)\|_X^q \\
&\leq \|u_n - u_m\|_X^q + \varepsilon q \langle Fu_m - Fu_n + v_n - v_m, j_q(u_n - u_m) \rangle \\
&\quad + \varepsilon^q d_q \|Fu_n - Fu_m + v_n - v_m\|_X^q \\
&\leq \|u_n - u_m\|_X^q - \varepsilon q \langle Fu_n - Fu_m, j_q(u_n - u_m) \rangle \\
&\quad + \varepsilon q \langle v_n - v_m, j_q(u_n - u_m) \rangle + \varepsilon^q d_q \|v_n - v_m\|_X^q \\
&\quad + \varepsilon^q d_q^2 \|Fu_m - Fu_n\|_X^q + q\varepsilon^q d_q \langle Fu_m - Fu_n, j_q(v_n - v_m) \rangle.
\end{aligned}
$$

Using (13.8) and the Lipschitz condition, we obtain the following estimates:

$$
\begin{aligned}
\|u_{n+1} - u_{m+1}\|_X^q &\leq \|u_n - u_m\|_X^q - \alpha \varepsilon q \|u_n - u_m\|_X^q + \varepsilon^q d_q^2 L^q \|u_n - u_m\|_X^q \\
&\quad + \varepsilon^q d_q \|v_n - v_m\|_X^q + q\varepsilon^q d_q \langle Fu_m - Fu_n, j_q(v_n - v_m) \rangle \\
&\quad + q\varepsilon \langle v_n - v_m, j_q(u_n - u_m) \rangle.
\end{aligned}
$$

Using Schwartz and Minkowsky inequalities and the fact that $\|j_q(x)\| = \|x\|^{q-1}$ we obtain

$$\langle Fu_m - Fu_n, j_q(v_n - v_m) \rangle \leq L \|u_m - u_n\|_X \|v_n - v_m\|_X^{q-1},$$

which implies

$$\langle Fu_m - Fu_n, j_q(v_n - v_m) \rangle \leq \frac{L}{q} \|u_n - u_m\|_X^q + \frac{L}{q'} \|v_n - v_m\|_X^q,$$

where q' is the Holder's conjugate of q. Finally we have

$$\|u_{n+1} - u_{m+1}\|_X^q \le \left(1 - \alpha q\varepsilon + L^q d_q^2 \varepsilon^q + L d_q \varepsilon^q\right) \|u_n - u_m\|_X^q$$
$$+ \left(\frac{q}{q'} d_q L \varepsilon^q + d_q \varepsilon^q\right) \|v_n - v_m\|_X^q$$
$$+ q\varepsilon \langle v_n - v_m, j_q(u_n - u_m)\rangle.$$

Using the same arguments as above, we obtain the following estimate:

$$\|v_{n+1} - v_{m+1}\|_X^q \le \left(1 - \beta q\varepsilon + L^q d_q^2 \varepsilon^q + L d_q \varepsilon^q\right) \|v_n - v_m\|_X^q$$
$$+ \left(\frac{q}{q'} d_q L_k \varepsilon^q + d_q \varepsilon^q\right) \|u_n - u_m\|_X^q$$
$$+ q\varepsilon \langle u_n - u_m, j_q(v_m - v_n)\rangle,$$

which implies that

$$\|w_{n+1} - w_{m+1}\|_X^q \le \left(1 - \gamma q\varepsilon + \varepsilon^q \delta_q\right) \|w_n - w_m\|_X^q + q\varepsilon A_{nm}, \quad (13.11)$$

where $A_{nm} := \langle v_n - v_m, j_q(u_n - u_m)\rangle - \langle u_n - u_m, j_q(v_n - v_m)\rangle$. Therefore, using Lemma 13.17, we have

$$\|w_{n+1} - w_{m+1}\|_X^q \le \left[1 - \varepsilon\left(\frac{q\gamma(\lambda_0 - 1)}{\lambda_0} - \gamma_q\right)\right] \|w_n - w_m\|_X^q. \quad (13.12)$$

Observing that $\varepsilon\left(\frac{q\gamma(\lambda_0-1)}{\lambda_0} - \gamma_q\right) \in (0,1)$, it follows that $\{w_n\}$ is Cauchy in E and so it converges to (u^*, v^*) satisfying $v^* = Fu^*$ and $u^* + KFu^* = 0$. This completes the proof.   □

Corollary 13.19 *Suppose* $X = L_p$, $2 \le p < \gamma + \sqrt{\gamma^2 + 4}$. *Let* $F, K : X \to X$ *be Lipschitzian maps and* $D(K) = F(X) = X$ *with conditions* (13.8) *and* (13.9) *where* $\gamma := \min\{\alpha, \beta\}$. *Assume that* $u + KFu = 0$ *has a solution* u^*. *Then,* $\{u_n\}$ *defined by* (13.10) *converges to* u^* *in* X, $\{v_n\}$ *converges to* v^* *in* X *and* u^* *is the unique solution of* $u + KFu = 0$ *with* $v^* = Fu^*$.

Proof. Since L_p spaces, $2 \le p < \infty$, are q-uniformly smooth spaces with $q = 2$ and $c_q = d_q = p - 1$ then, $\gamma_q = \frac{1}{p}(p^2 - 4)$ so that $\gamma > \frac{\gamma_q}{q}$ becomes $\gamma > \frac{1}{2p}(p^2 - 4)$ which is clearly satisfied. Hence, conditions (iii) preceding Theorem 13.18 are easily satisfied. The proof follows from Theorem 13.18.

In a real Hilbert space, $c_p = d_q = 1$, $q = 2$ and so $(1 + c_q)(1 + d_q) \ge 2^q$ is satisfied. Moreover, $\gamma_q = 0$ so that $\gamma > \frac{\gamma_q}{q}$ reduces to $\gamma > 0$. We then have the following corollary.

Corollary 13.20 *Let* H *be a real Hilbert space and* $F, K : H \to H$ *be maps with* $D(K) = F(K) = H$ *such that the following conditions hold:* (i) F *is Lipschitzian with constant* L_F *and there exists a positive constant* $\alpha > 0$ *such that*

$$\langle Fu_1 - Fu_2, u_1 - u_2\rangle \ge \alpha\|u_1 - u_2\|_H^2 \quad \forall u_1, u_2 \in H; \quad (13.13)$$

(*ii*) K *is Lipschitzian with constant* L_K *and there exists a positive constant* $\beta > 0$ *such that*

$$\langle Ku_1 - Ku_2, u_1 - u_2 \rangle \geq \beta \|u_1 - u_2\|_H^2 \quad \forall \, u_1, u_2 \in H; \qquad (13.14)$$

Let $L := \max\{L_F, L_K\}$, $\gamma := \min\{\alpha, \beta\}$. *Let* $\{u_n\}$ *and* $\{v_n\}$ *be sequences in* H *defined as follows:*

$$
\begin{aligned}
u_0 \in X, \ u_{n+1} &= u_n - \varepsilon(Fu_n - v_n), \ n \geq 1; \\
v_0 \in X, \ v_{n+1} &= v_n - \varepsilon(Kv_n + u_n), \ n \geq 1;
\end{aligned}
\qquad (13.15)
$$

with $\varepsilon \in \left(0, \frac{\gamma}{(3L+1)}\right)$. *Assume that* $u + KFu = 0$ *has a solution* u^*. *Then,* $\{u_n\}$ *converges to* u^* *in* H, $\{v_n\}$ *converges to* v^* *in* H *and* u^* *is the unique solution of* $u + KFu = 0$ *with* $v^* = Fu^*$.

Proof. The proof follows from Corollary 13.19 with $p = 2, \gamma_0 = 2$.

13.3.3 Convergence in $L_p, 1 < p \leq 2$

We observe that in L_p spaces, $1 < p < 2$, the condition $\gamma > \frac{\gamma_q}{q}$ is not easy to verify because c_q and d_q are not known precisely, so, in this section we use a different tool to obtain the conclusions of the last sub-section in these spaces. We begin with the following lemmas.

Lemma 13.21. *Let* $E = L_p$, $1 < p \leq 2$, *we have the following inequalities (see, e.g. Bynum [61]).*

$$(p - 1)\|x + y\|^2 \leq \|x\|^2 + 2\langle y, j(x) \rangle + \|y\|^2 \ \forall \, x, y \in E. \qquad (13.16)$$

$$\|x + y\|^2 + \|x - y\|^2 \geq 2\|x\|^2 + 2(p - 1)\|y\|^2 \ \forall \, x, y \in E. \qquad (13.17)$$

As a consequence of (13.16) and (13.17), following the method of the last sub-section, we have the following:

Lemma 13.22. *Let* $E = L_p$, $1 < p \leq 2$. *Let* x_1, x_2, y_1, y_2 *be points in* E. *Then we have the following estimate:*

$$
\begin{aligned}
A(x_1, x_2, y_1, y_2) &:= \langle y_1 - y_2, j(x_1 - x_2) \rangle + \langle x_1 - x_2, j(y_2 - y_1) \rangle \\
&\leq p(2 - p)\Big(\|x_1 - x_2\|_X^2 + \|y_1 - y_2\|_X^2\Big).
\end{aligned}
$$

Using Lemma 13.22, we now prove the following theorem.

Theorem 13.23. *Let* $X = L_p$, $1 + \sqrt{1-\gamma} < p \leq 2$, *and* $F, K : X \to X$ *with* $D(K) = F(K) = X$ *be maps such that the following conditions hold:* (i) F *is Lipschitzian with constant* L_F *and there exists a positive constant* $\alpha > 0$ *such that*

$$\langle Fu_1 - Fu_2, j_q(u_1 - u_2) \rangle \geq \alpha \|u_1 - u_2\|_X^2 \quad \forall \, u_1, u_2 \in D(F); \qquad (13.18)$$

(ii) K *is Lipschitzian with constant* L_K *and there exists a positive constant* $\beta > 0$ *such that*

$$\langle Ku_1 - Ku_2, j_q(u_1 - u_2) \rangle \geq \beta \|u_1 - u_2\|_X^2 \quad \forall \, u_1, u_2 \in D(K). \qquad (13.19)$$

(iii) *Let* $\gamma := \min\{\alpha, \beta\}$. *Let* $\{u_n\}$ *and* $\{v_n\}$ *be sequences in* X *defined as follows:*

$$\begin{aligned} u_0 \in X, \quad u_{n+1} = u_n - \varepsilon(Fu_n - v_n), \quad n \geq 1; \\ v_0 \in X, \quad v_{n+1} = v_n - \varepsilon(Kv_n + u_n), \quad n \geq 1; \end{aligned} \qquad (13.20)$$

with

$$0 < \varepsilon < \min\left\{\frac{1}{\Lambda_p}, \left(\frac{p\gamma}{\lambda_0 \delta_p}\right)^{\frac{1}{p-1}}\right\}, \quad \Lambda_p := \frac{1}{\frac{p\gamma(\lambda_0-1)}{\lambda_0} - p^2(2-p)},$$

where λ_0 *is any number such that* $\lambda_0 > \gamma/[\gamma - p(2-p)]$ *and* δ_p *is as in the definition preceding Theorem 13.18 (with q replaced by p). Assume that* $u + KFu = 0$ *has a solution* u^*. *Then,* $\{u_n\}$ *converges to* u^* *in* X, $\{v_n\}$ *converges to* v^* *in* X *and* u^* *is the unique solution of* $u + KFu = 0$ *with* $v^* = Fu^*$.

Proof. Using the same arguments as in the proof of Theorem 13.18, we have:

$$\|w_{n+1} - w_{m+1}\|_X^2 \leq (1 - \gamma p\varepsilon + \varepsilon^p \delta_p) \|w_n - w_m\|_X^2 + p\varepsilon \Lambda_{nm}, \quad (13.21)$$

which implies that

$$\|w_{n+1} - w_{m+1}\|_X^2 \leq \left[1 - \varepsilon\left(p\gamma - p^2(2-p) - \varepsilon^{p-1}\delta_p\right)\right] \times$$

$$\|w_n - w_m\|_X^2. \qquad (13.22)$$

Observing that $\varepsilon\left(p\gamma - p^2(2-p) - \varepsilon^{p-1}\delta_p\right) \in (0,1)$, it follows that $\{w_n\}$ is Cauchy in E and so it converges to (u^*, v^*). This completes the proof. $\square$

13.4 Coupled Scheme for the Case of Bounded Operators

We prove the following results.

13.4.1 Convergence Theorems

For $q > 1$, let X be a real q- uniformly smooth Banach space and $F, K : X \to X$ be maps with $D(K) = F(K) = X$ such that the following conditions hold: (i) F is bounded and there exists a positive constant $\alpha > 0$ such that

$$\langle Fu_1 - Fu_2, j_q(u_1 - u_2) \rangle \geq \alpha \|u_1 - u_2\|_X^q \quad \forall \ u_1, u_2 \in X; \qquad (13.23)$$

(ii) K is bounded and there exists a positive constant $\beta > 0$ such that

$$\langle Ku_1 - Ku_2, j_q(u_1 - u_2) \rangle \geq \beta \|u_1 - u_2\|_X^q \quad \forall \ u_1, u_2 \in X; \qquad (13.24)$$

(iii) $(1 + c_q)(1 + d_q) \geq 2^q$, $\gamma := \min\{\alpha, \beta\} > \frac{\delta_q}{q}$, where $\delta_q := \frac{\left[(1+c_q)(1+d_q)-2^q\right]}{1+c_q}$. With these assumptions, we prove the following theorem.

Theorem 13.24. *Let $\{u_n\}$ and $\{v_n\}$ be sequences in X defined iteratively from arbitrary points $u_1, v_1 \in X$ as follows:*

$$\begin{aligned} u_{n+1} &= u_n - \alpha_n(Fu_n - v_n), \ n \geq 1; \\ v_{n+1} &= v_n - \alpha_n(Kv_n + u_n), \ n \geq 1; \end{aligned} \qquad (13.25)$$

where $\{\alpha_n\}$ is a positive sequence satisfying $\lim \alpha_n = 0$, $\sum \alpha_n = \infty$ and $\sum \alpha_n^q < \infty$. Assume that $u + KFu = 0$ has a solution u^*. Then, there exists a constant $d_0 > 0$ such that if $0 < \alpha_n \leq d_0$, $\{u_n\}$ converges to u^* in X, $\{v_n\}$ converges to v^* in X and u^* is the unique solution of $u + KFu = 0$ with $v^* = Fu^*$.

Proof. Let $E := X \times X$ with the norm $\|z\|_E = (\|u\|_X^q + \|v\|_X^q)^{1/q}$ where $z = (u, v)$. Define the sequence $\{w_n\}$ in E by: $w_n := (u_n, v_n)$. Let $u^* \in X$ be the solution of $u + KFu = 0$, $v^* := Fu^*$ and $w^* := (u^*, v^*)$. We observe that $u^* = -Kv^*$. It suffices to show that $\{w_n\}$ converges to w^* in E. For this, let $r > 0$ be sufficient large such that $w_1 \in B(w^*, r)$. Define $B := \overline{B(w^*, r)}$. Since F and K are bounded, we set $M_1 := \sup\{\|Fx - y\|_X \ : \ (x, y) \in B\} < \infty$ and $M_2 := \sup\{\|Ky + x\|_X \ : \ (x, y) \in B\} < \infty$. Let $M := (M_1^q + M_2^q)^{1/q}$. Set

$$d_0 := \min \left\{ \frac{1}{q\gamma - \delta_q}, \left(\frac{q\gamma - \delta_q}{2M^2 d_q} r^q \right)^{\frac{1}{q-1}} \right\}.$$

Step1. The sequence $\{w_n\}$ is bounded in E. Indeed, it suffices to show that w_n is in B for all $n \geq 1$. The proof is by induction. By construction, $w_1 \in B$. Suppose that $w_n \in B$. We prove that $w_{n+1} \in B$. Assume for contradiction that $w_{n+1} \notin B$. Then, we have $\|w_{n+1} - w^*\|_E > r$. We compute as follows:

$$\|w_{n+1} - w^*\|^q = \|u_{n+1} - u^*\|_X^q + \|v_{n+1} - v^*\|_X^q.$$

From (5.10), it follows that

$$\begin{aligned}
\|u_{n+1} - u^*\|_X^q &= \|u_n - u^* - \alpha_n(Fu_n - v_n)\|^q \\
&\leq \|u_n - u^*\|_X^q - q\alpha_n \langle Fu_n - v_n, j_q(u_n - u^*) \rangle + d_q \alpha_n^q \|Fu_n - v_n\|_X^q \\
&\leq \|u_n - u^*\|_X^q - q\alpha_n \langle Fu_n - v_n, j_q(u_n - u^*) \rangle + d_q \alpha_n^q M_1^q.
\end{aligned}$$

Observing that

$$\langle Fu_n - v_n, j_q(u_n - u^*) \rangle = \langle Fu_n - Fu^*, j_q(u_n - u^*) \rangle - \langle v_n - v^*, j_q(u_n - u^*) \rangle$$

and using (13.23), we have the following estimate:

$$\begin{aligned}
\|u_{n+1} - u^*\|_X^q &\leq \left[1 - q\gamma\alpha_n\right] \|u_n - u^*\|_X^q + d_q \alpha_n^q M_1^q \\
&\quad + q\alpha_n \langle v_n - v^*, j_q(u_n - u^*) \rangle.
\end{aligned}$$

Following the same argument, we have

$$\begin{aligned}
\|v_{n+1} - v^*\|_X^q &\leq \left[1 - q\gamma\alpha_n\right] \|v_n - v^*\|_X^q + d_q \alpha_n^q M_2^q \\
&\quad + q\alpha_n \langle u^* - u_n, j_q(v_n - v^*) \rangle.
\end{aligned}$$

Thus, we obtain,

$$\begin{aligned}
\|w_{n+1} - w^*\|_X^q &\leq \left[1 - q\gamma\alpha_n\right] \|w_n - w^*\|_X^q + d_q M^q \alpha_n^q \\
&\quad + q\alpha_n \Big(\langle v_n - v^*, j_q(u_n - u^*) \rangle - \langle u_n - u^*, j_q(v_n - v^*) \rangle \Big).
\end{aligned}$$

Therefore, using Lemma 13.17, we have,

$$\begin{aligned}
\|w_{n+1} - w^*\|_X^q &\leq \left[1 - \alpha_n \Big(q\gamma - \delta_q \Big)\right] \|w_n - w^*\|_X^q + d_q M^q \alpha_n^q \\
&< \left[1 - \alpha_n \Big(q\gamma - \delta_q \Big)\right] r^q + d_q M^q \alpha_n^q \\
&< \left[1 - \alpha_n \Big(\frac{q\gamma - \delta_q}{2} \Big)\right] r^q.
\end{aligned}$$

Observing that $\alpha_n \Big(\frac{q\gamma - \delta_q}{2} \Big) \in (0,1)$, it follows that $\|w_{n+1} - w^*\|_X^q < r^q$, a contradiction. Therefore $w_{n+1} \in B$. Thus, by induction, $\{w_n\}$ is bounded and so are $\{u_n\}$ and $\{v_n\}$.

Step 2. We now prove that $\{w_n\}$ converges to w^* in E. Following the same method of computations as in step 1, we obtain the following inequality:

$$\|w_{n+1} - w^*\|_X^q \leq \left[1 - \alpha_n\left(q\gamma - \delta_q\right)\right]\|w_n - w^*\|_X^q + d_q M^q \alpha_n^q.$$

Observing that $\alpha_n\left(q\gamma - \delta_q\right) \in (0,1)$, $\sum \alpha_n = \infty$ and $\sum \alpha_n^q < \infty$, it follows from Lemma 6.34 that $\{w_n\}$ converge to w^* in E. This completes the proof. $\square$

Corollary 13.25 *Suppose $X = L_p$, $2 \leq p \leq \gamma + \sqrt{\gamma^2 + 4}$. Let $F, K : X \to X$ be bounded maps and $D(K) = F(X) = X$ with conditions (13.23) and (13.24) where $\gamma := \min\{\alpha, \beta\}$. Assume that $u + KFu = 0$ has a solution u^*. Then, there exists a constant $d_0 > 0$ such that if $0 < \alpha_n \leq d_0$, $\{u_n\}$ defined by (13.25) converges to u^* in X, $\{v_n\}$ converges to v^* in X and u^* is the unique solution of $u + KFu = 0$ with $v^* = Fu^*$.*

Proof. Since L_p spaces, $2 \leq p \leq \gamma + \sqrt{\gamma^2 + 4}$, are q-uniformly smooth spaces with $q = 2$ and $c_q = d_q = p - 1$ then, conditions (iii) preceding Theorem 13.24 are easily satisfied. The proof follows from Theorem 13.24. $\square$

In a real Hilbert space, $c_p = d_q = 1$, $q = 2$ and so $(1 + c_q)(1 + d_q) \geq 2^q$ is satisfied. Moreover, $\delta_q = 0$ so that $\gamma > \frac{\delta_q}{q}$ reduces to $\gamma > 0$. We then have the following corollary.

Corollary 13.26 *Let H be a real Hilbert space and $F, K : H \to H$ be maps with $D(K) = F(K) = H$ such that the following conditions hold: (i) F is bounded and there exists a positive constant $\alpha > 0$ such that*

$$\langle Fu_1 - Fu_2, u_1 - u_2 \rangle \geq \alpha \|u_1 - u_2\|_H^2 \quad \forall \, u_1, u_2 \in H; \tag{13.26}$$

(ii) K is bounded and there exists a positive constant $\beta > 0$ such that

$$\langle Ku_1 - Ku_2, u_1 - u_2 \rangle \geq \beta \|u_1 - u_2\|_H^2 \quad \forall \, u_1, u_2 \in H; \tag{13.27}$$

Let $\{u_n\}$ and $\{v_n\}$ be sequences in H defined iteratively from arbitrary points $u_1, v_1 \in H$ as follows:

$$\begin{aligned} u_{n+1} &= u_n - \alpha_n(Fu_n - v_n), \quad n \geq 1; \\ v_{n+1} &= v_n - \alpha_n(Kv_n + u_n), \quad n \geq 1; \end{aligned} \tag{13.28}$$

where $\{\alpha_n\}$ is a positive sequence satisfying $\lim \alpha_n = 0$, $\sum \alpha_n = \infty$ and $\sum \alpha_n^2 < \infty$, where $\gamma := \min\{\alpha, \beta\}$. Assume that $u + KFu = 0$ has a solution u^*. Then there exists a constant $d_0 > 0$ such that if $0 < \alpha_n \leq d_0$, $\{u_n\}$ converges to u^* in H, $\{v_n\}$ converges to v^* in H and u^* is the unique solution of $u + KFu = 0$ with $v^* = Fu^*$.

Proof. The proof follows from Corollary 13.25 with $p = 2$.

13.4.2 Convergence for Bounded Operators in L_p Spaces, $1 < p \leq 2$

We observe that in L_p spaces, $1 < p < 2$, the condition $\gamma > \frac{\delta_q}{q}$ is not easy to verify since c_q and d_q are not know precisely. So, in this section we use a different tool to obtain the conclusions of the last sub-section in these spaces. We begin with the following theorem.

Theorem 13.27. Let $X = L_p$, $1 + \sqrt{1 - \gamma} < p \leq 2$ and $F, K : X \to X$ with $D(K) = F(K) = X$ be maps such that the following conditions hold:

(i) F is bounded and there exists a positive constant $\alpha > 0$ such that

$$\langle Fu_1 - Fu_2, j(u_1 - u_2) \rangle \geq \alpha \|u_1 - u_2\|_X^2 \quad \forall \, u_1, u_2 \in D(F); \quad (13.29)$$

(ii) K is bounded and there exists a positive constant $\beta > 0$ such that

$$\langle Ku_1 - Ku_2, j(u_1 - u_2) \rangle \geq \beta \|u_1 - u_2\|_X^2 \quad \forall \, u_1, u_2 \in D(K). \quad (13.30)$$

(iii) Let $\gamma := \min\{\alpha, \beta\}$. Let $\{u_n\}$ and $\{v_n\}$ be sequences in X defined iteratively from arbitrary points $u_1, v_1 \in X$ as follows:

$$\begin{aligned} u_{n+1} &= u_n - \alpha_n(Fu_n - v_n), \ n \geq 1; \\ v_{n+1} &= v_n - \alpha_n(Kv_n + u_n), \ n \geq 1; \end{aligned} \quad (13.31)$$

where $\{\alpha_n\}$ is a positive sequence satisfying $\lim \alpha_n = 0$, $\sum \alpha_n = \infty$ and $\sum \alpha_n^2 < \infty$. Assume that $u + KFu = 0$ has a solution u^*. Then, there exists a constant $d_0 > 0$ such that if $0 < \alpha_n \leq d_0$, $\{u_n\}$ converges to u^* in X, $\{v_n\}$ converges to v^* in X and u^* is the unique solution of $u + KFu = 0$ with $v^* = Fu^*$.

Proof. Using the same arguments as in the proof of Theorem 13.24, we obtain the following estimate:

$$\begin{aligned} \|w_{n+1} - w^*\|_X^2 &\leq \left[1 - \alpha_n\left(2\gamma - 2p(2 - p)\right)\right]\|w_n - w^*\|_X^2 \\ &\quad + M\alpha_n^2, \end{aligned} \quad (13.32)$$

for some constant $M > 0$. It follows from Lemma 6.34 that $\{w_n\}$ converges to w^*. This completes the proof. $\qquad \square$

13.4.3 Convergence Theorems for Generalized Lipschitz Maps

Clearly, every generalized Lipschitz map is bounded. So, we obtain the following corollaries.

Corollary 13.28 *Let X be a q-uniformly smooth Bancah space, $q > 1$ and $F, K : X \rightarrow X$ with $D(K) = F(K) = X$ be maps such that the following conditions hold:*
(i) F is generalized Lipschitz and there exists a positive constant $\alpha > 0$ such that

$$\langle Fu_1 - Fu_2, j_q(u_1 - u_2) \rangle \geq \alpha \|u_1 - u_2\|_X^q \quad \forall\, u_1, u_2 \in D(F); \qquad (13.33)$$

(ii) K is generalized Lipschitz and there exists a positive constant $\beta > 0$ such that

$$\langle Ku_1 - Ku_2, j_q(u_1 - u_2) \rangle \geq \beta \|u_1 - u_2\|_X^q \quad \forall\, u_1, u_2 \in D(K). \qquad (13.34)$$

(iii) Assume $\gamma := \min\{\alpha, \beta\} > \frac{\delta_q}{q}$.
Let $\{u_n\}$ and $\{v_n\}$ be sequences in X defined iteratively from arbitrary points $u_1, v_1 \in X$ as follows:

$$\begin{aligned} u_{n+1} &= u_n - \alpha_n(Fu_n - v_n), \quad n \geq 1; \\ v_{n+1} &= v_n - \alpha_n(Kv_n + u_n), \quad n \geq 1; \end{aligned} \qquad (13.35)$$

where $\{\alpha_n\}$ is a positive sequence satisfying $\lim \alpha_n = 0$, $\sum \alpha_n = \infty$ and $\sum \alpha_n^q < \infty$. Assume that $u + KFu = 0$ has a solution u^*. Then, there exists a constant $d_0 > 0$ such that if $0 < \alpha_n \leq d_0$, $\{u_n\}$ converges to u^* in X, $\{v_n\}$ converges to v^* in X and u^* is the unique solution of $u + KFu = 0$ with $v^* = Fu^*$.

Corollary 13.29 *Let $X = L_p$, $1 + \sqrt{1 - \gamma} < p \leq 2$, and $F, K : X \rightarrow X$ with $D(K) = F(K) = X$ be maps such that the following conditions hold:*
(i) F is generalized Lipschitz and there exists a positive constant $\alpha > 0$ such that

$$\langle Fu_1 - Fu_2, j(u_1 - u_2) \rangle \geq \alpha \|u_1 - u_2\|_X^2 \quad \forall\, u_1, u_2 \in D(F); \qquad (13.36)$$

(ii) K is generalized Lipschitz and there exists a positive constant $\beta > 0$ such that

$$\langle Ku_1 - Ku_2, j(u_1 - u_2) \rangle \geq \beta \|u_1 - u_2\|_X^2 \quad \forall\, u_1, u_2 \in D(K). \qquad (13.37)$$

(iii) Let $\gamma := \min\{\alpha, \beta\}$. Let $\{u_n\}$ and $\{v_n\}$ be sequences in H defined iteratively from arbitrary points $u_1, v_1 \in H$ as follows:

$$\begin{aligned} u_{n+1} &= u_n - \alpha_n(Fu_n - v_n), \quad n \geq 1; \\ v_{n+1} &= v_n - \alpha_n(Kv_n + u_n), \quad n \geq 1; \end{aligned} \qquad (13.38)$$

where $\{\alpha_n\}$ is a positive sequence satisfying $\lim \alpha_n = 0$, $\sum \alpha_n = \infty$ and $\sum \alpha_n^2 < \infty$. Assume that $u + KFu = 0$ has a solution u^*. Then there exists a constant $d_0 > 0$ such that if $0 < \alpha_n \leq d_0$, $\{u_n\}$ converges to u^* in H, $\{v_n\}$

converges to v^* in H and u^* is the unique solution of $u + KFu = 0$ with $v^* = Fu^*$.

Corollary 13.30 *Let $X = L_p$, $2 \leq p < \gamma + \sqrt{\gamma^2 + 4}$, and $F, K : X \to X$ with $D(K) = F(K) = X$ be maps such that the following conditions hold:*
(i) F is generalized Lipschitz and there exists a positive constant $\alpha > 0$ such that

$$\langle Fu_1 - Fu_2, j(u_1 - u_2) \rangle \geq \alpha \|u_1 - u_2\|_X^2 \quad \forall \, u_1, u_2 \in D(F); \qquad (13.39)$$

(ii) K is generalized Lipschitz and there exists a positive constant $\beta > 0$ such that

$$\langle Ku_1 - Ku_2, j(u_1 - u_2) \rangle \geq \beta \|u_1 - u_2\|_X^2 \quad \forall \, u_1, u_2 \in D(K). \qquad (13.40)$$

(iii)Let $\gamma := \min\{\alpha, \beta\}$. Let $\{u_n\}$ and $\{v_n\}$ be sequences in H defined iteratively from arbitrary points $u_1, v_1 \in H$ as follows:

$$\begin{aligned} u_{n+1} &= u_n - \alpha_n(Fu_n - v_n), \quad n \geq 1; \\ v_{n+1} &= v_n - \alpha_n(Kv_n + u_n), \quad n \geq 1; \end{aligned} \qquad (13.41)$$

where $\{\alpha_n\}$ is a positive sequence satisfying $\lim \alpha_n = 0$, $\sum \alpha_n = \infty$ and $\sum \alpha_n^2 < \infty$. Assume that $u + KFu = 0$ has a solution u^*. Then there exists a constant $d_0 > 0$ such that if $0 < \alpha_n \leq d_0$, $\{u_n\}$ converges to u^* in H, $\{v_n\}$ converges to v^* in H and u^* is the unique solution of $u + KFu = 0$ with $v^* = Fu^*$.

Remark 13.31. A mapping $A : E \to E$ is called *strongly quasi-accretive* if $N(A) := \{x \in E : Ax = 0\} \neq \emptyset$ and

$$\langle Ax - Ap, j_q(x - p) \rangle \geq k\|x - p\|^q \ \forall \, x \in E, \ p \in N(A).$$

It is trivial to see that all the theorems of this chapter carry over to this class of mappings. Furthermore, the addition of *bounded* error terms to any of the recursion formulas of this chapter leads to no generalization.

13.5 Remarks and Open Questions

Remark 13.32. Theorem 13.27 and Corollary 13.29 are valid in L_p spaces where $1 + \sqrt{1 - \gamma} < p \leq 2$. Without loss of generality, we may assume that $\gamma \in (0, 1]$. Observe that if $\gamma = 1$ then the theorem and the corollary hold in all L_p spaces, $1 < p \leq 2$.

Open question 1. Do Theorem 13.27 and Corollary 13.29 hold in L_p spaces for all p such that $1 < p \leq 2$?

Open question 2. Do Corollaries 13.25 and 13.30 hold in L_p spaces for all p such that $2 \leq p < \infty$?

EXERCISES 13.1

1. Find an example to show that the composition of two monotone operators need not be monotone.

13.6 Historical Remarks

All the results of Sections 13.1 and 13.2 are due to Chidume and Zegeye [170]. The results of Sections 13.3 and 13.4 are due to Chidume and Djitte [137], [138].

Chapter 14
Iterative Methods for Some Generalizations of Nonexpansive Maps

14.1 Introduction

Some generalizations of nonexpansive mappings which have been studied extensively include the *(i) quasi-nonexpansive mappings; (ii) asymptotically nonexpansive mappings; (iii) asymptotically quasi-nonexpansive mappings* .

For the past 30 years or so, iterative algorithms for approximating fixed points of operators belonging to subclasses of these classes of nonlinear mappings and defined in appropriate Banach spaces have been flourishing areas of research for many mathematicians. For the classes of mappings mentioned here in *(i)* to *(iii)*, we show in this chapter that modifications of the Mann iteration algorithm and of the Halpern-type iteration process studied in chapter 6 can be used to approximate fixed points (when they exist).

14.2 Iteration Methods for Asymptotically Nonexpansive Mappings

14.2.1 Modified Mann Process

In this section, we examine the class of nonlinear *asymptotically nonexpansive mappings* which was introduced by Goebel and Kirk [231] (see also Kirk [282], Bruck *et al.* [60]) and which has been studied extensively by various authors. We begin with the following definition.

Definition 14.1. Let K be a nonempty subset of a normed linear space E. A mapping $T : K \to K$ is called *asymptotically nonexpansive* (Goebel and Kirk, [231]) if there exists a sequence $\{k_n\}, k_n \geq 1$, such that $\lim_{n \to \infty} k_n = 1$, and $||T^n x - T^n y|| \leq k_n ||x - y||$ holds for each $x, y \in K$ and for each integer

C. Chidume, *Geometric Properties of Banach Spaces and Nonlinear Iterations,*
Lecture Notes in Mathematics 1965,
© Springer-Verlag London Limited 2009

$n \geq 1$. The mapping $T : K \rightarrow K$ is called *asymptotically nonexpansive in the intermediate sense* (Bruck, *et al.* [60]) provided T is uniformly continuous and

$$\limsup_{n \rightarrow \infty} \left\{ \sup_{x,y \in K} (\|T^n x - T^n y\| - \|x - y\|) \right\} \leq 0.$$

It is clear that every nonexpansive mapping is asymptotically nonexpansive. The following example shows that the inclusion is proper.

Example 14.2. (Goebel and Kirk, [230, 231]) Let B denote the unit ball in the Hilbert space l^2 and let U be defined as follows:

$$U : (x_1, x_2, x_3, ...) \rightarrow (0, x_1^2, a_2 x_2, a_3 x_3, ...),$$

where $\{a_i\}$ is a sequence of numbers such that $0 < a_i < 1$ and $\prod_{i=2}^{\infty} a_i = \frac{1}{2}$. Then, U is Lipschitzian and $\|Ux - Uy\| \leq 2\|x - y\|, x, y \in B$; moreover,

$$\|U^i x - U^i y\| \leq 2 \prod_{j=2}^{i} a_j \|x - y\| \ \forall \ i = 2, 3, ...$$

Thus, $\lim_{i \rightarrow \infty} k_i = \lim_{i \rightarrow \infty} 2 \prod_{j=2}^{i} a_j = 1$. Clearly, U is not nonexpansive.

Two other definitions of asymptotically nonexpansive maps have also appeared in the literature. A definition weaker than Definition 14.1 was introduced in 1974 by Kirk [282] and requires that

$$\limsup_{n \rightarrow \infty} \sup_{y \in K} (\|T^n x - T^n y\| - \|x - y\|) \leq 0,$$

for every $x \in K$, and that T^N be continuous for some integer $N > 1$. The definition of mappings which are asymptotically nonexpansive in the intermediate sense is somewhat between this definition of Kirk and that of Goebel and Kirk, [231]. The other definition which has appeared required that

$$\limsup_{n \rightarrow \infty} (\|T^n x - T^n y\| - \|x - y\|) \leq 0 \ \forall \ x, y \in K.$$

This, however, has been shown to be unsatisfactory from the point of view of fixed point theory. Tingly [489] constructed an example of a closed bounded convex set K in a Hilbert space and a continuous map $T : K \rightarrow K$ which actually satisfies the following condition $\lim_{n \rightarrow \infty} \|T^n x - T^n y\| = 0, \ \forall \ x, y \in K$ but which has no fixed point. The averaging iteration process $x_{n+1} := (1 - \alpha_n) x_n + \alpha_n T^n x_n$ where $T : K \rightarrow K$ is asymptotically nonexpansive in the sense of Definition 14.1, K is a closed convex and bounded subset of a Hilbert space was introduced by Schu [435]. He considered the following iteration scheme: $E = H$, a Hilbert space, K is a nonempty closed convex and bounded subset of H, $T : K \to K$ is completely continuous and asymptotically nonexpansive; for each $x_0 \in K$, $x_{n+1} := (1 - \alpha_n)x_n + \alpha_n T^n x_n, n \geq 0$, where $\sum(k_n^2 - 1) < \infty$ and $\{\alpha_n\}$ is a real sequence satisfying appropriate conditions. He proved that $\{x_n\}$ converges strongly to a fixed point of T. This result has been extended to uniformly convex Banach spaces in the following theorems.

Theorem 14.3. *(Rhoades, [421]) Let E be uniformly convex and K be a nonempty closed convex and bounded subset of E. Suppose $T : K \to K$ is completely continuous and asymptotically nonexpansive; for each $x_0 \in K$, let $\{x_n\}$ be defined as follows:*

$$x_{n+1} = (1 - \alpha_n)x_n + \alpha_n T^n x_n, n \geq 0,$$

where $\sum(k_n^r - 1) < \infty$, for some $r > 1$, $\epsilon \leq 1 - \alpha_n \leq 1 - \epsilon$ for all positive integer n and some $\epsilon > 0$. Then $\lim \|x_n - Tx_n\| = 0$.

Proof. Set $d = [dim(K)]^p$. From [231], Theorem 1, T has a fixed point x^* in K, with r such that $K \subseteq B_r$, it follows from inequality (4.32) that

$$
\begin{aligned}
\|x_{n+1} - x^*\|^p &\leq \alpha_n \|T^n x_n - x^*\|^p + (1 - \alpha_n)\|x_n - x^*\|^p \\
&\quad - W_p(\alpha_n)g(\|x_n - T^n x_n\|) \\
&\leq \alpha_n k_n^p \|x_n - x^*\|^p + (1 - \alpha_n)\|x_n - x^*\|^p \\
&\quad - W_p(\alpha_n)g(\|x_n - T^n x_n\|) \\
&\leq \|x_n - x^*\|^p + \alpha_n(k_n^p - 1)d \\
&\quad - W_p(\alpha_n)g(\|x_n - T^n x_n\|).
\end{aligned}
\tag{14.1}
$$

Note that $W_p(\alpha_n) \geq 2\epsilon^{p+1}$. Thus, from (14.1),

$$0 \leq 2\epsilon^{p+1} g(\|x_n - T^n x_n\|) \leq \|x_n - x^*\|^p - \|x_{n+1} - x^*\| + \alpha_n d(k_n^p - 1),$$

and hence

$$2\epsilon^{p+1} \sum_{n=0}^{m} g(\|x_n - T^n x_n\|) \leq \|x_0 - x^*\|^p + (1 - \epsilon)d \sum_{n=0}^{m}(k_n^p - 1),$$

which implies, from the hypothesis on $\{k_n\}$, that the series in the left converges. Therefore $\lim g(\|x_n - T^n x_n\|) = 0$ which, since g is continuous at 0 and strictly increasing, implies that $\lim \|x_n - T^n x_n\| = 0$. That $\lim \|x_n - Tx_n\| = 0$ now follows from [435], Lemma 1.2. $\square$

Theorem 14.4. *(Rhoades, [421]) Let E be uniformly convex and K be a nonempty closed convex and bounded subset of E. Suppose $T : K \to K$ is completely continuous and asymptotically nonexpansive; for each $x_0 \in K$, let $\{x_n\}$ be defined as follows:*

$$x_{n+1} = (1 - \alpha_n)x_n + \alpha_n T^n x_n, n \geq 0,$$

where $\sum(k_n^r - 1) < \infty$, $r := \max\{2, p\}$ and $\epsilon \leq \alpha_n \leq 1 - \epsilon$ for all positive integer n and some $\epsilon > 0$. Then, $\{x_n\}$ converges strongly to some fixed point of T.

Proof. From the assumption it follows that $\Omega := \overline{co}(\{x_1\} \bigcup T(K))$ is a compact subset of K containing $\{x_n\}$ where $\overline{co}$ denotes *closed convex hull*. Hence, there exist an $x \in K$ and some subsequence $\{x_{n_i}\}$ of $\{x_n\}$ which converges strongly to x. But T is continuous and $\lim \|x_n - Tx_n\| = 0$ by Theorem 14.3. Thus $Tx = x$. This implies that

$$\|x_{n+1} - x\| \leq \alpha_n \|T^n(x_n) - T^n(x)\| + (1 - \alpha_n)\|x_n - x\|$$

$$\leq (\alpha_n k_n + (1 - \alpha_n))\|x_n - x\|$$

$$\leq k_n \|x_n - x\|$$

for all $n \in \mathbb{N}$. Since $\prod_{n=1}^{\infty} k_n$ converges and $\lim x_{n_i} = x$, this is easily seen to ensure that the whole sequence $\{x_n\}$ converges to x. $\qquad\square$

The following theorem was proved by Chang. (It is a special case of the "sufficiency" part of Theorem 2.2 of [71]).

Theorem 14.5. *(Chang, [71]). Let E be a real uniformly smooth Banach space, K be a nonempty bounded closed convex subset of E, $T : K \to K$ be an asymptotically nonexpansive mapping with a sequence $\{k_n\} \subset (0, \infty)$, $\lim k_n = 1$, and let $F(T) := \{x \in K : Tx = x\} \neq \emptyset$. Let $\{\alpha_n\}$ and $\{\beta_n\}$ be real sequences in $[0, 1]$ satisfying the following conditions: (i) $\alpha_n \to 0, \beta_n \to 0$, $(n \to \infty)$; (ii) $\sum \alpha_n = \infty$. Let $x_0 \in K$ be arbitrary and let $\{x_n\}, \{y_n\}$ be defined iteratively by*

$$x_{n+1} := (1 - \alpha_n)x_n + \alpha_n T^n y_n,$$

$$y_n = (1 - \beta_n)x_n + \beta_n T^n x_n, n \geq 0.$$

Suppose there exists a strictly increasing function $\phi : [0, \infty) \to [0, \infty)$, $\phi(0) = 0$ such that

$$\langle Ty_n - x^*, j(y_n - x^*) \rangle \leq k_n \|y_n - x^*\|^2 - \phi(\|y_n - x^*\|)$$

for all integers $n \geq 0$. Then, $x_n \to x^* \in F(T)$.

For other results concerning the *strong convergence* of successive approximations to some fixed point of asymptotically nonexpansive mappings, we refer the reader to Chidume([117, 119]), Chidume and Ofoedu ([164]), Chang et al. [79], Chidume and Zegeye ([172], [177]), Chidume et al. ([180], [178]), Kaczor et al. [282], Oka [222], Osilike and Aniagboso [373], Passty [380], Tan and Xu ([484], [485], [488]), and the references contained therein.

Weak convergence theorems have also been proved for asymptotically nonexpansive mappings. In particular, we have the following theorem.

Theorem 14.6. *(Lim and Xu, [307]) Suppose E is a Banach space with a weakly sequentially continuous duality mapping, K is a weakly compact convex subset of E, and $T : K \to K$ is an asymptotically nonexpansive mapping. Then, the following conclusions hold: (i) T has fixed point in K, and (ii) if T is weakly asymptotically regular at some $x \in K$, that is, $w - \lim(T^n x - T^{n+1}x) = 0$, then $\{T^n x\}$ converges weakly to a fixed point of T.*

Remark 14.7. For other results concerning the *weak* convergence of successive approximations to fixed points of asymptotically nonexpansive mappings, we refer the reader to Gornicki [239, 240], Qihou [390, 391], and the references contained therein.

Bruck *et al.* [60] proved a convergence theorem for mappings which are asymptotically nonexpansive in the intermediate sense (see also, Chidume *et al.*, [164]).

14.2.2 Iteration Method of Schu

Our next algorithm for approximating fixed points of asymptotically nonexpansive mappings deals with the convergence of *almost fixed points, $x_n := \mu_n T^n x_n$* of an asymptotically nonexpansive mappings T. Schu [440] proved the convergence of this sequence $\{x_n\}$ to some fixed point of T, under the additional assumption that *T is uniformly asymptotically regular and $(I - T)$ is demiclosed.* These assumptions had actually been made by Vijayaraju [497] to ensure the existence of a fixed point for T. By strengthening the asymptotic regularity condition on T, Schu established the convergence of *an explicit iteration scheme, $z_{n+1} := \mu_{n+1} T^n z_n$* to some fixed point of T.

Definition 14.8. Let $\{\epsilon_n\} \subset (0, \infty)$ and $\{\mu_n\} \subset (0, 1)$. Then $(\{\epsilon_n\}, \{\mu_n\})$ is called *admissible* ([434, 439]) if and only if (1)$\{\epsilon_n\}$ is decreasing; (2)$\{\mu_n\}$ is strictly increasing with $\lim \mu_n = 1$; (3) there exists $\{\beta_n\} \subset I\!N$ such that $(i)\{\beta_n\}$ is increasing; $(ii) \lim \beta_n(1 - \mu_n) = \infty$; $(iii) \lim \frac{(1 - \mu_{n+\beta_n})}{(1 - \mu_n)} = 1$ and $(iv) \lim \frac{\epsilon_n \beta_n \mu_{n+\beta_n}}{(1 - \mu_n)} = 0$.

Theorem 14.9. *(Schu, [439]) Let E be a reflexive smooth Banach space possessing a duality map that is weakly sequentially continuous at 0; $\emptyset \neq K \subset E$ closed bounded and starshaped with respect to 0; $T : K \to K$ asymptotically nonexpansive with sequence $\{k_n\} \subset [1, \infty)$; $(I - T)$ demiclosed and $\{\lambda_n\} \subset (\frac{1}{2}, 1)$;*

$$x_0 \in K, \quad z_{n+1} := \left(\frac{\lambda_{n+1}}{k_{n+1}}\right) T^n z_n \quad \text{for} \quad n \geq N_0, \quad \text{some } N_0 > 0;$$

and (a) $\lim \lambda_n = 1, \mu_n := \frac{\lambda_n}{k_n}$ for all integers $n \geq N$, $k_n \leq \frac{\lambda_n^2}{2\lambda_n - 1}$; $\frac{(1-\mu_n)}{(1-\lambda_n)}$ bounded; $\epsilon_n \in (0, \infty)$ such that $(\{\epsilon_n\}, \{\mu_n\})$ is admissible. (b) $\|T^n x - T^{n+1} x\| \leq \epsilon_n$ for all integers $n \geq 1$ and for all $x \in K$. Then, $\{x_n\}$ converges strongly to some fixed point of T.

14.2.3 Halpern-type Process

Let E be a real uniformly smooth Banach space, K be a nonempty bounded closed convex subset of E, and $T : K \to K$ be a *nonexpansive mapping*. Then, for a fixed $u \in K$ and each integer $n \geq 1$, by the Banach Contraction Mapping Principle, there exists a unique $x_n \in K$ such that

$$x_n = \frac{1}{n} u + (1 - \frac{1}{n}) T x_n. \tag{14.2}$$

It follows immediately from this equation that $\lim \|x_n - T x_n\| = 0$. One of the most useful results concerning algorithms for approximating fixed points of nonexpansive mappings *in real uniformly smooth Banach spaces* is the celebrated convergence theorem of Reich [401] who proved that the implicit sequence $\{x_n\}$ defined in equation (14.2) actually converges strongly to a fixed point of T. Several authors have tried to obtain a result analogous to that of Reich [401] for *asymptotically nonexpansive mappings*. Suppose K is a nonempty bounded closed convex subset of a real uniformly smooth Banach space E and $T : K \to K$ is an *asymptotically nonexpansive mapping* with sequence $k_n \geq 1$ for all $n \geq 1$. Fix $u \in K$ and define, for each integer $n \geq 1$, the contraction mapping $S_n : K \to K$ by

$$S_n(x) = \left(1 - \frac{t_n}{k_n}\right) u + \frac{t_n}{k_n} T^n x,$$

where $\{t_n\} \subset [0, 1)$ is any sequence such that $t_n \to 1$. Then, by the Banach Contraction Mapping Principle, there exists a unique point x_n fixed by S_n, i.e., there exists x_n such that

$$x_n = \left(1 - \frac{t_n}{k_n}\right) u + \frac{t_n}{k_n} T^n x_n. \tag{14.3}$$

The question now arises as to whether or not this sequence converges to a fixed point of T. A partial answer is given in the following theorem.

Theorem LX (Lim and Xu, [307]) *Let E be a uniformly smooth Banach space, K be a nonempty closed convex and bounded subset of E, $T : K \to K$ be an asymptotically nonexpansive mapping with sequence $\{k_n\} \subset [1, \infty)$. Fix $u \in K$ and let $\{t_n\} \subset [0, 1)$ be chosen such that $\lim_{n \to \infty} t_n = 1$ and*

$\lim_{n\to\infty} \dfrac{k_n - 1}{k_n - t_n} = 0$. *Then, (i) for each integer $n \geq 0$, there is a unique $x_n \in K$ such that*

$$x_n = \left(1 - \frac{t_n}{k_n}\right)u + \frac{t_n}{k_n}T^n x_n;$$

suppose in addition that $\lim_{n\to\infty} \|x_n - Tx_n\| = 0$, then, (ii) the sequence $\{x_n\}_n$ converges strongly to a fixed point of T.

Chidume *et al.* [143] extended Theorem LX to reflexive Banach spaces with uniformly Gâteaux differentiable norms. As an application, they proved that the *explicit sequence* $\{z_n\}_{n\geq 0}$ iteratively generated by

$$z_1 \in K, \quad z_{n+1} = \left(1 - \frac{t_n}{k_n}\right)u + \frac{t_n}{k_n}T^n z_n, \quad n \geq 1,$$

converges strongly to a fixed point of the asymptotically nonexpansive mapping T.

We now have the following theorems.

Theorem 14.10. *(Chidume, Li and Udomene, [143]) Let E be a real Banach space with uniformly Gâteaux differentiable norm possessing uniform normal structure, K be a nonempty closed convex and bounded subset of E, $T : K \to K$ be an asymptotically nonexpansive mapping with sequence $\{k_n\} \subset [1, \infty)$. Let $u \in K$ be fixed and let $\{t_n\} \subset (0,1)$ be such that $\lim_{n\to\infty} t_n = 1$ and $\lim_{n\to\infty} \dfrac{k_n - 1}{k_n - t_n} = 0$. Then, (i) for each integer $n \geq 1$, there is a unique $x_n \in K$ such that*

$$x_n = \left(1 - \frac{t_n}{k_n}\right)u + \frac{t_n}{k_n}T^n x_n;$$

and if in addition $\lim_{n\to\infty} \|x_n - Tx_n\| = 0$, then, (ii) the sequence $\{x_n\}_n$ converges strongly to a fixed point of T.

Theorem 14.11. *(Chidume, Li and Udomene, [143]) Let E be a real Banach space with uniformly Gâteaux differentiable norm possessing uniform normal structure, K be a nonempty closed convex and bounded subset of E, $T : K \to K$ be an asymptotically nonexpansive mapping with sequence $\{k_n\} \subset [1, \infty)$. Let $u \in K$ be fixed and let $\{t_n\} \subset (0,1)$ be such that $\lim_{n\to\infty} t_n = 1$ and $\lim_{n\to\infty} \dfrac{k_n - 1}{k_n - t_n} = 0$. Define the sequence $\{z_n\}_n$ iteratively by $z_1 \in K$,*

$$z_{n+1} = \left(1 - \frac{t_n}{k_n}\right)u + \frac{t_n}{k_n}T^n z_n, \quad n \geq 1.$$

Then, (i) for each integer $n \geq 1$, there is a unique $x_n \in K$ such that

$$x_n = \left(1 - \frac{t_n}{k_n}\right)u + \frac{t_n}{k_n}T^n x_n;$$

and if in addition $\lim_{n\to\infty} \|x_n - Tx_n\| = 0$, $\|z_n - T^n z_n\| = o\left(1 - \frac{t_n}{k_n}\right)$ then, (ii) the sequence $\{z_n\}_n$ converges strongly to a fixed point of T.

In [453], the authors proved the following theorems.

Theorem 14.12. *(Shioji and Takahashi, [453], Theorem 1). Let K be a closed convex subset of a uniformly convex Banach space whose norm is uniformly Gâteaux differentiable and let T be an asymptotically nonexpansive mapping from K into itself such that the set $F(T)$ of fixed points of T is nonempty. Then, $F(T)$ is a sunny nonexpansive retract of K.*

Theorem 14.13. *(Shioji and Takahashi, [453], Theorem 2). Let K be a closed convex subset of a uniformly convex Banach space whose norm is uniformly Gâteaux differentiable and let T be an asymptotically nonexpansive mapping from K into itself with Lipschitz constants $\{k_n\}$ such that the set of fixed points of T is nonempty and let P be the sunny nonexpansive retraction from K onto $F(T)$. Let $\{\alpha_n\}$ be a real sequence such that*

$0 < \alpha_n < 1$; $\lim \alpha_n = 0$, $\lim \frac{b_n - 1}{\alpha_n} < 1$, *where* $b_n := \frac{1}{n+1}\sum_{j=0}^{n} k_j$, $n = 0, 1.2, ...$.

Let u be an element of K and let $\{x_n\}$ be the unique point of K which satisfies

$$x_n = \alpha_n u + (1 - \alpha_n)\frac{1}{n+1}\sum_{j=0}^{n}T^j x_n, \text{ for } n \geq N_0,$$

where N_0 is a sufficiently large number. Then, $\{x_n\}$ converges strongly to $Pu \in F(T)$.

14.3 Asymptotically Quasi-nonexpansive Mappings

In Chapter 6, we introduced the class of quasi-nonexpansive mappings which includes the class of nonexpansive maps with fixed points as a proper subclass.

In this section, we introduce a class of nonlinear mappings more general than the class of quasi-nonexpansive mappings and more general than the class of asymptotically nonexpansive mappings with fixed points and prove convergence theorems.

Definition 14.14. Let K be a nonempty subset of a normed linear space, E. A mapping $T : K \to K$ is called *asymptotically quasi-nonexpansive* if $F(T) \neq \emptyset$ and there exists $u_n \in [0, \infty)$, $\lim u_n = 0$ such that for all integers $n \geq 0$, the following inequality holds:

$$\|T^n x - x^*\| \leq (1 + u_n)\|x - x^*\| \ \forall x \in K, \ x^* \in F(T).$$

It is clear from this definition that every asymptotically nonexpansive mapping with a fixed point is asymptotically quasi-nonexpansive.

Petryshyn and Williamson [382] proved a necessary and sufficient condition for a Mann iteration algorithm to converge strongly to a fixed point of a quasi-nonexpansive mapping. In [118], Chidume proved general convergence theorems for asymptotically quasi-nonexpansive mappings.

We begin by proving the following lemmas. The method of proof is basically that of Qihou [393].

Lemma 14.15. *Let K be a nonempty closed convex subset of a real Banach space, and let $T : K \to K$ be an asymptotically quasi-nonexpansive mapping of K with sequence $\{u_n\}$. Let $\{x_n\}$ be defined iteratively by*

$$x_0 \in K, \ x_{n+1} := (1 - \alpha_n)x_n + \alpha_n T^n x_n, \tag{14.4}$$

where $(i)\, 0 < \alpha_n < 1$, $(ii)\, \sum_{n=1}^{\infty} \alpha_n < \infty$. Then, there exist an integer $N_1 > 0$ and some constant $M > 0$ such that for all $n \geq N_1$, the following inequalities hold:
(a) $\|x_{n+1} - x^*\| \leq (1 + D\alpha_n)\|x_n - x^*\|$, $D = \sup u_n$,
(b) $\|x_{n+m} - x^*\| \leq M\|x_n - x^*\|$.

Proof. For (a): Since T is asymptotically quasi-nonexpansive, $F(T) \neq \emptyset$. From (14.4),

$$\begin{aligned}
\|x_{n+1} - x^*\| &= \|(1 - \alpha_n)(x_n - x^*) + \alpha_n(T^n x_n - T^n x^*)\| \\
&\leq \left[1 - \alpha_n + \alpha_n(1 + u_n)\right]\|x_n - x^*\| \\
&\leq \left[1 + D\alpha_n\right]\|x_n - x^*\|.
\end{aligned}$$

For (b): Let $n, m \geq N_1$ and replace n in (a) by $n + m - 1$ to get the following estimates:

$$\begin{aligned}
\|x_{n+m} - x^*\| &\leq (1 + D\alpha_{n+m-1})\|x_{n+m-1} - x^*\| \\
&\leq \exp\left(D \sum_{j=n}^{n+m-1} \alpha_j\right)\|x_n - x^*\| \\
&\leq M\|x_n - x^*\|,
\end{aligned}$$

where $M := \exp(D \sum_{j=1}^{\infty} \alpha_j) > 0$. This completes the proof. $\qquad \square$

Lemma 14.16. *Let $a_n \geq 0, \sigma_n \geq 0$ be such that*

$$a_{n+1} \leq (1 + \alpha_n)a_n + \sigma_n.$$

If $(a) \sum_{n=1}^{\infty} \alpha_n < \infty$; $(b) \sum_{n=1}^{\infty} \sigma_n < \infty$; and $(c) \liminf_{n \to \infty} a_n = 0$, then $\lim_{n \to \infty} a_n = 0$.

We now prove the following theorems.

Theorem 14.17. *Let K be a nonempty closed convex subset of a real Banach space E, and let $T : K \to K$ be an asymptotically quasi-nonexpansive self-mapping of K. For arbitrary $x_0 \in K$, let $\{x_n\}$ be iteratively defined as follows:*

$$x_{n+1} = (1 - \alpha_n)x_n + \alpha_n T^n x_n,$$

where $(i) 0 < \alpha_n < 1$; $(ii) \sum_{n=1}^{\infty} \alpha_n < \infty$. Then, $\{x_n\}$ converges strongly to a fixed point of T if and only if $\liminf_{n\to\infty} d(x_n, F(T)) = 0$, where $d(y, K)$ denotes the distance of y to the set K, i.e. $d(y, K) := \inf_{x \in K} d(y, x)$.

Proof. Necessity. We basically use the technique of Qihou [393]. Let $x^* \in F(T)$. By Lemma 14.15(a) we have that, (using $D = \sup u_n$), $||x_{n+1} - x^*|| \leq (1 + D\alpha_n)||x_n - x^*||$. This implies that, $d(x_{n+1}, F(T)) \leq (1 + D\alpha_n)d(x_n, F(T))$. But $\liminf_{n\to\infty} d(x_n, F(T)) = 0$. So, by Lemma 14.15, $\lim_{n\to\infty} d(x_n, F(T)) = 0$. This completes the proof of the necessity.

Sufficiency. First we prove that the sequence is Cauchy. By condition (ii) and $\lim_{n\to\infty} d(x_n, F(T)) = 0$, given any $\epsilon > 0$, there exists an integer $N^* > N_1$ such that $d(x_n, F(T)) \leq \frac{\epsilon}{2(M+1)} \ \forall n \geq N^*$. This implies, there exists, $x^* \in F(T)$ such that $||x_n - x^*|| \leq \frac{\epsilon}{2(M+1)} \ \forall n \geq N^*$. Then $\forall n \geq N^*$ and integers $m \geq 1$, using Lemma 14.15(b), we obtain that

$$||x_{n+m} - x_n|| \leq ||x_{n+m} - x^*|| + ||x_n - x^*|| \leq M||x_n - x^*|| + ||x_n - x^*||$$

$$< \frac{M\epsilon}{2(M + 1)} + \frac{\epsilon}{2(M + 1)} < \epsilon.$$

Hence, $\{x_n\}$ is Cauchy. Completeness of E implies $x_n \to y^*$ for some $y^* \in E$. But clearly $y^* \in K$. We now prove $y^* \in F(T)$. Let $\tilde{\epsilon} > 0$ be given. Since $x_n \to y^*$, there exists an integer $N_2 > 0$ such that $||x_n - y^*|| < \frac{\tilde{\epsilon}}{2(2+u_1)}, \ \forall n \geq N_2$. Also, $\lim_{n\to\infty} d(x_n, F(T)) = 0$ implies there exists an integer $N_1^* \geq N^* \geq N_2 > 0$ such that for all $n \geq N_1^*$, $d(x_n, F(T)) < \frac{\tilde{\epsilon}}{2(2+u_1)}$. In particular, $d(x_{N_1^*}, F(T)) < \frac{\tilde{\epsilon}}{2(2+u_1)}$. Hence there exists $p^* \in F(T)$ such that $||x_{N_1^*} - p^*|| \leq \frac{\tilde{\epsilon}}{2(2+u_1)}$. Then, for all $n \geq N_1^*$,

$$||Ty^* - y^*|| \leq ||Ty^* - p^*|| + ||x_{N_1^*} - p^*||$$
$$+ ||x_{N_1^*} - y^*||$$
$$\leq (1 + u_1)||y^* - p^*|| + ||x_{N_1^*} - p^*|| + ||x_{N_1^*} - y^*||$$
$$\leq (1 + u_1)||y^* - x_{N_1^*}|| + (1 + u_1)||x_{N_1^*} - p^*||$$
$$+ ||x_{N_1^*} - p^*|| + ||x_{N_1^*} - y^*||$$

$$\leq (2+u_1)||y^* - x_{N_1^*}|| + (2+u_1)||x_{N_1^*} - p^*||$$

$$\leq (2+u_1)\frac{\tilde{\epsilon}}{2(2+u_1)} + (2+u_1)\frac{\tilde{\epsilon}}{2(2+u_1)} = \tilde{\epsilon}.$$

Hence $Ty^* = y^*$. The proof is complete.   □

Corollary 14.18 *Let K be a nonempty closed convex subset of a real Banach space E, and let $T : K \to K$ be a quasi-nonexpansive mapping of K. For arbitrary $x_0 \in K$, let $\{x_n\}$ be iteratively defined as follows:*

$$x_{n+1} := (1-\alpha_n)x_n + \alpha_n T^n x_n,$$

where $\{\alpha_n\}$ satisfies (i) and (ii) of Theorem 14.17. Then, $\{x_n\}$ converges strongly to a fixed point of T if and only if $\liminf_{n\to\infty} d(x_n, F(T)) = 0$.

Corollary 14.19 *Let K be a nonempty closed convex subset of a real Banach space E, and let $T : K \to K$ be an asymptotically nonexpansive mapping of K. For arbitrary $x_0 \in K$, let $\{x_n\}$ be iteratively defined as follows:*

$$x_{n+1} := (1-\alpha_n)x_n + \alpha_n T^n x_n,$$

where $\{\alpha_n\}$ satisfies (i) and (ii) of Theorem 14.17. Then, $\{x_n\}$ converges strongly to a fixed point of T if and only if $\liminf_{n\to\infty} d(x_n, F(T)) = 0$.

Corollary 14.20 *Let K be a nonempty closed convex subset of a real Banach space E, and let $T : K \to K$ be an asymptotically quasi-nonexpansive mapping of K. For arbitrary $x_0 \in K$, let $\{x_n\}$ be iteratively defined as follows:*

$$x_{n+1} := (1-\alpha_n)x_n + \alpha_n T^n x_n,$$

where $\{\alpha_n\}$ satisfies (i) and (ii) of Theorem 14.17. Suppose T satisfies the following conditions: (i) T is asymptotically regular at x_0 i.e., $\lim ||x_n - Tx_n|| = 0$; (ii) $\lim_{n\to\infty} ||z_n - Tz_n|| = 0$ implies $\liminf_{n\to\infty} d(z_n, F(T)) = 0$ for any sequence $\{z_n\}$ in K. Then, $\{x_n\}$ converges strongly to a fixed point of T.

Proof. Conditions (i) and (ii) imply $\liminf_{n\to\infty} d(z_n, F(T)) = 0$. The result follows from Theorem 14.17.   □

EXERCISES 14.1

1. Prove that the sequence $\{\alpha_n\}$ defined by

$$\alpha_n := \begin{cases} n^{-\frac{1}{2}}, & \text{if } n \text{ is odd}, \\ (n^{-\frac{1}{2}} - 1)^{-1}, & \text{if } n \text{ is even} \end{cases}$$

satisfies the condition $\lim\limits_{n\to\infty} \frac{1}{\alpha_n}(|\alpha_{n-1} - \alpha_n|) = 0$, but does not satisfy $\sum\limits_{n=0}^{\infty} |\alpha_{n+1} - \alpha_n| < \infty$.

2. Show that the map U defined in Example 14.2 is *not* nonexpansive.
3. Prove Lemma 14.16.

14.4 Historical Remarks

The theorems of Section 14.3 are due to Chidume [118]. Further results on approximation of fixed points of mappings in subclasses of the class of asymptotically quasi-nonexpansive mappings can be found in the numerous papers that have been published on methods of approximating fixed points of such mappings (see e.g., Al'ber *et al.* [6], Bose [34], Chang *et al.* [81, 84, 73, 82], Chidume and Zegeye [171], Cho *et al.* [187], Eshita *et al* [220], Falset *et al.* [222], Gu *et al.* [243], Kaczor [267], Nishiura *et al.* [354], Oka [362], Rhoades [429], Shahzad and Udomene [444, 445], Shioji and Takahashi [451, 452, 453], Takahashi and Zembayashi [479, 480], Sun [458], Tan and Xu [484, 485, 488], Wang [498] and the references contained therein). However, examples of asymptotically nonexpansive mappings which *are not* nonexpansive are still difficult to find in applications. It is certainly of interest to find or construct such examples.

Chapter 15
Common Fixed Points for Finite Families of Nonexpansive Mappings

15.1 Introduction

Markov ([320])(see also Kakutani [270]) showed that if a *commuting* family of bounded *linear* transformations $T_\alpha, \alpha \in \Delta$ (Δ an arbitrary index set) of a normed linear space E into itself leaves some nonempty compact convex subset K of E invariant, then the family has at least one common fixed point. (The actual result of Markov is more general than this but this version is adequate for our purposes).

Motivated by this result, De Marr studied the problem of the existence of a common fixed point for a family of *nonlinear* maps, and proved the following theorem.

Theorem DM (De Marr [200], p.1139) *Let E be a Banach space and K be a nonempty compact convex subset of E. If $\mathcal{F}$ is a nonempty commuting family of nonexpansive mappings of K into itself, then the family $\mathcal{F}$ has a common fixed point in K.*

Browder proved the result of De Marr in a uniformly convex Banach space, requiring that K be only nonempty closed bounded and convex.

Theorem B (Browder [42], Theorem 1) *Let E be a uniformly convex Banach space, K a nonempty closed convex and bounded subset of E, $\{T_\lambda\}$ a commuting family of nonexpansive self-mappings of K. Then, the family $\{T_\lambda\}$ has a common fixed point in K.*

For other fixed point theorems for families of nonexpansive mappings the reader may consult any of the following references: Belluce and Kirk [27]; Bruck [57]; and Lim [304].

Recently, considerable research efforts have been devoted to developing iterative methods for approximating a common fixed point (when it exists) for a family of nonexpansive mappings (see e.g., Atsushiba *et al.* [18], Chang *et al.* ([84]), Chen *et al.* [85]), Chidume and Ali [125], Chidume *et al.* ([164], [183],

C. Chidume, *Geometric Properties of Banach Spaces and Nonlinear Iterations,* 205
Lecture Notes in Mathematics 1965,
© Springer-Verlag London Limited 2009

[184], [153]), Jung et al. [265], Kang et al. [269], Nakajo et al. [349], Rhoades and Soltuz [423], Suzuki ([467], [461], [463], [464]), O'Hara et al.[361]), Sun [458], Takahashi et al. [481], Yao et al. [535], Zhou et al. ([559]), and the references contained therein).

In this chapter, we study a Halpern-type iteration process for approximating a common fixed point (assuming existence) for a family of nonexpansive mappings.

Bauschke [25] was the first to introduce a Halpern-type iterative process for approximating a common fixed point for a finite family of r nonexpansive self-mappings. He proved the following theorem in which, for an operator A, $Fix(A) := \{x \in D(A) : Ax = x\}$.

Theorem BSK (Bauschke, [25], Theorem 3.1) *Let K be a nonempty closed convex subset of a Hilbert space H and $T_1, T_2, ..., T_r$ be a finite family of nonexpansive mappings of K into itself with $F := \bigcap_{i=1}^{r} Fix(T_i) \neq \emptyset$ and $F = Fix(T_r T_{r-1}...T_1) = Fix(T_1 T_r...T_2) = ... = Fix(T_{r-1} T_{r-2}...T_1 T_r)$. Let $\{\lambda_n\}$ be a real sequence in $[0,1]$ which satisfies $C1 : \lim \lambda_n = 0$; $C2 : \sum \lambda_n = \infty$ and $C3: \sum_n |\lambda_n - \lambda_{n+r}| < \infty$. Given points $u, x_0 \in K$, let $\{x_n\}$ be generated by*

$$x_{n+1} = \lambda_{n+1} u + (1 - \lambda_{n+1}) T_{n+1} x_n, n \geq 0, \qquad (15.1)$$

where $T_n = T_{n \bmod r}$. Then, $\{x_n\}$ converges strongly to $P_F u$, where $P_F : H \to F$ is the metric projection.

Takahashi et al. [481] extended this result to uniformly convex Banach spaces. O'Hara et al. [361] proved a complementary result to that of Bauschke, still in the framework of Hilbert spaces, replacing C3 with the following new condition: $C4 : \lim_{n \to \infty} \frac{\lambda_n}{\lambda_{n+r}} = 1$, *or equivalently,* $\lim_{n \to \infty} \frac{\lambda_n - \lambda_{n+r}}{\lambda_{n+r}} = 0$. Their main theorems are the following.

Theorem OPH1 (O'Hara et al., [361], Theorem 3.3) *Let $\{\lambda_n\} \subset (0,1)$ satisfy $\lim \lambda_n = 0$ and $\sum \lambda_n = \infty$. Let K be a nonempty closed and convex subset of a Hilbert space H and let $T_n : K \to K$, $n = 1, 2, ...$ be nonexpansive mappings such that $F := \bigcap_{i=1}^{\infty} Fix(T_i) \neq \emptyset$. Assume that $V_1, V_2, ..., V_n : K \to K$ are nonexpansive mappings with the property: for all $k = 1, 2, ..., N$ and for any bounded subset C of K, there holds $\lim_{n \to \infty} \sup_{x \in C} ||T_n x - V_k(T_n x)|| = 0$. For $x_0, u \in K$ define*

$$x_{n+1} = \lambda_{n+1} u + (1 - \lambda_{n+1}) T_{n+1} x_n, n \geq 0.$$

Then, $x_n \to Pu$, where P is the projection from H.

Theorem OPH2 (O'Hara *et al.*, [361], Theorem 4.1) *Let K be a nonempty closed convex subset of a Hilbert space H and $T_1, T_2, ..., T_N$ be nonexpansive self-mappings of K with $F := \bigcap_{i=1}^{N} Fix(T_i) \neq \emptyset$. Assume that $F = Fix(T_N...T_1) = Fix(T_1 T_N...T_2) = ... = Fix(T_{N-1} T_{N-2}...T_N)$. Let $\{\lambda_n\} \subset (0,1)$ satisfy the following conditions: (i) $\lim \lambda_n = 0$; (ii) $\sum \lambda_n = \infty$, and (iii) $\lim \frac{\lambda_n}{\lambda_{n+N}} = 1$. Given points $x_0, u \in K$, the sequence $\{x_n\} \subset K$ is defined by*

$$x_{n+1} = \lambda_{n+1} u + (1 - \lambda_{n+1}) T_{n+1} x_n, \ n \geq 0.$$

Then, $x_n \to P_F u$, where P_F is the projection of u onto F.

Jung [264] extended the results of O'Hara *et al.* to uniformly smooth Banach spaces. Recently, Jung *et al.* [265] studied the iteration scheme (15.1), where the iteration parameter $\{\lambda_n\}$ satisfies the following conditions: $C1 : \lim \lambda_n = 0$; $C2 : \sum \lambda_n = \infty$ and $C5 : |\lambda_{n+N} - \lambda_n| \leq o(\lambda_{n+N}) + \sigma_n$, where $\sum \sigma_n < \infty$.

They proved the following theorem.

Theorem JCA (Jung *et al.*, [265], Theorem 3.1) *Let E be a uniformly smooth Banach space with a weakly sequentially continuous duality mapping $J : E \to E^*$ and let K be a nonempty closed convex subset of E. Let $T_1, T_2, ..., T_N$ be nonexpansive mappings from K into itself with $F := \bigcap_{i=1}^{N} Fix(T_i) \neq \emptyset$ and $F = Fix(T_N T_{N-1}...T_1) = Fix(T_1 T_N...T_3 T_2) = ... = Fix(T_{N-1} T_{N-2}...T_1 T_N)$. Let $\{\lambda_n\}$ be a sequence in $(0,1)$ satisfying the conditions : (i) $\lim \lambda_n = 0$, (ii) $\sum \lambda_n = \infty$ and (iii) $|\lambda_{n+N} - \lambda_n| \leq o(\lambda_{n+N}) + \sigma_n$ where $\sum \sigma_n < \infty$. Then, the iterative sequence $\{x_n\}$ defined by (15.1) converges strongly to $Q_F u$, where Q_F is a sunny nonexpansive retraction of K onto F.*

In the sequel, we shall say that the sequence generated by (15.1) is *weakly asymptotically regular* if $\omega - \lim_{n \to \infty} (x_{n+j} - x_n) = 0$, and is *asymptotically regular* if $s - \lim_{n \to \infty} (x_{n+j} - x_n) = 0, j = 1, 2, 3, ...$.

Zhou *et al.* [559] proved that the conditions: (C1) and (C2) are indeed sufficient to guarantee the strong convergence of the iteration sequence of (15.1) in each of the following situations: (a) E is a Hilbert space; (b) E is a Banach space with weakly sequentially continuous duality map and the sequence $\{x_n\}$ of (15.1) is weakly asymptotically regular; (c) E is a reflexive Banach space whose norm is uniformly Gâteaux differentiable and in which every weakly compact convex subset of E has the fixed point property for nonexpansive mappings, and the sequence $\{x_n\}$ of (15.1) is asymptotically regular.

Their main results are the following theorems.

Theorem ZWC1 (Zhou *et al.* [559], Theorem 6) *Let E be a reflexive Banach space with a uniformly Gâteaux differentiable norm and a weakly continuous duality mapping J_φ for some gauge function φ. Let K be a nonempty closed convex subset of E. Assume that every weakly compact convex subset of E has the fixed point property for nonexpansive mappings. Let $T_1, T_2, ..., T_r$ be nonexpansive mappings of K into itself such that $F := \bigcap_{i=1}^{r} Fix(T_i) \neq \emptyset$. Assume also that $F = Fix(T_r T_{r-1}...T_1) = Fix(T_1 T_r...T_2) = ... = Fix(T_{r-1}T_{r-2}...T_1 T_r)$. Let $\{\lambda_n\}$ be a sequence in $(0,1)$ which satisfies (C1) and (C2). Let $\{x_n\}$ be the sequence defined by (15.1) and assume that $\{x_n\}$ is weakly asymptotically regular, then the sequence $\{x_n\}$ converges strongly to a common fixed point of $T_1, T_2, ..., T_r$.*

Theorem ZWC2 (Zhou *et al.* [559], Theorem 10) *Let E be a reflexive Banach space with a uniformly Gâteaux differentiable norm, and let K be a nonempty closed convex subset of E. Assume that every weakly compact convex subset of E has the fixed point property for nonexpansive mappings. Let $T_1, T_2, ..., T_r$ be nonexpansive mappings of K into itself such that $F := \bigcap_{i=1}^{r} Fix(T_i) \neq \emptyset$. Assume also that*

$$F = Fix(T_r T_{r-1}...T_1) = Fix(T_1 T_r...T_2)... = Fix(T_{r-1}T_{r-2}...T_1 T_r).$$

Let $\{\lambda_n\}$ be a sequence in $(0,1)$ which satisfies (C1) and (C2). Let $\{x_n\}$ be the sequence generated by (15.1) and assume that $\{x_n\}$ is asymptotically regular. Then, $\{x_n\}$ converges strongly to a common fixed point of $T_1, T_2, ..., T_r$.

Remark 15.1. In their proofs of theorems ZWC1 and ZWC2, the authors use the concept of Banach limits, proving in the process two results involving these limits.

In all the above discussion, $T_1, T_2, ..., T_N$ remain self-mappings of a nonempty subset of the Banach space E. If, however, the domain of $T_1, T_2, ..., T_N$, $D(T_i) \equiv K, i = 1, 2, ..., N$, is a *proper* subset of E and T_i maps K into E for each i, then the recursion formula (15.1) may fail to be well defined. To overcome this, an algorithm for non-self mappings was defined for the scheme (15.1) by Chidume *et al.* [184]. Using this algorithm, Chidume *et al.* [184] proved a strong convergence theorem to a common fixed point of the family $T_1, T_2, ..., T_N$ of *non-self* nonexpansive mappings in a reflexive Banach space with uniformly Gâteaux differentiable norm when the parameter $\{\lambda_n\}$ satisfies conditions C1, C2 and C3.

We remark that the requirement that the underlying space E be a Hilbert space, or satisfy Opial's condition, or admit weak sequential continuous duality map imposed in several theorems, in particular, in the theorems of Bauschke [25], in Theorems JCA, ZWC1, OPH1 and OPH2 excludes the application of any of these theorems in, for example, L_p spaces, $1 < p < \infty$,

$p \neq 2$ because it is well known that these spaces do not admit weak sequentially continuous duality mappings and do not satisfy Opial's condition.

In this chapter, we study an iteration scheme introduced by Chidume and Ali [125] with respect to which these strong conditions on the underlying space are dispensed with and conditions C1 and C2 are sufficient to guarantee the *strong convergence* of the sequence generated by the recursion formula of the iterative scheme to a common fixed point of $T_1, T_2, ..., T_r$. Furthermore, this iteration process can be used, for example, in the cases when E is a reflexive Banach space with uniformly Gâteaux differentiable norm, and in which every weakly compact convex subset of E has the fixed point property for nonexpansive mappings, and T_n satisfies a mild condition. Moreover, the underlying space will not be required to admit weak sequential continuous duality maps or to satisfy Opial's condition. In addition, the sequence $\{x_n\}$ will not be assumed to be asymptotically regular and the method of proof, which is of independent interest, will not involve the use of Banach limits. Finally, convergence theorem for non-self maps which complement the results of Chidume *et al.* [184] are proved. All the theorems proved in this chapter hold, in particular, in L_p spaces, $1 < p < \infty$.

In the sequel, we shall make use of the following theorem.

Theorem MJ (Morales and Jung, [341], Theorem 1) *Let E be a real reflexive Banach space with uniformly Gâteaux differentiable norm. Suppose K is a nonempty closed convex subset of E and $T : K \to E$ a continuous pseudo-contractive mapping satisfying the weakly inward condition. Suppose that every nonempty closed convex bounded subset of K has the fixed point property for nonexpansive mappings. If there exists $u \in K$ such that the set $B := \{x \in K : Tx = \gamma x + (1 - \gamma)u, \ for \ some \ \gamma > 1\}$ is bounded, then the path $t \to z_t \ 0 < t < 1$, defined by $z_t = tu + (1 - t)Tz_t$, converges strongly to a fixed point of T.*

We note that if K is bounded, then the set B is clearly bounded.

15.2 Convergence Theorems for a Family of Nonexpansive Mappings

In the next theorem, we shall assume that every nonempty closed convex bounded subset of E has the fixed point property for nonexpansive mappings. We now prove the following theorem

Theorem 15.2. *Let E be a real reflexive Banach space with uniformly Gâteaux differentiable norm. Let K be a nonempty closed convex subset of E. Let $T_1, T_2, ..., T_N$ be a family of nonexpansive self-mappings of K, with $F := \cap_{i=1}^{N} Fix(T_i)$ and*

$$F = Fix(T_N T_{N-1}...T_1) = Fix(T_1 T_N...T_2)$$
$$= ... = Fix(T_{N-1} T_{N-2}...T_1 T_N) \neq \emptyset.$$

Let $\{\lambda_n\}$ be a sequence in $(0,1)$ satisfying the following conditions: $C1$: $\lim \lambda_n = 0$; $C2 : \sum \lambda_n = \infty$. *For a fixed $\delta \in (0,1)$, define $S_n : K \to K$ by* $S_n x := (1-\delta)x + \delta T_n x \ \forall \ x \in K$ *where $T_n = T_{n \ mod \ N}$. For arbitrary fixed* $u, x_0 \in K$, *let $B := \{x \in K : T_N T_{N-1}...T_1 x = \gamma x + (1-\gamma)u, \ for \ some \ \gamma > 1\}$* *be bounded and let*

$$x_{n+1} = \lambda_{n+1} u + (1 - \lambda_{n+1}) S_{n+1} x_n, \ for \ n \geq 0. \qquad (15.2)$$

Assume $\lim_{n \to \infty} ||T_n x_n - T_{n+1} x_n|| = 0$. Then, $\{x_n\}$ converges strongly to a common fixed point of the family $T_1, T_2, ..., T_N$.

Proof. Observe that S_n is nonexpansive and $Fix(S_n) = Fix(T_n)$ for each $n \in \mathbb{N}$. We prove the theorem in stages. Let $x^* \in F$.

Step 1 $||x_n - x^*|| \leq \max\{||u - x^*||, ||x_0 - x^*||\}$ for every $n \geq 0$.
We prove this by induction. It is clear that the statement is true for $n = 0$.
Assume it is true for $n = k$. Then,

$$||x_{k+1} - x^*|| \leq \lambda_{k+1}||u - x^*|| + (1 - \lambda_{k+1})||S_{k+1} x_k - x^*||$$
$$\leq \max\{||u - x^*||, ||x_0 - x^*||\}.$$

So, $\{x_n\}$ is bounded and it follows that $\{S_{n+1} x_n\}$ and $\{T_{n+1} x_n\}$ are bounded.

Step 2 $\lim_{n \to \infty} ||x_{n+N} - S_{n+N} x_{n+N-1}|| = 0$.
From step 1 , and the recursion (15.2), we have,
$x_{n+N} - S_{n+N} x_{n+N-1} = \lambda_{n+N}(u - S_{n+N} x_{n+N-1}) \to 0$ as $n \to \infty$.

Step 3 $\lim_{n \to \infty} ||x_{n+N} - x_n|| = 0$.
To prove this, define $\{\beta_n\}$ by $\beta_n := (1 - \delta)\lambda_{n+N+1} + \delta$ and a sequence $\{y_n\}$ by

$$y_n := \frac{x_{n+N+1} - x_{n+N} + \beta_n x_{n+N}}{\beta_n}.$$

Then, using the definition of $\{\beta_n\}$ and $\{S_n\}$ we obtained that

$$y_n = \frac{\lambda_{n+N+1} u + (1 - \lambda_{n+N+1})\delta T_{n+N+1} x_{n+N}}{\beta_n}.$$

It is clear from this that $\{y_n\}$ is bounded. Moreover,

$$||y_{n+1} - y_n|| \leq \left| \frac{\lambda_{n+N+2}}{\beta_{n+1}} - \frac{\lambda_{n+N+1}}{\beta_n} \right| ||u||$$
$$+ \left| \frac{(1 - \lambda_{n+N+2})}{\beta_{n+1}} - \frac{(1 - \lambda_{n+N+1})}{\beta_n} \right| \delta ||T_{n+N+2} x_{n+N+1}||$$

$$+ \ \frac{(1 - \lambda_{n+N+1})}{\beta_n} \ \delta || x_{n+N+1} - x_{n+N} ||$$

$$+ \ \frac{(1 - \lambda_{n+N+1})}{\beta_n} \ \delta || T_{n+N+2} x_{n+N} - T_{n+N+1} x_{n+N} ||,$$

so that

$$|| y_{n+1} - y_n || - || x_{n+N+1} - x_{n+N} || \leq \left| \frac{\lambda_{n+N+2}}{\beta_{n+1}} - \frac{\lambda_{n+N+1}}{\beta_n} \right| || u ||$$

$$+ \left| \frac{(1 - \lambda_{n+N+2})}{\beta_{n+1}} - \frac{(1 - \lambda_{n+N+1})}{\beta_n} \right| \delta || T_{n+N+2} x_{n+N+1} ||$$

$$+ \left| \frac{(1 - \lambda_{n+N+1})}{\beta_n} \ \delta - 1 \right| || x_{n+N+1} - x_{n+N} ||$$

$$+ \ \frac{(1 - \lambda_{n+N+1})}{\beta_n} \ \delta || T_{n+2} x_{n+N} - T_{n+1} x_{n+N} ||.$$

This implies, $\lim\sup_{n \to \infty}(|| y_{n+1} - y_n || - || x_{n+N+1} - x_{n+N} ||) \leq 0$, and by Lemma 6.33, $\lim_{n \to \infty} || y_n - x_{n+N} || = 0$. Hence,

$$|| x_{n+N+1} - x_{n+N} || = \beta_n || y_n - x_{n+N} || \to 0 \ as \ n \to \infty. \tag{15.3}$$

But,

$$|| x_{n+N} - x_n || \leq || x_{n+N} - x_{n+N-1} || + || x_{n+N-1} - x_{n+N-2} || + ... + || x_{n+1} - x_n || \to 0 \tag{15.4}$$

as $n \to \infty$, completing step 3.

Step 4 $\lim_{n \to \infty} || x_n - T_{n+N} T_{n+N-1} ... T_{n+1} x_n || = 0.$

From step 2 we have $x_{n+N} - S_{n+N} x_{n+N-1} \to 0$ as $n \to \infty$, which implies $x_{n+N} - [(1 - \delta) x_{n+N-1} + \delta T_{n+N} x_{n+N-1}] \to 0$, that is,

$$(x_{n+N} - x_{n+N-1}) + \delta[x_{n+N-1} - T_{n+N} x_{n+N-1}] \to 0$$

as $n \to \infty$. Thus, using this and (15.3) we have $x_{n+N-1} - T_{n+N} x_{n+N-1} \to 0$ as $n \to \infty$ and so,

$$|| x_{n+N} - T_{n+N} x_{n+N-1} || \leq || x_{n+N} - x_{n+N-1} || + || x_{n+N-1} - T_{n+N} x_{n+N-1} || \to 0$$

as $n \to \infty$. Using the fact that T_i is nonexpansive for each i we obtain the following finite table:

$$x_{n+N} - T_{n+N} x_{n+N-1} \to 0 \quad as \quad n \to \infty$$

$$T_{n+N}x_{n+N-1} - T_{n+N}T_{n+N-1}x_{n+N-2} \to 0 \quad as \quad n \to \infty$$

$$\vdots$$

$$T_{n+N}T_{n+N-1}...T_{n+2}x_{n+1} - T_{n+N}T_{n+N-1}...T_{n+2}T_{n+1}x_n \to 0 \quad as \quad n \to \infty$$

and adding the table yields $x_{n+N} - T_{n+N}T_{n+N-1}...T_{n+1}x_n \to 0$ as $n \to \infty$. Using this and (15.4), we get that $\lim_{n\to\infty} ||x_n - T_{n+N}T_{n+N-1}...T_{n+1}x_n|| = 0$.

Step 5. Let $x_{t_n} \in K$ be a continuous path satisfying

$$x_{t_n} = t_n u + (1 - t_n)T_N T_{N-1}...T_1 x_{t_n} \tag{15.5}$$

and $lim\ x_{t_n} = x^* \in Fix(T_N T_{N-1}...T_1)$ as $t_n \to 0^+$ (guaranteed by Theorem MJ) which implies (by hypothesis) that $x^* \in F := \overset{N}{\underset{i=1}{\cap}} Fix(T_i)$. From step 4 we have that $x_n - T_{n+N}T_{n+N-1}...T_{n+1}x_n \to 0$ as $n \to \infty$. Let $\{t_n\}$ be a real sequence in $(0, 1)$ such that $t_n \to 0$ as $n \to \infty$ and

$$\lim_{n\to\infty} \frac{||x_n - T_{n+N}T_{n+N-1}...T_{n+1}x_n||}{t_n} = 0.$$

Let $\{x_{t_n}\}$ be a sequence satisfying (15.5) such that $x_{t_n} \to x^* \in F$ as $t_n \to 0$. Then, using inequality (4.4) we have,

$$||x_{t_n} - x_n||^2 \leq (1 - t_n)^2||x_n - T_N T_{N-1}...T_1 x_{t_n}||^2 + 2t_n \langle u - x_n, j(x_{t_n} - x_n) \rangle$$

$$\leq (1 - t_n)^2 \Big[||T_N T_{N-1}...T_1 x_n - T_N T_{N-1}...T_1 x_{t_n}||$$

$$+ ||x_n - T_N T_{N-1}...T_1 x_n|| \Big]^2 + 2t_n \langle u - x_n, j(x_{t_n} - x_n) \rangle,$$

which implies

$$\langle u - x_{t_n}, j(x_n - x_{t_n}) \rangle \leq \left(\frac{(1 - t_n)^2 + 2t_n - 1}{2t_n} \right) ||x_{t_n} - x_n||^2$$

$$+ (1 - t_n)^2 ||x_{t_n} - x_n|| \frac{||x_n - T_N T_{N-1}...T_1 x_n||}{t_n}$$

$$+ \frac{(1 - t_n)^2 ||x_n - T_N T_{N-1}...T_1 x_n||^2}{2t_n}$$

and hence $\limsup \langle u - x_{t_n}, j(x_n - x_{t_n}) \rangle \leq 0$. Moreover,

$$\langle u - x_{t_n}, j(x_n - x_{t_n}) \rangle = \langle u - x^*, j(x_n - x^*) \rangle$$

$$+ \langle u - x^*, j(x_n - x_{t_n}) - j(x_n - x^*) \rangle$$

$$+ \langle x^* - x_{t_n}, j(x_n - x_{t_n}) \rangle,$$

and since j is norm-to-$weak^*$ uniformly continuous on bounded sets, we have $\limsup \langle u - x^*, j(x_n - x^*) \rangle \leq 0$.

Step 6 From the recursion formula (15.2) and inequality (4.4), we have the following;

$$||x_{n+1} - x^*||^2 \leq (1 - \lambda_{n+1})^2 ||S_{n+1}x_n - x^*||^2 + 2\lambda_{n+1}\langle u - x^*, j(x_{n+1} - x^*)\rangle$$

$$\leq (1 - \lambda_{n+1})||x_n - x^*||^2 + 2\lambda_{n+1}\langle u - x^*, j(x_{n+1} - x^*)\rangle,$$

and by Lemma 6.34, we have $\{x_n\}$ converges strongly to some fixed point of the family $\{T_i\}_{i=1}^N$. $\square$

15.3 Non-self Mappings

For non-self mappings, we prove the following theorem. In the theorem, we shall assume that K is a sunny nonexpansive retract of E with Q as the sunny nonexpansive retraction and that every nonempty closed bounded convex subset of E has the fixed point property for nonexpansive mappings.

Theorem 15.3. *Let K be a nonempty closed convex subset of a real reflexive Banach space E which has a uniformly Gâteaux differentiable norm. Let $T_i : K \to E, i = 1, 2, ..., N$ be a family of nonexpansive mappings which are weakly inward with $F := \overset{N}{\underset{i=1}{\cap}} Fix(T_i) = Fix(QT_NQT_{N-1}...QT_1) = Fix(QT_1QT_N...QT_2) = ... = Fix(QT_{N-1}QT_{N-2}...QT_1QT_N) \neq \emptyset$. For a fixed $\delta \in (0,1)$, define $S_n : K \to K$ by $S_nx := (1 - \delta)x + \delta QT_nx \ \forall \ x \in E$. For arbitrary fixed $u, x_0 \in K$, let $B := \{x \in K : T_NT_{N-1}...T_1x = \gamma x + (1 - \gamma)u$, for some $\gamma > 1\}$ be bounded and let the sequence $\{x_n\}$ be generated iteratively by*

$$x_{n+1} = \lambda_{n+1}u + (1 - \lambda_{n+1})S_{n+1}x_n, \ for \ n \geq 0, \tag{15.6}$$

where $T_n = T_{n \bmod N}$ and $\{\lambda_n\}$ is a real sequence which satisfies (C1) and (C2). Assume $\lim_{n\to\infty} ||QT_nx_n - QT_{n+1}x_n|| = 0$. Then, $\{x_n\}$ converges strongly to a common fixed point of $T_1, T_2, ..., T_N$. Further, if $Pu = \lim_{n\to\infty} x_n$ for each $u \in K$, then P is sunny nonexpansive retraction of K onto F.

Proof. The boundedness of $\{x_n\}$, $\{S_{n+1}x_n\}$ and $\{QT_{n+1}x_n\}$ follows as in the proof of step 1 of Theorem 15.2. From the recursion formula (15.6), we have $x_{n+N} - S_{n+N}x_{n+N-1} = \lambda_{n+N}(u - S_{n+N}x_{n+N-1}) \to 0$ as $n \to \infty$. Now, define $\{\beta_n\}$ by $\beta_n := (1 - \delta)\lambda_{n+N+1} + \delta$ and a sequence $\{w_n\}$ by

$$w_n := \frac{x_{n+N+1} - x_{n+N} + \beta_nx_{n+N}}{\beta_n}, \quad n \geq 1.$$

Then, using the definitions of $\{\beta_n\}$ and $\{S_n\}$ we obtain that

$$w_n = \frac{\lambda_{n+N+1}u + (1 - \lambda_{n+N+1})\delta Q T_{n+N+1} x_{n+N}}{\beta_n}.$$

Following the line of the proof in step 3 of Theorem 15.2, we get $\lim_{n \to \infty} \|x_{n+N} - x_n\| = 0$. The rest of the proof now follows as in the proof of Theorem 3.1 of [184]. □

Let K be a nonempty closed convex subset of a real Banach space with uniformly Gâteaux differentiable norm and $T : K \to K$ a nonexpansive map. For arbitrary $x_0, u \in K$, let $\{x_n\}$ be defined by

$$x_{n+1} = \lambda_n u + (1 - \lambda_n) S x_n \tag{15.7}$$

where $\{\lambda_n\}$ is a real sequence in $[0, 1)$ satisfying some conditions and $S := (1 - \delta)I + \delta T$ for some fixed real number $\delta \in (0, 1)$. It is proved in Chidume and Chidume [132] that $\{x_n\}$ is an approximate fixed point sequence of the nonexpansive map S (i.e., $\|x_n - S x_n\| \to 0$ as $n \to \infty$).

An example of a finite family of maps satisfying the limiting condition in Theorem 15.2 is given as follows.

Example 15.4. Let K be a closed convex nonempty subset of a real Banach space E with uniformly Gâteaux differentiable norm. Let $S : K \to K$ be a nonexpansive mapping. Let $\{x_n\}$ be defined by (15.7). From [132], $\{x_n\}$ is an approximate fixed point sequence of S. Now, let $\{S_i\}_{i=1}^{m}$ be a finite family of maps defined by $S_n := S^n$, $n = n \mod m$. Write $S^m = S^{m-1} \circ S$. Then, it is easy to see that

$$\|S_n x_n - S_{n+1} x_n\| = \begin{cases} \|x_n - S x_n\|, & \text{if } n \neq km, \quad k \in \mathbb{N}; \\ (m-1)\|x_n - S x_n\|, & \text{if } n = km, \quad k \in \mathbb{N}, \end{cases}$$

so that, in all cases, $\|S_n x_n - S_{n+1} x_n\| \to 0$ as $n \to \infty$.

Historical remarks. More convergence theorems on iterative methods for common fixed points for families of nonexpansive mappings can be found in, for example, Chidume and Shahzad [164], Kim and Takahashi [280], Shahzad and Al-Dubiban [443], Suzuki [466], Takahashi and Tamura [477], Zegeye and Shahzad [543], [544] and the references contained therein. Theorems 15.2, 15.3 and Example 15.4 are due to Chidume and Ali [125].

Chapter 16
Common Fixed Points for Countable Families of Nonexpansive Mappings

16.1 Introduction

Various authors have studied iterative schemes similar to that of Bauschke (Theorem BSK, Chapter 15) in more general Banach spaces on one hand and using various conditions on the sequence $\{\lambda_n\}$ on the other hand (see, for example, Colao *et al.* [192], Yao [532], Takahashi and Takahashi [482], Plubtieng and Punpaeng [385], Ceng *et al.* [193], Chidume and Ali [125], Jung [264], Jung *et al.* [265], O'Hara *et al.* [361], Zhou *et al.* [559]). Most of the results in these references are proved for *finite* families of nonexpansive mappings defined in Hilbert spaces.

Convergence theorems have also been proved for common fixed points of *countable infinite* families of nonexpansive mappings. Before we proceed, we first state the following important theorem.

Theorem 16.1. *(Bruck, [55]) Let K be a nonempty closed convex subset of a strictly convex Banach space E. Let $\{S_k\}$ be a sequence of nonexpansive mappings of K into E and $\{\beta_k\}$ be a sequence of positive real numbers such that $\sum\limits_{k=1}^{\infty} \beta_k = 1$. If $\mathcal{F} := \bigcap\limits_{k=1}^{\infty} F(S_k) \neq \emptyset$, then the mapping $T := \sum\limits_{k=1}^{\infty} \beta_k S_k$ is well defined and $F(T) = \bigcap\limits_{k=1}^{\infty} F(S_k)$, where $F(T)$ denotes the set of fixed points of T.*

Remark 16.2. It is easy to see from Theorem 16.1 that T is also nonexpansive. Therefore, we obtain that the set of common fixed points of a countable family of nonexpansive maps is a sunny nonexpansive retract of E, if E is uniformly convex and has a uniformly Gâteaux differentiable norm.

A recent convergence result for approximating a common fixed point for a *countable infinite* family of nonexpansive maps in a Banach space even more general than Hilbert spaces is the following theorem.

C. Chidume, *Geometric Properties of Banach Spaces and Nonlinear Iterations*, Lecture Notes in Mathematics 1965,
© Springer-Verlag London Limited 2009

Theorem 16.3. *(Aoyama et al., [16]) Let E be a real uniformly convex Banach space whose norm is uniformly Gâteaux differentiable and K be a nonempty closed convex subset of E. Let $\{T_n\}$ be a sequence of nonexpansive mappings of K into itself such that $\bigcap_{n=1}^{\infty} F(T_n) \neq \emptyset$ and $\{\alpha_n\}$ be a sequence in $[0, 1]$ such that (i) $\lim \alpha_n = 0$; and (ii) $\sum_{n=1}^{\infty} \alpha_n = \infty$. Let $\{x_n\}$ be a sequence in K defined as follows: $x_1 = u \in K$, and $x_{n+1} = \alpha_n u + (1 - \alpha_n) T_n x_n \ \forall \ n \in \mathbb{N}$. Suppose that*

$$\sum_{n=1}^{\infty} \sup\{\|T_{n+1}z - T_n z\| : z \in C\} < \infty, \tag{16.1}$$

for any bounded subset C of K. Let T be a mapping of K into itself defined by $Tz := \lim T_n z \ \forall \ z \in K$, and suppose $F(T) = \bigcap_{n=1}^{\infty} F(T_n)$. If either

(i) $\sum_{n=1}^{\infty} |\alpha_{n+1} - \alpha_n| < \infty$ or,

(ii) $\alpha_n \in (0, 1) \ \forall \ n$ and $\lim \frac{\alpha_n}{\alpha_{n+1}} = 1$, then, $\{x_n\}$ converges strongly to Qu, where Q is the sunny nonexpansive retraction of E onto $F(T) = \bigcap_{n=1}^{\infty} F(T_n)$.

Nilsrakoo and Saejung [353] used the condition (16.1) to establish weak and strong convergence theorems for finding common fixed points for a countable family of certain Lipschitzian mappings *in a real Hilbert space*. They also imposed additional conditions on the Lipschitzan mappings.

Yao *et al.* [533] studied the following problem. Let H be a real Hilbert space. Consider the iterative sequence

$$x_{n+1} = \alpha_n \gamma f(x_n) + \beta_n x_n + [(1 - \beta_n)I - \alpha_n A]W_n x_n, \tag{16.2}$$

where $\gamma > 0$ is some constant, $f : H \to H$ is a given contractive mapping, A is a strongly positive bounded linear operator on H and W_n is the so called $W-$mapping generated by an countable infinite family of nonexpansive mappings $T_1, T_2, ..., T_n, ...$ and scalars $\lambda_1, \lambda_2, ..., \lambda_n, ...$ such that the common fixed points set $\mathcal{F} := \bigcap_{n=1}^{\infty} Fix(T_n) \neq \emptyset$. Under appropriate conditions on the iteration parameters, Yao *et al.* proved that $\{x_n\}$ converges strongly to $x^* \in \mathcal{F}$, where x^* is the unique solution of the variational inequality $\langle (A - \gamma f)x^*, x^* - y \rangle \geq 0 \ \forall \ y \in \mathcal{F}$.

The operator W_n in (16.2) is defined as follows:

$$U_{n,n+1} = I,$$
$$U_{n,n} = \lambda_n T_n U_{n,n+1} + (1 - \lambda_n)I,$$
$$U_{n,n-1} = \lambda_{n-1} T_{n-1} U_{n,n} + (1 - \lambda_{n-1})I,$$

$$\vdots$$

$$U_{n,k} = \lambda_k T_k U_{n,k+1} + (1 - \lambda_k)I,$$

$$\vdots$$

$$U_{n,2} = \lambda_2 T_2 U_{n,3} + (1 - \lambda_2)I,$$

$$W_n = U_{n,1} = \lambda_1 T_1 U_{n,2} + (1 - \lambda_1)I. \tag{16.3}$$

Yao [532] studied the iteration process (16.3) and "proved" strong convergence theorems. Colao *et al.* [192], however, pointed out ([192], p.341) that there is a gap in the proof of Yao, and that the same gap also appears in Atsushiba and Takahashi [19], where the technique used by Yao was first introduced. Colao *et al.* then proved strong convergence theorems in a Hilbert space, using (16.3).

Suzuki [463] proved convergence theorems for common fixed points for a countable infinite family of nonexpansive mappings in a real Banach space under the assumption that *the family is commuting* and *the domain of the operators is compact and convex*. We shall discuss this in the next chapter.

Maingé [318] studied the Halpern-type scheme for approximation of a common fixed point for a *countable infinite* family of nonexpansive mappings in a Hilbert space. Define $\mathcal{N}_{\mathcal{I}} := \{i \in \mathbb{N} : T_i \neq I\}$ (I being the identity mapping on a real normed space E).

He proved the following theorems.

Theorem 16.4. *(Maingé,[318]) Let K be a nonempty closed convex subset of a real Hilbert space H. Let $\{T_i\}$ be a countable family of nonexpansive self-mappings of K, $\{t_n\}$ and $\{\sigma_{i,t_n}\}$ be sequences in $(0,1)$ satisfying the following conditions: (i) $\lim t_n = 0$, (ii) $\sum_{i \geq 1} \sigma_{i,t_n} = (1-t_n)$, (iii) $\forall i \in \mathcal{N}_{\mathcal{I}}$, $\lim_{n \to \infty} \frac{t_n}{\sigma_{i,t_n}} = 0$.*
Define a fixed point sequence $\{x_{t_n}\}$ by

$$x_{t_n} = t_n C x_{t_n} + \sum_{i \geq 1} \sigma_{i,t_n} T_i x_{t_n} \tag{16.4}$$

where $C : K \to K$ is a strict contraction. Assume $F := \cap_i F(T_i) \neq \emptyset$, then $\{x_{t_n}\}$ converges strongly to a unique fixed point of the contraction $P_F \circ C$, where P_F is the metric projection from H onto F.

Theorem 16.5. *(Maingé,[318]) Let K be a nonempty closed convex subset of a real Hilbert space H. Let $\{T_i\}$ be a countable family of nonexpansive self-mappings of K, $\{\alpha_n\}$ and $\{\sigma_{i,n}\}$ be sequences in $(0,1)$ satisfying the following conditions: (i) $\sum \alpha_n = \infty$, $\sum_{i \geq 1} \sigma_{i,n} = (1 - \alpha_n)$,*

$$(ii) \begin{cases} \frac{1}{\sigma_{i,n}}\left|1 - \frac{\alpha_{n-1}}{\alpha_n}\right| \to 0, \text{ or } \sum_n \frac{1}{\sigma_{i,n}}|\alpha_{n-1} - \alpha_n| < \infty, \\[2mm] \frac{1}{\alpha_n}\left|\frac{1}{\sigma_{i,n}} - \frac{1}{\sigma_{i,n-1}}\right| \to 0, \text{ or } \sum_n \left|\frac{1}{\sigma_{i,n}} - \frac{1}{\sigma_{i,n-1}}\right| < \infty, \\[2mm] \frac{1}{\sigma_{i,n}\alpha_n}\sum_{k\geq 0}|\sigma_{k,n} - \sigma_{k,n-1}| \to 0, \text{ or } \sum_n \frac{1}{\sigma_{i,n}}\sum_{k\geq 0}|\sigma_{k,n} - \sigma_{k,n-1}| < \infty. \end{cases}$$

(iii) $\forall i \in \mathcal{N_I}$, $\lim\limits_{n\to\infty} \frac{\alpha_n}{\sigma_{i,n}} = 0$. Then, the sequence $\{x_n\}$ defined iteratively by $x_1 \in K$,

$$x_{n+1} = \alpha_n C x_n + \sum_{i\geq 1} \sigma_{i,n} T_i x_n \qquad (16.5)$$

converges strongly to the unique fixed point of $P_F \circ C$, where P_F is the metric projection from H onto F.

Theorem 16.5, from the point of view of applications, seems much better than Theorem 16.3 and the theorems of Nilsrakoo and Saejung, both of which impose the condition (16.1), a condition that is clearly very difficult to verify in any application. The recursion formula of Maingé is also better than that of Yao *et al.* which involves the recursion formula (16.2) with W_n defined by the cumbersome (16.3).

Chidume *et al.* [135] proved theorems, with recursion formulas simpler than (16.4) and (16.5), that extended Theorems 16.4 and 16.5 to l_p spaces, $1 < p < \infty$. Furthermore, they also proved convergence theorems which are applicable in L_p spaces, $1 < p < \infty$. Moreover, in their more general setting, some of the conditions on the sequences $\{\alpha_n\}$ and $\{\sigma_{i,n}\}$ imposed in Theorem 16.5 were dispensed with or weakened.

In this chapter, we present theorems of Chidume and Chidume [130] on iterative approximation for common fixed points for a *countable infinite family* of nonexpansive mappings in *real uniformly convex Banach spaces with uniformly Gâteaux differentiable norms*. Furthermore, in this more general setting, the recursion formula studied is much simpler than that studied by Yao *et al.* [533] *in a Hilbert space*. In the special case that $E = l_p, 1 < p < \infty$, the condition (16.1) of Aoyama *et al.* (which is also imposed in Yao *et al.*) is dispensed with.

16.2 Path Convergence Theorems

Let K be a nonempty closed and convex subset of a real Banach space E. Let $\{T_i\}_{i=1}^\infty$ be a family of nonexpansive self-mappings of K. For a fixed $\delta \in (0,1)$, define a family of mappings, $S_i : K \to K$ by $S_i x := (1-\delta)x + \delta T_i x \ \forall\, x \in K$, and $i \in \mathbb{N}$. For $t \in (0,1)$, let $\{\sigma_{i,t}\}_{i=1}^\infty$ be a sequence in $(0,1)$ such that $\sum_{i\geq 1}\sigma_{i,t} = (1-t)$. For arbitrary fixed $u \in K$, define a map $T_t : K \to K$ by

$$T_t x = tu + \sum_{i \geq 1} \sigma_{i,t} S_i x \quad \forall x \in K. \tag{16.6}$$

Then, T_t is a strict contraction on K. For, if $x, y \in K$, we have

$$\|T_t x - T_t y\| = \left\| \sum_{i \geq 1} \sigma_{i,t} \Big((1 - \delta)(x - y) + \delta(T_i x - T_i y) \Big) \right\|$$

$$\leq \sum_{i \geq 1} \sigma_{i,t} \Big((1 - \delta)\|x - y\| + \delta\|T_i x - T_i y\| \Big)$$

$$= (1 - t)\|x - y\|.$$

Thus, for each $t \in (0,1)$, there is a unique $z_t \in K$ satisfying

$$z_t = tu + \sum_{i \geq 1} \sigma_{i,t} S_i z_t. \tag{16.7}$$

Lemma 16.6. *Let E be a real Banach space. Let K be a closed, convex and nonempty subset of E. Let $\{T_i\}_{i=1}^{\infty}$ be a family of nonexpansive self-mappings of K. For $t \in (0,1)$, let $\{z_t\}$ be a sequence satisfying (16.7) and assume $\mathcal{F} := \bigcap_{i=1}^{\infty} F(T_i) \neq \emptyset$. Then, $\{z_t\}$ is bounded and admits a unique accumulation point as $t \to 0$.*

Proof. Let $x^* \in \mathcal{F}$. Then, using (16.7), we have

$$\|z_t - x^*\|^2 = \left\langle t(u - x^*) + \sum_{i \geq 1} \sigma_{i,t}(S_i z_t - x^*), j(z_t - x^*) \right\rangle$$

$$\leq t\langle u - x^*, j(z_t - x^*)\rangle + \sum_{i \geq 1} \sigma_{i,t}\|z_t - x^*\|^2$$

$$= t\langle u - x^*, j(z_t - x^*)\rangle + (1 - t)\|z_t - x^*\|^2$$

which implies $\|z_t - x^*\| \leq \|u - x^*\|$. Thus $\{z_t\}$ is bounded. Assume for contradiction that x' and x^* are two distinct accumulation points of $\{z_t\}$. Then, it is clear from the above argument that

$$\|x' - x^*\|^2 \leq \left\langle t(u - x^*) + \sum_{i \geq 1} \sigma_{i,t}(S_i x' - x^*), j(x' - x^*) \right\rangle$$

$$= t\langle u - x^*, j(x' - x^*)\rangle + (1 - t)\|x' - x^*\|^2$$

so that $\|x' - x^*\|^2 \leq \langle u - x^*, j(x' - x^*)\rangle$. Similarly, $\|x^* - x'\|^2 \leq \langle u - x', j(x^* - x')\rangle$. These inequalities imply $2\|x^* - x'\|^2 \leq \|x^* - x'\|^2$, a contradiction, and thus $x' = x^*$. $\qquad\square$

Lemma 16.7. *Let E be a real uniformly convex Banach space. For arbitrary $r > 0$, let $B_r(0) := \{x \in E : \|x\| \leq r\}$. Then, there exists a continuous* *strictly increasing function* $g : [0, \infty) \to [0, \infty)$, $g(0) = 0$ *such that for every* $x, y \in B_r(0)$ *and* $p \in (1, \infty)$, *the following inequality holds:*

$$4.2^p g\left(\frac{1}{2}\|x + y\|\right) \le (p.2^p - 4)\|x\|^p + p.2^p\langle y, j_p(x)\rangle + 4\|y\|^p. \quad (16.8)$$

Proof. Under the hypotheses of the lemma, the following inequality holds (inequality (4.31) of Chapter 4): For all $x, y \in B_r(0)$,

$$\|x + y\|^p \ge \|x\|^p + p\langle y, j_p(x)\rangle + g(\|y\|). \quad (16.9)$$

Replacing y by $\frac{1}{2}(x + y)$ and x by $(-\frac{1}{2}x)$ in (16.9), we obtain the following:

$$g\left(\frac{1}{2}\|x + y\|\right) \le \left(\frac{1}{4.2^p}(p.2^p - 4)\right)\|x\|^p + \frac{1}{4}p\langle y, j_p(x)\rangle + \frac{1}{2^p}\|y\|^p, \quad (16.10)$$

which completes the proof. $\square$

16.3 Path Convergence in Uniformly Convex Real Banach Spaces

Lemma 16.8. *Let E be a real uniformly convex Banach space. Let K be a closed, convex and nonempty subset of E. Let $\{T_i\}_{i=1}^{\infty}$ be a family of nonexpansive self-mappings of K. Let $\{t_n\}$ be a sequence in $(0, 1)$ such that $\lim\limits_{n \to \infty} t_n = 0$ and $\lim\limits_{n \to \infty} \frac{t_n}{\sigma_{i,t_n}} = 0 \ \forall \ i \in \mathcal{N}_{\mathcal{I}}$, $\sum\limits_{i \ge 1} \sigma_{i,t_n} = (1 - t_n)$. Let $\{z_{t_n}\}$ be a sequence satisfying (16.7) and let $\mathcal{F} := \bigcap\limits_{i=1}^{\infty} F(T_i) \ne \emptyset$. Then, $\lim\limits_{n \to \infty} \|z_{t_n} - T_i z_{t_n}\| = 0 \ \forall \ i \in \mathbb{N}$.*

Proof. For $i \in \mathbb{N}$ and $x^* \in \mathcal{F}$, we have the following estimates, using inequality (16.8),

$$4.2^p g\left(\frac{1}{2}\|S_i z_{t_n} - z_{t_n}\|\right)$$

$$= 4.2^p g\left(\frac{1}{2}\|S_i z_{t_n} - x^* + x^* - z_{t_n}\|\right)$$

$$\le (p.2^p - 4)\|x^* - z_{t_n}\|^p + p.2^p\langle S_i z_{t_n} - x^*, j_p(x^* - z_{t_n})\rangle + 4\|S_i z_{t_n} - x^*\|^p$$

$$\le (p.2^p - 4)\|x^* - z_{t_n}\|^p + p.2^p\langle S_i z_{t_n} - z_{t_n} + z_{t_n} - x^*, j_p(x^* - z_{t_n})\rangle$$
$$+ 4\|x^* - z_{t_n}\|^p$$

$$= (p.2^p - 4)\|x^* - z_{t_n}\|^p + p.2^p\langle S_i z_{t_n} - z_{t_n}, j_p(x^* - z_{t_n})\rangle$$
$$- p.2^p\langle z_{t_n} - x^*, j_p(z_{t_n} - x^*)\rangle + 4\|x^* - z_{t_n}\|^p$$

$$= p.2^p\left\langle z_{t_n} - S_i z_{t_n}, j_p(z_{t_n} - x^*)\right\rangle.$$

Hence,

$$\frac{4}{p}g\left(\frac{1}{2}\|S_i z_{t_n} - z_{t_n}\|\right) \le \left\langle z_{t_n} - S_i z_{t_n}, j_p(z_{t_n} - x^*)\right\rangle,$$

and so,

$$\frac{4}{p}\sum_{i=1}^{\infty}\sigma_{i,t_n}g\left(\frac{1}{2}\|S_i z_{t_n} - z_{t_n}\|\right) \le \sum_{i=1}^{\infty}\sigma_{i,t_n}\left\langle z_{t_n} - S_i z_{t_n}, j_p(z_{t_n} - x^*)\right\rangle. \quad (16.11)$$

Using (16.7), we have

$$\langle z_{t_n} - x^*, j_p(z_{t_n} - x^*)\rangle = t_n\langle u - x^*, j_p(z_{t_n} - x^*)\rangle$$

$$+ \sum_{i=1}^{\infty}\sigma_{i,t_n}\left\langle S_i z_{t_n} - z_{t_n} + z_{t_n} - x^*, j_p(z_{t_n} - x^*)\right\rangle$$

$$= t_n\langle u - x^*, j_p(z_{t_n} - x^*)\rangle + \sum_{i=1}^{\infty}\sigma_{i,t_n}\left\langle S_i z_{t_n} - z_{t_n}, j_p(z_{t_n} - x^*)\right\rangle$$

$$+ (1 - t_n)\langle z_{t_n} - x^*, j_p(z_{t_n} - x^*)\rangle$$

which implies

$$\sum_{i=1}^{\infty}\sigma_{i,t_n}\left\langle z_{t_n} - S_i z_{t_n}, j_p(z_{t_n} - x^*)\right\rangle = t_n\langle u - z_{t_n}, j_p(z_{t_n} - x^*)\rangle.$$

Using this and (16.11), we get,

$$\frac{4}{p}\sum_{i=1}^{\infty}\sigma_{i,t_n}g\left(\frac{1}{2}\|S_i z_{t_n} - z_{t_n}\|\right) \le t_n\langle u - z_{t_n}, j_p(z_{t_n} - x^*)\rangle.$$

Since $\{z_{t_n}\}$ is bounded, we have that

$$\sum_{i=1}^{\infty}\sigma_{i,t_n}g\left(\frac{1}{2}\|S_i z_{t_n} - z_{t_n}\|\right) \le t_n M,$$

for some constant $M > 0$. This yields, $\forall\, i \in \mathcal{N}_{\mathcal{I}}$,

$$g\left(\frac{1}{2}\|S_i z_{t_n} - z_{t_n}\|\right) \le \frac{t_n}{\sigma_{i,t_n}}M,$$

which yields, since g is continuous, strictly increasing, $g(0) = 0$ and $\lim_{n\to\infty}\frac{t_n}{\sigma_{i,t_n}} = 0 \,\forall\, i \in \mathcal{N}_{\mathcal{I}}$, that

$$\|S_i z_{t_n} - z_{t_n}\| \le 2g^{-1}\left(\frac{t_n}{\sigma_{i,t_n}}M\right) \to 0,$$

as $n \to \infty$. Hence, $\lim_{n\to\infty}\|S_i z_{t_n} - z_{t_n}\| = 0$, establishing the lemma. □

Theorem 16.9. *Let $E = l_p$, $1 < p < \infty$. Let K be a closed, convex and nonempty subset of E. Let $\{T_i\}_{i=1}^\infty$ be a family of nonexpansive self-mappings of K. Let $\{t_n\}$ be a sequence in $(0,1)$ such that $\lim\limits_{n\to\infty} t_n = 0$ and $\lim\limits_{n\to\infty} \frac{t_n}{\sigma_{i,t_n}} = 0 \ \forall \, i \in \mathcal{N}_\mathcal{I}$, $\sum\limits_{i\geq 1}\sigma_{i,t_n} = (1 - t_n)$. Let $\{z_{t_n}\}$ be a sequence satisfying (16.7) and let $\mathcal{F} := \overset{\infty}{\underset{i=1}{\cap}} F(T_i) \neq \emptyset$. Then, $\{z_{t_n}\}$ converges strongly to an element in $\mathcal{F}$.*

Proof. Since $\{z_{t_n}\}$ is bounded, there exists a subsequence say $\{z_{t_{n_k}}\}$ of $\{z_{t_n}\}$ that converges weakly to some point $z \in K$. Using the demiclosedness property of $(I - T_i)$ for each $i \in \mathbb{N}$, and the fact that $\lim\limits_{k\to\infty} \|T_i z_{t_{n_k}} - z_{t_{n_k}}\| = 0$, we get that z is a point in $\mathcal{F}$. We also observe from (16.7) that

$$\|z_{t_{n_k}} - z\|^p = \Big\langle t_{n_k}(u - z) + \sum_{i\geq 1}\sigma_{i,t_{n_k}}(S_i z_{t_{n_k}} - z), j_p(z_{t_{n_k}} - z)\Big\rangle$$

$$\leq t_{n_k}\langle u - z, j_p(z_{t_{n_k}} - z)\rangle + \sum_{i\geq 1}\sigma_{i,t_{n_k}}\|z_{t_{n_k}} - z\|^p$$

$$= t_{n_k}\langle u - z, j_p(z_{t_{n_k}} - z)\rangle + (1 - t_{n_k})\|z_{t_{n_k}} - x^*\|^p,$$

which implies, $\|z_{t_{n_k}} - z\|^p \leq \langle u - z, j_p(z_{t_{n_k}} - z)\rangle$. Since j_p admits weak sequential continuity, this inequality implies that the subsequence $\{z_{t_{n_k}}\}$ converges strongly to z, and since z_{t_n} admits unique accumulation point, z_{t_n} converges strongly to z. $\qquad\square$

The following corollary follows from Theorem 16.9.

Corollary 16.10 *Let H be a real Hilbert space. Let K be a closed, convex and nonempty subset of H. Let $\{T_i\}_{i=1}^\infty$ be a family of nonexpansive self-mappings of K. Let $\{t_n\}$ be a sequence in $(0,1)$ such that $\lim\limits_{n\to\infty} t_n = 0$ and $\lim\limits_{n\to\infty} \frac{t_n}{\sigma_{i,t_n}} = 0 \ \forall \, i \in \mathcal{N}_\mathcal{I}$, $\sum\limits_{i\geq 1}\sigma_{i,t_n} = (1 - t_n)$. Let $\{z_{t_n}\}$ be a sequence satisfying (16.7) and let $\mathcal{F} := \overset{\infty}{\underset{i=1}{\cap}} F(T_i) \neq \emptyset$. Then, $\{z_{t_n}\}$ converges strongly to an element of $\mathcal{F}$.*

We also prove the following theorem.

Theorem 16.11. *Let E be a uniformly convex real Banach space. Let K be a closed, convex and nonempty subset of E. Let $\{T_i\}_{i=1}^\infty$ be a family of nonexpansive self-mappings of K. Let $\{t_n\}$ be a sequence in $(0,1)$ such that $\lim\limits_{n\to\infty} t_n = 0$ and $\lim\limits_{n\to\infty} \frac{t_n}{\sigma_{i,t_n}} = 0 \ \forall \, i \in \mathcal{N}_\mathcal{I}$, $\sum\limits_{i\geq 1}\sigma_{i,t_n} = (1 - t_n)$. Let $\{z_{t_n}\}$ be a sequence satisfying (16.7) and let $\mathcal{F} := \overset{\infty}{\underset{i=1}{\cap}} F(T_i) \neq \emptyset$. If at least one of the maps T_i is demicompact, then $\{z_{t_n}\}$ converges strongly to an element of $\mathcal{F}$.*

Proof. For some fixed $s \in \mathbb{N}$, let T_s be demicompact. Since $\lim\limits_{n\to\infty} \|T_s z_{t_n} - z_{t_n}\| = 0$, there exists a subsequence say $\{z_{t_{n_k}}\}$ of $\{z_{t_n}\}$ that converges strongly to some point $z \in K$. By the continuity of T_i $\forall i \in \mathbb{N}$, we have that $z \in \mathcal{F}$. But z_{t_n} admits a unique accumulation point, so z_{t_n} converges strongly to z.   □

The following corollary follows from Theorem 16.11.

Corollary 16.12 *Let E be a uniformly convex real Banach space. Let K be a compact, convex and nonempty subset of E. Let $\{T_i\}_{i=1}^{\infty}$ be a family of nonexpansive self-mappings of K. Let $\{t_n\}$ be a sequence in $(0, 1)$ such that* $\lim_{n \to \infty} t_n = 0$ *and* $\lim_{n \to \infty} \frac{t_n}{\sigma_{i,t_n}} = 0$ $\forall$ $i \in \mathcal{N}_{\mathcal{I}}$, $\sum_{i \geq 1} \sigma_{i,t_n} = (1 - t_n)$. *Let $\{z_{t_n}\}$ be a sequence satisfying (16.7) and let $\mathcal{F} := \overset{\infty}{\underset{i=1}{\cap}} F(T_i) \neq \emptyset$. Then, $\{z_{t_n}\}$ converges strongly to an element of $\mathcal{F}$.*

Proof. Compactness of K implies $\{z_{t_n}\}$ has a subsequence $\{z_{t_{n_k}}\}$ which converges strongly to some z in K. The rest follows as in the proof of Theorem 16.11.   □

16.4 Iterative Convergence in Uniformly Convex Real Banach Spaces

Theorem 16.13. *Let E be a uniformly convex real Banach space with a uniformly Gâteaux differentiable norm. Let K be a closed, convex and nonempty subset of E. Let $\{T_i\}_{i=1}^{\infty}$ be a family of nonexpansive self-mappings of K. For arbitrary fixed $\delta \in (0, 1)$, define a family of nonexpansive maps $\{S_i\}_{i=1}^{\infty}$ by $S_i := (1 - \delta)I + \delta T_i$ $\forall i \in \mathbb{N}$ where I is an identity map of K. Assume $\mathcal{F} := \overset{\infty}{\underset{i=1}{\cap}} F(T_i) \neq \emptyset$. Let $\{\alpha_n\}$ and $\{\sigma_{i,n}\}$ be sequences in $(0, 1)$ satisfying the following conditions: (i) $\lim_{n \to \infty} \alpha_n = 0$, (ii) $\sum \alpha_n = \infty$, (iii) $\sum_{i \geq 1} \sigma_{i,n} := (1 - \alpha_n)$ and (iv) $\lim_{n \to \infty} \sum_{i \geq 1} |\sigma_{i,n+1} - \sigma_{i,n}| = 0$. Define a sequence $\{x_n\}$ iteratively by $x_1, u \in K$,*

$$x_{n+1} = \alpha_n u + \sum_{i=1}^{n} \sigma_{i,n} S_i x_n, \ n \geq 1. \tag{16.12}$$

If at least one of the $T_i's$ is demicompact, then, $\{x_n\}$ converges strongly to an element of $\mathcal{F}$.

Proof. Let $x^* \in \mathcal{F}$ be arbitrary. Then, the sequence $\{x_n\}$ defined by (16.12) satisfies $\|x_n - x^*\| \leq \max\{\|u - x^*\|, \|x_1 - x^*\|\}$ for all $n \in \mathbb{N}$. We verify this by induction. It is clear that this is true for $n = 1$. Assume it is true for $n = k$ for some $k > 1$, $k \in \mathbb{N}$. Then, using (16.12) we have

$$\|x_{k+1} - x^*\| \le \alpha_k \|u - x^*\|$$
$$+ \sum_{i=1}^{n} \sigma_{i,k} \left[(1 - \delta) \|x_k - x^*\| + \delta \|T_i x_k - x^*\| \right]$$
$$\le \alpha_k \|u - x^*\| + (1 - \alpha_k) \|x_k - x^*\|$$
$$\le \max\{\|u - x^*\|, \|x_1 - x^*\|\},$$

and the result follows by induction. So, $\{x_n\}$ is bounded and so are $\{T_i x_n\}$ and $\{S_i x_n\}$. Define two sequences $\{\beta_n\}$ and $\{y_n\}$ by $\beta_n := (1 - \delta)\alpha_n + \delta$ and $y_n := \frac{x_{n+1} - x_n + \beta_n x_n}{\beta_n}$. Then,

$$y_n = \frac{\alpha_n u + \delta \sum_{i=1}^{n} \sigma_{i,n} T_i x_n}{\beta_n}.$$

Observe that $\{y_n\}$ is bounded and that

$$\|y_{n+1} - y_n\| - \|x_{n+1} - x_n\| \le \left| \frac{\alpha_{n+1}}{\beta_{n+1}} - \frac{\alpha_n}{\beta_n} \right| \|u\|$$
$$+ \left| \frac{\delta(1 - \alpha_{n+1})}{\beta_{n+1}} - 1 \right| \|x_{n+1} - x_n\|$$
$$+ \frac{\delta M}{\beta_{n+1} \beta_n} \sum_{i=1}^{n} |\sigma_{i,n+1} - \sigma_{i,n}| + \frac{\delta M}{\beta_{n+1} \beta_n} |\beta_n - \beta_{n+1}|,$$

for some $M > 0$. This implies,

$$\limsup_{n \to \infty} (\|y_{n+1} - y_n\| - \|x_{n+1} - x_n\|) \le 0,$$

and by Lemma 6.33, $\lim_{n \to \infty} \|y_n - x_n\| = 0$. Hence,

$$\|x_{n+1} - x_n\| = \beta_n \|y_n - x_n\| \to 0 \quad \text{as } n \to \infty. \tag{16.13}$$

From (16.12) we have $x_{n+1} - x_n = \alpha_n(u - x_n) + \sum_{i=1}^{n} \sigma_{i,n}(S_i x_n - x_n)$ which implies, $\left\| \sum_{i=1}^{n} \sigma_{i,n}(S_i x_n - x_n) \right\| \le \|x_{n+1} - x_n\| + \alpha_n \|u - x_n\|$ and thus $\lim_{n \to \infty} \|\sum_{i=1}^{n} \sigma_{i,n}(S_i x_n - x_n)\| = 0$. Let $\{t_n\}$ be a real sequence in $(0,1)$ satisfying the following conditions:

$$\lim_{n \to \infty} t_n = 0, \quad \sum_{i=1}^{\infty} \sigma_{i,t_n} = (1 - t_n) \quad \text{and} \quad \lim_{n \to \infty} \frac{\left\| \sum_{i=1}^{n} \sigma_{i,n}(S_i x_n - x_n) \right\|}{t_n} = 0.$$

Let $z_{t_n} \in K$ be the unique fixed point satisfying (16.7) for each $n \in \mathbb{N}$ and let $z_{t_n} \to z \in \mathcal{F}$ as $n \to \infty$ (this is guaranteed by Theorem 16.11). Using (16.7) and inequality (4.4), we have the following estimates

$$\|z_{t_n} - x_n\|^2 \leq \left\| \sum_{i=1}^{n} \sigma_{i,t_n} \left(S_i z_{t_n} - S_i x_n + S_i x_n - x_n \right) \right\|^2$$
$$+ 2t_n \langle u - x_n, j_p(z_{t_n} - x_n) \rangle$$
$$\leq \left((1 - t_n) \|z_{t_n} - x_n\| + \left\| \sum_{i=1}^{n} \sigma_{i,n}(S_i x_n - x_n) \right\| \right)^2$$
$$+ 2t_n \langle u - x_n, j_p(z_{t_n} - x_n) \rangle.$$

This implies,

$$\langle u - z_{t_n}, j_p(x_n - z_{t_n}) \rangle \leq \frac{t_n}{2} \|z_{t_n} - x_n\|^2$$
$$+ (1 - t_n) \|z_{t_n} - x_n\| \left(\frac{\left\| \sum_{i=1}^{n} \sigma_{i,n}(S_i x_n - x_n) \right\|}{t_n} \right)$$
$$+ \frac{\left\| \sum_{i=1}^{n} \sigma_{i,n}(S_i x_n - x_n) \right\|^2}{2t_n}$$

and hence $\limsup \langle u - z_{t_n}, j_p(x_n - z_{t_n}) \rangle \leq 0$. Moreover,

$$\langle u - z_{t_n}, j_p(x_n - z_{t_n}) \rangle = \langle u - z, j_p(x_n - z) \rangle$$
$$+ \langle u - z, j_p(x_n - z_{t_n}) - j_p(x_n - z) \rangle$$
$$+ \langle z - z_{t_n}, j_p(x_n - z_{t_n}) \rangle,$$

and since j_p is norm-to-weak* uniformly continuous on bounded sets, we have $\limsup \langle u - z, j_p(x_n - z) \rangle \leq 0$. From the recursion formula (16.12) and inequality (4.4), we have the following.

$$\|x_{n+1} - z\|^2 \leq \left\| \sum_{i=1}^{n} \sigma_{i,n}(S_i x_n - z) \right\|^2 + 2\alpha_n \langle u - z, j_p(x_{n+1} - z) \rangle$$
$$\leq (1 - \alpha_n) \|x_n - z\|^2 + 2\alpha_n \langle u - z, j_p(x_{n+1} - z) \rangle,$$

and by Lemma 6.34, we have $\{x_n\}$ converges strongly to $z \in \mathcal{F}$. This completes the proof. $\square$

The following is an immediate corollary of Theorem 16.13.

Corollary 16.14 *Let E be a uniformly convex real Banach space with a uniformly Gâteaux differentiable norm. Let K be a compact, convex and nonempty subset of E. Let $\{T_i\}_{i=1}^{\infty}$ be a family of nonexpansive self-mappings* *of K. For arbitrary fixed $\delta \in (0,1)$, define a family of nonexpansive maps $\{S_i\}_{i=1}^\infty$ by $S_i := (1-\delta)I + \delta T_i \ \forall i \in \mathbb{N}$ where I is an identity map of K. Let $\mathcal{F} := \bigcap_{i=1}^\infty F(T_i) \neq \emptyset$. Let $\{\alpha_n\}$ and $\{\sigma_{i,n}\}$ be sequences as in Theorem 16.13. Let $\{x_n\}$ be a sequence defined by (16.12). Then, $\{x_n\}$ converges strongly to an element of $\mathcal{F}$.*

Following the method of proof of Theorem 16.13, and using Theorem 16.9, the following theorem is easily proved.

Theorem 16.15. *Let $E = l_p$, $1 < p < \infty$. Let K be a closed, convex and nonempty subset of E. Let $\{T_i\}_{i=1}^\infty$ be a family of nonexpansive self-mappings of K. For arbitrary fixed $\delta \in (0,1)$, define a family of nonexpansive maps $\{S_i\}_{i=1}^\infty$ by $S_i := (1-\delta)I + \delta T_i \ \forall i \in \mathbb{N}$ where I is an identity map of K. Assume $\mathcal{F} := \bigcap_{i=1}^\infty F(T_i) \neq \emptyset$. Let $\{\alpha_n\}$ and $\{\sigma_{i,n}\}$ be sequences in $(0,1)$ satisfying the following conditions:*

$$(i) \ \lim_{n \to \infty} \alpha_n = 0, \quad (ii) \ \sum \alpha_n = \infty, \quad (iii) \sum_{i \geq 1} \sigma_{i,n} := (1 - \alpha_n)$$

and $(iv) \lim_{n \to \infty} \sum_{i \geq 1} |\sigma_{i,n+1} - \sigma_{i,n}| = 0$. Define a sequence $\{x_n\}$ iteratively by $x_1, u \in K$,

$$x_{n+1} = \alpha_n u + \sum_{i=1}^n \sigma_{i,n} S_i x_n, \ n \geq 1. \tag{16.14}$$

Then, $\{x_n\}$ converges strongly to an element of $\mathcal{F}$.

We also obtain the following corollary.

Corollary 16.16 *Let H be a real Hilbert space. Let K be a closed, convex and nonempty subset of E. Let $\{T_i\}_{i=1}^\infty$ be a family of nonexpansive self-mappings of K. For arbitrary fixed $\delta \in (0,1)$, define a family of nonexpansive maps $\{S_i\}_{i=1}^\infty$ by $S_i := (1 - \delta)I + \delta T_i \ \forall i \in \mathbb{N}$ where I is an identity map of K. Assume $\mathcal{F} := \bigcap_{i=1}^\infty F(T_i) \neq \emptyset$. Let $\{\alpha_n\}$ and $\{\sigma_{i,n}\}$ be sequences in $(0,1)$ satisfying the following conditions: (i) $\lim_{n \to \infty} \alpha_n = 0$, (ii) $\sum \alpha_n = \infty$, (iii) $\sum_{i \geq 1} \sigma_{i,n} := (1 - \alpha_n)$ and (iv) $\lim_{n \to \infty} \sum_{i=1}^\infty |\sigma_{i,n+1} - \sigma_{i,n}| = 0$. Define a sequence $\{x_n\}$ iteratively by $x_1, u \in K$,*

$$x_{n+1} = \alpha_n u + \sum_{i=1}^n \sigma_{i,n} S_i x_n, \ n \geq 1. \tag{16.15}$$

Then, $\{x_n\}$ converges strongly to an element of $\mathcal{F}$.

16.5 Non-self Mappings

The following theorems are now easily proved.

Theorem 16.17. *Let E be a uniformly convex real Banach space with a uniformly Gâteaux differentiable norm. Let K be a closed, convex and nonempty sunny nonexpansive retract of E with Q as the sunny nonexpansive retraction. Let $T_i : K \to E$, $i \in \mathbb{N}$ be a family of nonexpansive mappings of K into E. For arbitrary fixed $\delta \in (0,1)$, define a family of nonexpansive maps $\{S_i\}_{i=1}^{\infty}$ by $S_i := (1 - \delta)I + \delta Q T_i$ $\forall i \in \mathbb{N}$ where I is the identity map of K. Assume $\mathcal{F} := \bigcap_{i=1}^{\infty} F(T_i) \neq \emptyset$. Let $\{\alpha_n\}$ and $\{\sigma_{i,n}\}$ be sequences in $(0,1)$ satisfying the following conditions: (i) $\lim_{n\to\infty} \alpha_n = 0$, (ii) $\sum \alpha_n = \infty$, (iii) $\sum_{i\geq 0} \sigma_{i,n} = 1 - \alpha_n$ and (iv) $\lim_{n\to\infty} \sum_{i\geq 1} |\sigma_{i,n+1} - \sigma_{i,n}| = 0$. Define a sequence $\{x_n\}$ iteratively by $x_1, u \in K$*

$$x_{n+1} = \alpha_n u + \sum_{i=1}^{n} \sigma_{i,n} S_i x_n. \tag{16.16}$$

If at least one of the $T_i's$ is demicompact, then, $\{x_n\}$ converges strongly to an element of $\mathcal{F}$.

Theorem 16.18. *Let E be a uniformly convex real Banach space with a uniformly Gâteaux differentiable norm. Let K be a compact, convex and nonempty sunny nonexpansive retract of E with Q as the sunny nonexpansive retraction. Let $T_i : K \to E$, $i \in \mathbb{N}$ be a family of nonexpansive mappings of K into E. For arbitrary fixed $\delta \in (0,1)$, define a family of nonexpansive maps $\{S_i\}_{i=1}^{\infty}$ by $S_i := (1 - \delta)I + \delta Q T_i$ $\forall i \in \mathbb{N}$ where I is the identity map of K. Assume $\mathcal{F} := \bigcap_{i=1}^{\infty} F(T_i) \neq \emptyset$. Let $\{\alpha_n\}$ and $\{\sigma_{i,n}\}$ be sequences in $(0,1)$ satisfying the following conditions: (i) $\lim_{n\to\infty} \alpha_n = 0$, (ii) $\sum \alpha_n = \infty$, (iii) $\sum_{i\geq 0} \sigma_{i,n} = 1 - \alpha_n$ and (iv) $\lim_{n\to\infty} \sum_{i\geq 1} |\sigma_{i,n+1} - \sigma_{i,n}| = 0$. Define a sequence $\{x_n\}$ iteratively by $x_1, u \in K$*

$$x_{n+1} = \alpha_n u + \sum_{i=1}^{n} \sigma_{i,n} S_i x_n. \tag{16.17}$$

Then, $\{x_n\}$ converges strongly to an element of $\mathcal{F}$.

Theorem 16.19. *Let $E = l_p, 1 < p < \infty$. Let K be a closed, convex and nonempty sunny nonexpansive retract of E with Q as the sunny nonexpansive retraction. Let $T_i : K \to E$, $i \in \mathbb{N}$ be a family of nonexpansive mappings of K into E. For arbitrary fixed $\delta \in (0,1)$, define a family of nonexpansive maps $\{S_i\}_{i=1}^{\infty}$ by $S_i := (1 - \delta)I + \delta Q T_i$ $\forall i \in \mathbb{N}$ where I is the identity map of K. Assume $\mathcal{F} := \bigcap_{i=1}^{\infty} F(T_i) \neq \emptyset$. Let $\{\alpha_n\}$ and $\{\sigma_{i,n}\}$ be sequences in*

$(0,1)$ satisfying the following conditions: (i) $\lim\limits_{n\to\infty} \alpha_n = 0$, (ii) $\sum \alpha_n = \infty$, $(iii) \sum\limits_{i\geq 0} \sigma_{i,n} = 1 - \alpha_n$ and $(iv) \lim\limits_{n\to\infty} \sum\limits_{i\geq 1} |\sigma_{i,n+1} - \sigma_{i,n}| = 0$. Define a sequence $\{x_n\}$ iteratively by $x_1, u \in K$

$$x_{n+1} = \alpha_n u + \sum_{i=1}^{n} \sigma_{i,n} S_i x_n. \tag{16.18}$$

Then, $\{x_n\}$ converges strongly to an element of $\mathcal{F}$.

Theorem 16.20. *Let H be a real Hilbert space. Let K be a closed, convex and nonempty sunny nonexpansive retract of E with Q as the sunny nonexpansive retraction. Let $T_i : K \to E$, $i \in \mathbb{N}$ be a family of nonexpansive mappings of K into E. For arbitrary fixed $\delta \in (0,1)$, define a family of nonexpansive maps $\{S_i\}_{i=1}^{\infty}$ by $S_i := (1 - \delta)I + \delta Q T_i$ $\forall i \in \mathbb{N}$ where I is the identity map of K. Assume $\mathcal{F} := \bigcap\limits_{i=1}^{\infty} F(T_i) \neq \emptyset$. Let $\{\alpha_n\}$ and $\{\sigma_{i,n}\}$ be sequences in $(0,1)$ satisfying the following conditions: (i) $\lim\limits_{n\to\infty} \alpha_n = 0$, (ii) $\sum \alpha_n = \infty$, $(iii) \sum\limits_{i\geq 0} \sigma_{i,n} = 1 - \alpha_n$ and $(iv) \lim\limits_{n\to\infty} \sum\limits_{i\geq 1} |\sigma_{i,n+1} - \sigma_{i,n}| = 0$. Define a sequence $\{x_n\}$ iteratively by $x_1, u \in K$*

$$x_{n+1} = \alpha_n u + \sum_{i=1}^{n} \sigma_{i,n} S_i x_n. \tag{16.19}$$

Then, $\{x_n\}$ converges strongly to an element of $\mathcal{F}$.

Remark 16.21. Prototypes of the sequences $\{\alpha_n\}$ and $\{\sigma_{i,n}\}$ in the theorems of this chapter are the following: $\alpha_n := \frac{1}{n+1}$, $\sigma_{i,n} := \frac{n}{2^i(n+1)}$ $\forall i \in \mathbb{N}$. For these choices, the recursion formulas (16.15) and (16.19) become $x_1, u \in K$,

$$x_{n+1} = \left(\frac{1}{n+1}\right) u + \frac{n}{n+1} \sum_{i=1}^{n} \frac{1}{2^i} S_i x_n, \quad n \geq 1. \tag{16.20}$$

Remark 16.22. We first observe that the function C introduced in Theorems 16.4 and 16.5 makes the computation of each iterate using the recursion formula (16.5) more involved (and therefore less efficient) than using the recursion formula (16.12). With this in mind, we observe that Theorems 16.9 and 16.15 extend Theorems 16.4 and 16.5, respectively, to l_p spaces, $1 < p < \infty$. Furthermore, the following conditions in (ii) :

$$\begin{cases} \frac{1}{\sigma_{i,n}}\left|1 - \frac{\alpha_{n-1}}{\alpha_n}\right| \to 0, \text{ or } \sum\limits_{n} \frac{1}{\sigma_{i,n}} |\alpha_{n-1} - \alpha_n| < \infty \\ \frac{1}{\alpha_n}\left|\frac{1}{\sigma_{i,n}} - \frac{1}{\sigma_{i,n-1}}\right| \to 0, \text{ or } \sum\limits_{n}\left|\frac{1}{\sigma_{i,n}} - \frac{1}{\sigma_{i,n-1}}\right| < \infty \end{cases}$$

imposed in Theorem 16.5 are dispensed with even in the more general settings of these theorems. In addition, the requirement,

$$\frac{1}{\sigma_{i,n}\alpha_n} \sum_{k\geq 0} |\sigma_{k,n} - \sigma_{k,n-1}| \to 0, \quad \text{or} \quad \sum_n \frac{1}{\sigma_{i,n}} \sum_{k\geq 0} |\sigma_{k,n} - \sigma_{k,n-1}| < \infty$$

also imposed in (ii) of Theorem 16.5 is weakened to $\lim_{n\to\infty} \sum_{i\geq 1} |\sigma_{i,n+1} - \sigma_{i,n}| = 0$.

All the other theorems of this chapter are of independent interest.

16.6 Historical Remarks

Lemma 16.8 and all the results of Sections 16.3, 16.4 and 16.5 are due to Chidume and Chidume [130].

Chapter 17
Common Fixed Points for Families of Commuting Nonexpansive Mappings

17.1 Introduction

In this chapter, we present an iteration process which has been studied for approximating common fixed points for families of *commuting* nonexpansive mappings defined on a *compact* convex subset of a Banach space.

We first prove the following lemmas which are connected with real numbers.

Lemma 17.1. *Let $\{\alpha_n\}$ be a real sequence with $\lim_n(\alpha_{n+1} - \alpha_n) = 0$. Then, every $\tau \in \mathbb{R}$ satisfying $\liminf \alpha_n < \tau < \limsup \alpha_n$ is a cluster point of $\{\alpha_n\}$.*

Proof. We assume that there exists $\tau \in (\liminf \alpha_n, \limsup \alpha_n)$ such that τ is not a cluster point of $\{\alpha_n\}$. Then, there exist $\epsilon > 0$ and $n_1 \in \mathbb{N}$ such that $\liminf_{n\to\infty} \alpha_n < \tau - \epsilon < \tau < \tau + \epsilon < \limsup_{n\to\infty} \alpha_n$ and $\alpha_n \in (-\infty, \tau - \epsilon] \cup [\tau + \epsilon, \infty)$ for all $n \geq n_1$. We choose $n_2 \geq n_1$ such that $|\alpha_{n+1} - \alpha_n| < \epsilon$ for all $n \geq n_2$. Then, there exist $n_3, n_4 \in \mathbb{N}$ such that $n_4 \geq n_3 \geq n_2$, $\alpha_{n_3} \in (-\infty, \tau - \epsilon]$ and $\alpha_{n_4} \in [\tau + \epsilon, \infty)$. We put $n_5 = \max\{n : n < n_4, \alpha_n \leq \tau - \epsilon\} \geq n_3$. Then, we have $\alpha_{n_5} \leq \tau - \epsilon < \tau + \epsilon \leq \alpha_{n_5+1}$ and hence $\epsilon \leq 2\epsilon \leq \alpha_{n_5+1} - \alpha_{n_5} = |\alpha_{n_5+1} - \alpha_{n_5}| < \epsilon$. This is a contradiction. Therefore we obtain the desired result. $\square$

Lemma 17.2. *For $\alpha, \beta \in (0, \frac{1}{2})$ and $n \in \mathbb{N}$, $|\alpha^n - \beta^n| \leq |\alpha - \beta|$ and*

$$\sum_{k=1}^{\infty} |\alpha^k - \beta^k| \leq 4|\alpha - \beta|$$

hold.

Proof. We assume that $n \geq 2$ because the conclusion is obvious in the case of $n = 1$. Since $\alpha^n - \beta^n = (\alpha - \beta) \sum_{k=0}^{n-1} \alpha^{n-1-k} \beta^k$, we have

C. Chidume, *Geometric Properties of Banach Spaces and Nonlinear Iterations*,
Lecture Notes in Mathematics 1965,
© Springer-Verlag London Limited 2009

$$|\alpha^n - \beta^n| = |\alpha - \beta| \sum_{k=0}^{n-1} \alpha^{n-1-k} \beta^k \leq |\alpha - \beta| \sum_{k=1}^{n-1} \frac{1}{2^{n-1}}$$

$$= |\alpha - \beta| \frac{n}{2^{n-1}} \leq |\alpha - \beta|.$$

We also have

$$\sum_{k=1}^{\infty} |\alpha^k - \beta^k| = \left| \sum_{k=1}^{\infty} (\alpha^k - \beta^k) \right| = \left| \frac{\alpha}{1-\alpha} - \frac{\beta}{1-\beta} \right|$$

$$= \left| \frac{\alpha - \beta}{(1-\alpha)(1-\beta)} \right| \leq 4|\alpha - \beta|.$$

This completes the proof. $\qquad\qquad\qquad\qquad\qquad\qquad\qquad\qquad\qquad$ □

We know the following.

Lemma 17.3. *Let C be a subset of a Banach space E and $\{T_n\}$ be a sequence of nonexpansive mappings on C with a common fixed point $w \in C$. Let $x_1 \in C$ and define a sequence $\{x_n\}$ in C by $x_{n+1} = T_n x_n$ for $n \in \mathbb{N}$. Then, $\{\|x_n - w\|\}$ is a nonincreasing sequence in $\mathbb{R}$.*

Proof. We have $\|x_{n+1} - w\| = \|T_n x_n - T_n w\| \leq \|x_n - w\|$ for all $n \in \mathbb{N}$. □

17.2 Three Commuting Nonexpansive Mappings

In this section, we prove a convergence theorem for three commuting nonexpansive mappings. The purpose of this is to give an idea of the proof of the main results. We begin with the following lemma.

Lemma 17.4. *Let C be a closed convex subset of a Banach space E. Let T_1 and T_2 be nonexpansive mappings on C with $T_1 \circ T_2 = T_2 \circ T_1$. Let $\{t_n\}$ be a sequence in $(0,1)$ converging to 0, and let $\{z_n\}$ be a sequence in C such that $\{z_n\}$ converges strongly to some $w \in C$ and $\lim\limits_{n\to\infty} \frac{\|(1-t_n)T_1 z_n + t_n T_2 z_n - z_n\|}{t_n} = 0$. Then, w is a common fixed point of T_1 and T_2.*

Proof. It is obvious that $\sup\limits_{m,n\in\mathbb{N}} \|T_1 z_m - T_1 z_n\| \leq \sup\limits_{m,n\in\mathbb{N}} \|z_m - z_n\|$. So, $\{T_1 z_n\}$ is bounded because $\{z_n\}$ is bounded. Similarly, we have that $\{T_2 z_n\}$ is also bounded. Since $\lim\limits_{n\to\infty} \|(1 - t_n)T_1 z_n + t_n T_2 z_n - z_n\| = 0$, we have

$$\|T_1 w - w\| \leq \limsup_{n\to\infty} \Big(\|T_1 w - T_1 z_n\| + \|T_1 z_n - (1 - t_n)T_1 z_n - t_n T_2 z_n\|$$

$$+ \|(1 - t_n)T_1 z_n + t_n T_2 z_n - z_n\| + \|z_n - w\| \Big)$$

$$\leq \limsup_{n\to\infty} \Big(2\|w - z_n\| + t_n\|T_1 z_n - T_2 z_n\|$$

$$+ \|(1 - t_n)T_1 z_n + t_n T_2 z_n - z_n\| \Big) = 0,$$

and hence w is a fixed point of T_1. We note that $T_1 \circ T_2 w = T_2 \circ T_1 w = T_2 w$. We assume that w is not a fixed point of T_2. Put $\epsilon = \frac{\|T_2 w - w\|}{3} > 0$. Then, there exists $m \in \mathbb{N}$ such that $\|z_m - w\| < \epsilon$ and $\frac{\|(1 - t_m)T_1 z_m + t_m T_2 z_m - z_m\|}{t_m} < \epsilon$. Since $3\epsilon = \|T_2 w - w\| \leq \|T_2 w - z_m\| + \|z_m - w\| < \|T_2 w - z_m\| + \epsilon$, we have $2\epsilon < \|T_2 w - z_m\|$. So, we obtain

$$\|T_2 w - z_m\| \leq \|T_2 w - (1 - t_m)T_1 z_m - t_m T_2 z_m\|$$

$$+ \|(1 - t_m)T_1 z_m + t_m T_2 z_m - z_m\|$$

$$\leq (1 - t_m)\|T_2 w - T_1 z_m\| + t_m\|T_2 w - T_2 z_m\|$$

$$+ \|(1 - t_m)T_1 z_m + t_m T_2 z_m - z_m\|$$

$$= (1 - t_m)\|T_1 \circ T_2 w - T_1 z_m\| + t_m\|T_2 w - T_2 z_m\|$$

$$+ \|(1 - t_m)T_1 z_m + t_m T_2 z_m - z_m\|$$

$$\leq (1 - t_m)\|T_2 w - z_m\| + t_m\|w - z_m\|$$

$$+ \|(1 - t_m)T_1 z_m + t_m T_2 z_m - z_m\|$$

$$< (1 - t_m)\|T_2 w - z_m\| + 2t_m \epsilon$$

$$< (1 - t_m)\|T_2 w - z_m\| + t_m\|T_2 w - z_m\|$$

$$= \|T_2 w - z_m\|.$$

This is a contradiction. Hence, w is a common fixed point of T_1 and T_2. □

Lemma 17.5. *Let C be a closed convex subset of a Banach space E. Let T_1, T_2 and T_3 be commuting nonexpansive mappings on C. Let $\{t_n\}$ be a sequence in $(0, \frac{1}{2})$ converging to 0, and let $\{z_n\}$ be a sequence in C such that $\{z_n\}$ converges strongly to some $w \in C$, and*

$$\lim_{n\to\infty} \frac{\|(1 - t_n - t_n^2)T_1 z_n + t_n T_2 z_n + t_n^2 T_3 z_n - z_n\|}{t_n^2} = 0.$$

Then, w is a common fixed point of T_1, T_2 and T_3.

Proof. We note that $\{T_1 z_n\}$, $\{T_2 z_n\}$ and $\{T_3 z_n\}$ are bounded sequences in C because $\{z_n\}$ is bounded. We have

$$\limsup_{n \to \infty} \frac{\|(1 - t_n)T_1 z_n + t_n T_2 z_n - z_n\|}{t_n}$$

$$\leq \limsup_{n \to \infty} \frac{\|(1 - t_n - t_n^2)T_1 z_n + t_n T_2 z_n + t_n^2 T_3 z_n - z_n\| + t_n^2 \|T_1 z_n - T_3 z_n\|}{t_n}$$

$$\leq \lim_{n \to \infty} \left(t_n \frac{\|(1 - t_n - t_n^2)T_1 z_n + t_n T_2 z_n + t_n^2 T_3 z_n - z_n\|}{t_n^2} + t_n \|T_1 z_n - T_3 z_n\| \right)$$

$$= 0.$$

So, by Lemma 17.4, we have that w is a common fixed point of T_1 and T_2. We note that $T_1 \circ T_3 w = T_3 \circ T_1 w = T_3 w$, and $T_2 \circ T_3 w = T_3 \circ T_2 w = T_3 w$. We assume that w is not a fixed point of T_3. Put $\epsilon = \frac{\|T_3 w - w\|}{3} > 0$. Then, there exists $m \in \mathbb{N}$ such that $\|z_m - w\| < \epsilon$ and $\frac{\|(1 - t_m - t_m^2)T_1 z_m + t_m T_2 z_m + t_m^2 T_3 z_m - z_m\|}{t_m^2} < \epsilon$. Since $3\epsilon = \|T_3 w - w\| \leq \|T_3 w - z_m\| + \|z_m - w\| < \|T_3 w - z_m\| + \epsilon$, we have $2\epsilon < \|T_3 w - z_m\|$. So, we obtain

$$
\begin{aligned}
\|T_3 w - z_m\| &\leq \|T_3 w - (1 - t_m - t_m^2)T_1 z_m - t_m T_2 z_m - t_m^2 T_3 z_m\| \\
&\quad + \|(1 - t_m - t_m^2)T_1 z_m + t_m T_2 z_m + t_m^2 T_3 z_m - z_m\| \\
&\leq (1 - t_m - t_m^2)\|T_3 w - T_1 z_m\| + t_m\|T_3 w - T_2 z_m\| \\
&\quad + t_m^2\|T_3 w - T_3 z_m\| \\
&\quad + \|(1 - t_m - t_m^2)T_1 z_m + t_m T_2 z_m + t_m^2 T_3 z_m - z_m\| \\
&= (1 - t_m - t_m^2)\|T_1 \circ T_3 w - T_1 z_m\| + t_m\|T_2 \circ T_3 w - T_2 z_m\| \\
&\quad + t_m^2\|T_3 w - T_3 z_m\| \\
&\quad + \|(1 - t_m - t_m^2)T_1 z_m + t_m T_2 z_m + t_m^2 T_3 z_m - z_m\| \\
&\leq (1 - t_m - t_m^2)\|T_3 w - z_m\| + t_m\|T_3 w - z_m\| + t_m^2\|w - z_m\| \\
&\quad + \|(1 - t_m - t_m^2)T_1 z_m + t_m T_2 z_m + t_m^2 T_3 z_m - z_m\| \\
&< (1 - t_m^2)\|T_3 w - z_m\| + 2t_m^2 \epsilon \\
&< (1 - t_m^2)\|T_3 w - z_m\| + t_m^2\|T_3 w - z_m\| = \|T_3 w - z_m\|.
\end{aligned}
$$

This is a contradiction. Hence, w is common fixed point of T_1, T_2 and T_3. $\square$

We now prove the following convergence theorem.

Theorem 17.6. *Let C be a compact convex subset of a Banach space E. Let T_1, T_2 and T_3 be commuting nonexpansive mappings on C. Fix $\lambda \in (0, 1)$. Let $\{\alpha_n\}$ be a sequence in $[0, \frac{1}{2}]$ satisfying $\liminf_{n \to \infty} \alpha_n = 0$, $\limsup_{n \to \infty} \alpha_n > 0$, and $\lim_{n \to \infty} (\alpha_{n+1} - \alpha_n) = 0$. Define a sequence $\{x_n\}$ in C by $x_1 \in C$ and*

$$x_{n+1} = \lambda(1 - \alpha_n - \alpha_n^2)T_1 x_n + \lambda \alpha_n T_2 x_n + \lambda \alpha_n^2 T_3 x_n + (1 - \lambda)x_n$$

for $n \in \mathbb{N}$. Then, $\{x_n\}$ converges strongly to a common fixed point of T_1, T_2 and T_3.

Proof. Put $\alpha = \limsup\limits_{n\to\infty} \alpha_n > 0$, $M = \sup\limits_{x\in C} \|x\| < \infty$, and

$$y_n = (1 - \alpha_n - \alpha_n^2)T_1 x_n + \alpha_n T_2 x_n + \alpha_n^2 T_3 x_n$$

for $n \in \mathbb{N}$. We note that $x_{n+1} = \lambda y_n + (1 - \lambda)x_n$ for all $n \in \mathbb{N}$. Since

$$\|y_{n+1} - y_n\| = \|(1 - \alpha_{n+1} - \alpha_{n+1}^2)T_1 x_{n+1} + \alpha_{n+1}T_2 x_{n+1}$$
$$+ \alpha_{n+1}^2 T_3 x_{n+1} - (1 - \alpha_n - \alpha_n^2)T_1 x_n - \alpha_n T_2 x_n - \alpha_n^2 T_3 x_n\|$$
$$\leq (1 - \alpha_{n+1} - \alpha_{n+1}^2)\|T_1 x_{n+1} - T_1 x_n\| + \alpha_{n+1}\|T_2 x_{n+1} - T_2 x_n\|$$
$$+ \alpha_{n+1}^2\|T_3 x_{n+1} - T_3 x_n\| + |\alpha_{n+1} + \alpha_{n+1}^2 - \alpha_n - \alpha_n^2|.\| - T_1 x_n\|$$
$$+ |\alpha_{n+1} - \alpha_n|.\|T_2 x_n\| + |\alpha_{n+1}^2 - \alpha_n^2|.\|T_3 x_n\|$$
$$\leq (1 - \alpha_{n+1} - \alpha_{n+}^2)\|x_{n+1} - x_n\| + \alpha_{n+1}\|x_{n+1} - x_n\|$$
$$+ \alpha_{n+1}^2\|x_{n+1} - x_n\| + |\alpha_{n+1} + \alpha_{n+1}^2 - \alpha_n - \alpha_n^2|M$$
$$+ |\alpha_{n+1} - \alpha_n|M + |\alpha_{n+1}^2 - \alpha_n^2|M$$
$$\leq \|x_{n+1} - x_n\| + 4|\alpha_{n+1} - \alpha_n|M$$

for some $M \in \mathbb{R}$, we have $\limsup\limits_{n\to\infty}(\|y_{n+1} - y_n\| - \|x_{n+1} - x_n\|) \leq 0$. So, by Lemma 6.33, we have $\lim\limits_{n\to\infty} \|x_n - y_n\| = 0$. Fix $t \in \mathbb{R}$ with $0 < t < \alpha$. Then, by Lemma 17.1, there exists a subsequence $\{\alpha_{n_k}\}$ of $\{\alpha_n\}$ converging to t. Since C is compact, there exists a subsequence $\{x_{n_{k_j}}\}$ of $\{x_{n_k}\}$ converging to some point $z_t \in C$. We have

$$\|(1 - t - t^2)T_1 z_t + tT_2 z_t + t^2 T_3 z_t - z_t\|$$
$$\leq \|(1 - t - t^2)T_1 z_t + tT_2 z_t + t^2 T_3 z_t - y_n\|$$
$$+ \|y_n - x_n\| + \|x_n - z_t\|$$
$$= \|(1 - t - t^2)T_1 z_t + tT_2 z_t + t^2 T_3 z_t - (1 - \alpha_n - \alpha_n^2)T_1 x_n$$
$$- \alpha_n T_2 x_n - \alpha_n^2 T_3 x_n\| + \|y_n - x_n\| + \|x_n - z_t\|$$
$$\leq (1 - t - t^2)\|T_1 z_t - T_1 x_n\| + t\|T_2 z_t - T_2 x_n\|$$
$$+ t^2\|T_3 z_t - T_3 x_n\| + |t + t^2 - \alpha_n - \alpha_n^2|.\| - T_1 x_n\|$$
$$+ |t - \alpha_n|.\|T_2 x_n\| + |t^2 - \alpha_n^2|.\|T_3 x_n\|$$
$$+ \|y_n - x_n\| + \|x_n - z_t\|$$
$$\leq (1 - t - t^2)\|z_t - x_n\| + t\|z_t - x_n\| + t^2\|z_t - x_n\|$$
$$+ |t + t^2 - \alpha_n - \alpha_n^2|M + |t - \alpha_n|M$$
$$+ |t^2 - \alpha_n^2|M + \|y_n - x_n\| + \|x_n - z_t\|$$
$$\leq 2\|z_t - x_n\| + 4|t - \alpha_n|M + \|y_n - x_n\|$$

for $n \in \mathbb{N}$, and hence

$$\|(1 - t - t^2)T_1 z_t + t T_2 z_t + t^2 T_3 z_t - z_t\|$$

$$\leq \lim_{j \to \infty} (2\|z_t - x_{n_{k_j}}\| + 4|t - \alpha_{n_{k_j}}|M + \|y_{n_{k_j}} - x_{n_{k_j}}\|) = 0.$$

Therefore, we have $(1 - t - t^2)T_1 z_t + t T_2 z_t + t^2 T_3 z_t = z_t$ for all $t \in \mathbb{R}$ with $0 < t < \alpha$. Since C is compact, there exists a real sequence $\{t_n\}$ in $(0, \alpha)$ such that $\lim_{n \to \infty} t_n = 0$, and $\{z_{t_n}\}$ converges strongly to some point $w \in C$. By Lemma 17.5, we obtain that such a w is a common fixed point of T_1, T_2 and T_3. We note that w is a cluster point of $\{x_n\}$ because so are z_{t_n} for all $n \in \mathbb{N}$. Hence, $\liminf_{n \to \infty} \|x_n - w\| = 0$. We also have $\{\|x_n - w\|\}$ is non-increasing by Lemma 17.3. Thus, $\lim_{n \to \infty} \|x_n - w\| = 0$. This completes the proof. $\qquad\square$

We give an example concerning $\{\alpha_n\}$.

Example 17.7. Define a sequence $\{\beta_n\}$ in $[-\frac{1}{2}, \frac{1}{2}]$ by

$$\beta_n = \begin{cases} \frac{1}{2k}, & \text{if } 2\sum_{j=1}^{k-1} j < n \leq 2\sum_{j=1}^{k-1} j + k \text{ for some } k \in \mathbb{N}, \\ -\frac{1}{2k}, & \text{if } 2\sum_{j=1}^{k-1} j + k < n \leq 2\sum_{j=1}^{k} j \text{ for some } k \in \mathbb{N}. \end{cases}$$

Define a sequence $\{\alpha_n\}$ in $[0, \frac{1}{2}]$ by $\alpha_n = \sum_{k=1}^{n} \beta_k$ for $n \in \mathbb{N}$. Then, $\{\alpha_n\}$ satisfies the assumption of theorem 17.6.

Remark. The sequence $\{\alpha_n\}$ is as follows:

$$\alpha_1 = \frac{1}{2}, \quad \alpha_2 = 0,$$

$$\alpha_3 = \frac{1}{4}, \quad \alpha_4 = \frac{2}{4}, \quad \alpha_5 = \frac{1}{4}, \quad \alpha_6 = 0,$$

$$\alpha_7 = \frac{1}{6}, \quad \alpha_8 = \frac{2}{6}, \quad \alpha_9 = \frac{3}{6}, \quad \alpha_{10} = \frac{2}{6}, \quad \alpha_{11} = \frac{1}{6}, \quad \alpha_{12} = 0$$

$$\alpha_{13} = \frac{1}{8}, \quad \alpha_{14} = \frac{2}{8}, \quad \alpha_{15} = \frac{3}{8}, \quad \alpha_{16} = \frac{4}{8},$$

$$\alpha_{17} = \frac{3}{8}, \quad \alpha_{18} = \frac{2}{8}, \quad \alpha_{19} = \frac{1}{8}, \quad \alpha_{20} = 0,$$

$$\alpha_{21} = \frac{1}{10}, \quad \alpha_{22} = \frac{2}{10}, \quad \alpha_{23} = \frac{3}{10}, \quad \alpha_{24} = \frac{4}{10}, \quad \alpha_{25} = \frac{5}{10},$$

$$\alpha_{26} = \frac{4}{10}, \quad \alpha_{27} = \frac{3}{10}, \quad \alpha_{28} = \frac{2}{10}, \quad \dots.$$

17.3 Common Fixed Points for Family of Commuting Nonexpansive Mappings

In this section, we prove the main results. We begin with the following lemma.

Lemma 17.8. *Let C be a closed convex subset of a Banach space E. Let $l \in \mathbb{N}$ with $l \geq 2$ and let $T_1, T_2, ..., T_l$ be commuting nonexpansive mappings on C. Let $\{t_n\}$ be a sequence in $(0, \frac{1}{2})$ converging to 0, and let $\{z_n\}$ be a sequence in C such that $\{z_n\}$ converges strongly to some $w \in C$, and*

$$\lim_{n \to \infty} \frac{\left\|\left(1 - \sum_{k=1}^{l-1} t_n^k\right)T_1 z_n + \sum_{k=2}^{l} t_n^{k-1} T_k z_n - z_n\right\|}{t_n^{l-1}} = 0.$$

Then w is a common fixed point of $T_1, T_2, ..., T_l$.

Proof. We shall prove this lemma by induction. We have already proved the conclusion in the cases of $l = 2$ and 3. Fix $l \in \mathbb{N}$ with $l \geq 4$. We assume that the conclusion holds for any integer less than l and greater than 1. We note that $\{T_1 z_n\}, \{T_2 z_n\}, ..., \{T_l z_n\}$ are bounded sequences in C because $\{z_n\}$ is bounded. We have

$$\limsup_{n \to \infty} \frac{\left\|\left(1 - \sum_{k=1}^{l-2} t_n^k\right)T_1 z_n + \sum_{k=2}^{l-1} t_n^{k-1} T_k z_n - z_n\right\|}{t_n^{l-2}}$$

$$\leq \limsup_{n \to \infty} \frac{\left\|\left(1 - \sum_{k=1}^{l-1} t_n^k\right)T_1 z_n + \sum_{k=2}^{l} t_n^{k-1} T_k z_n - z_n\right\| + t_n^{l-1}\|T_1 z_n - T_l z_n\|}{t_n^{l-2}}$$

$$= \lim_{n \to \infty} \left(t_n \cdot \frac{\left\|\left(1 - \sum_{k=1}^{l-1} t_n^k\right)T_1 z_n + \sum_{k=2}^{l} t_n^{k-1} T_k z_n - z_n\right\|}{t_n^{l-1}} + t_n\|T_1 z_n - T_l z_n\|\right)$$

$$= 0.$$

So, by the assumption of induction, we have w is a common fixed point of $T_1, T_2, ..., T_{l-1}$. We note that $T_k \circ T_l w = T_l \circ T_k w = T_l w$ for all $k \in \mathbb{N}$ with $1 \leq k < l$. We assume that w is not a fixed point of T_l. Put $\epsilon = \frac{\|T_l w - w\|}{3} > 0$. Then there exists $m \in \mathbb{N}$ such that

$$\|z_m - w\| < \epsilon \text{ and } \frac{\left\|\left(1 - \sum_{k=1}^{l-1} t_m^k\right)T_1 z_m + \sum_{k=2}^{l} t_m^{k-1} T_k z_m - z_m\right\|}{t_m^{l-1}} < \epsilon.$$

Since $3\epsilon = \|T_l w - w\| \leq \|T_l w - z_m\| + \|z_m - w\| < \|T_l w - z_m\| + \epsilon$, we have $2\epsilon < \|T_l w - z_m\|$. So, we obtain

$$\|T_l w - z_m\| \leq \left\| T_l w - \left(1 - \sum_{k=1}^{l-1} t_m^k\right) T_1 z_m - \sum_{k=2}^{l} t_m^{k-1} T_k z_m \right\|$$

$$+ \left\| \left(1 - \sum_{k=1}^{l-1} t_m^k\right) T_1 z_m + \sum_{k=2}^{l} t_m^{k-1} T_k z_m - z_m \right\|$$

$$< \left(1 - \sum_{k=1}^{l-1} t_m^k\right) \|T_l w - T_1 z_m\| + \sum_{k=2}^{l} t_m^{k-1} \|T_l w - T_k z_m\| + t_m^{l-1} \epsilon$$

$$= \left(1 - \sum_{k=1}^{l-1} t_n^k\right) \|T_1 \circ T_l w - T_1 z_m\| + \sum_{k=2}^{l-1} t_m^{k-1} \|T_k \circ T_l w - T_k z_m\|$$

$$+ t_m^{l-1} \|T_l w - T_l z_m\| + t_m^{l-1} \epsilon$$

$$\leq \left(1 - \sum_{k=1}^{l-1} t_m^k\right) \|T_l w - z_m\| + \sum_{k=2}^{l-1} t_m^{k-1} \|T_l w - z_m\|$$

$$+ t_m^{l-1} \|w - z_m\| + t_m^{l-1} \epsilon$$

$$< (1 - t_m^{l-1}) \|T_l w - z_m\| + 2 t_m^{l-1} \epsilon$$

$$< (1 - t_m^{l-1}) \|T_l w - z_m\| + t_m^{l-1} \|T_l w - z_m\|$$

$$= \|T_l w - z_m\|.$$

This is a contradiction. Hence, w is a common fixed point of T_1, T_2, ..., T_l. By induction, we obtain the desired result. □

Lemma 17.9. *Let C be a bounded closed convex subset of a Banach space E. Let $\{T_n : n \in \mathbb{N}\}$ be an infinite family of commuting nonexpansive mappings on C. Let $\{t_n\}$ be a sequence in $(0, \frac{1}{2})$ converging to 0, and let $\{z_n\}$ be a sequence in C such that $\{z_n\}$ converges strongly to some $w \in C$, and*

$$\left(1 - \sum_{k=1}^{\infty} t_n^k\right) T_1 z_n + \sum_{k=2}^{\infty} t_n^{k-1} T_k z_n = z_n$$

for all $n \in \mathbb{N}$. Then, w is a common fixed point of $\{T_n : n \in \mathbb{N}\}$.

Proof. Fix $l \in \mathbb{N}$ with $l \geq 2$. We put $M = 2 \sup\{\|x\| : x \in C\} < \infty$. We have

$$\limsup_{n \to \infty} \frac{\left\| \left(1 - \sum_{k=1}^{l-1} t_n^k\right) T_1 z_n + \sum_{k=2}^{l} t_n^{k-1} T_k z_n - z_n \right\|}{t_n^{l-1}}$$

$$\leq \limsup_{n \to \infty} \left(\frac{\left\| \left(1 - \sum_{k=1}^{\infty} t_n^k\right) T_1 z_n + \sum_{k=2}^{\infty} t_n^{k-1} T_k z_n - z_n \right\|}{t_n^{l-1}} \right.$$

$$+ \frac{\sum\limits_{k=l+1}^{\infty} t_n^{k-1} \|T_1 z_n - T_k z_n\|}{t_n^{l-1}} \Bigg)$$

$$\leq \limsup_{n \to \infty} \sum_{k=l+1}^{\infty} t_n^{k-l} M = \lim_{n \to \infty} \frac{t_n}{1 - t_n} M = 0.$$

So, by Lemma 17.8, we have that w is a common fixed point of $T_1, T_2, ..., T_l$. Since $l \in \mathbb{N}$ is arbitrary, we obtain that w is a common fixed point of $\{T_n : n \in \mathbb{N}\}$. This completes the proof. $\qquad\square$

17.4 Convergence Theorems for Infinite Family of Commuting Nonexpansive Mappings

Theorem 17.10. *Let C be a compact convex subset of a Banach space E. Let $\{T_n : n \in \mathbb{N}\}$ be an infinite family of commuting nonexpansive mappings on C. Fix $\lambda \in (0, 1)$. Let $\{\alpha_n\}$ be an infinite sequence in $[0, \frac{1}{2}]$ satisfying $\liminf\limits_{n \to \infty} \alpha_n = 0$, $\limsup\limits_{n \to \infty} \alpha_n > 0$, and $\lim\limits_{n \to \infty} (\alpha_{n+1} - \alpha_n) = 0$. Define a sequence $\{x_n\}$ in C by $x_1 \in C$ and*

$$x_{n+1} = \lambda \Big(1 - \sum_{k=1}^{n-1} \alpha_n^k \Big) T_1 x_n + \lambda \sum_{k=2}^{n} \alpha_n^{k-1} T_k x_n + (1 - \lambda) x_n$$

for $n \in \mathbb{N}$. Then, $\{x_n\}$ converges strongly to a common fixed point of $\{T_n : n \in \mathbb{N}\}$.

Remark 17.11. We know that $\bigcap\limits_{n=1}^{\infty} F(T_n) \neq \emptyset$ by Theorem DM (stated in chapter 15). We define $\sum\limits_{k=1}^{0} \alpha_1^k = 0$ and $\sum\limits_{k=2}^{1} \alpha_1^{k-1} T_k x_1 = 0$.

Proof. Put $\alpha = \limsup\limits_{n \to \infty} \alpha_n > 0$, $M = \sup\limits_{x \in C} \|x\| < \infty$, and

$$y_n = \Big(1 - \sum_{k=1}^{n-1} \alpha_n^k \Big) T_1 x_n + \sum_{k=2}^{n} \alpha_n^{k-1} T_k x_n$$

for $n \in \mathbb{N}$. We note that $x_{n+1} = \lambda y_n + (1 - \lambda) x_n$ for all $n \in \mathbb{N}$. Since

$$\|y_{n+1} - y_n\| = \Big\| \Big(1 - \sum_{k=1}^{n} \alpha_{n+1}^k \Big) T_1 x_{n+1} + \sum_{k=2}^{n+1} \alpha_{n+1}^{k-1} T_k x_{n+1}$$

$$- \Big(1 - \sum_{k=1}^{n-1} \alpha_n^k\Big) T_1 x_n - \sum_{k=2}^{n} \alpha_n^{k-1} T_k x_n \Big\|$$

$$\le \Big(1 - \sum_{k=1}^{n} \alpha_{n+1}^k\Big) \|T_1 x_{n+1} - T_1 x_n\| + \sum_{k=2}^{n+1} \alpha_{n+1}^{k-1} \|T_k x_{n+1} - T_k x_n\|$$

$$+ \Big| \sum_{k=1}^{n} \alpha_{n+1}^k - \sum_{k=1}^{n} \alpha_n^k \Big| \| - T_1 x_n\| + \sum_{k=2}^{n+1} \Big| \alpha_{n+1}^{k-1} - \alpha_n^{k-1} \Big| \|T_k x_n\|$$

$$+ |\alpha_n^n| \cdot \| - T_1 x_n\| + |\alpha_n^n| \cdot \|T_{n+1} x_n\|$$

$$\le \Big(1 - \sum_{k=1}^{n} \alpha_{n+1}^k\Big) \|x_{n+1} - x_n\| + \sum_{k=2}^{n+1} \alpha_{n+1}^{k-1} \|x_{n+1} - x_n\|$$

$$+ 2M \sum_{k=1}^{n} |\alpha_{n+1}^k - \alpha_n^k| + 2M |\alpha_n^n|$$

$$\le \|x_{n+1} - x_n\| + 8M |\alpha_{n+1} - \alpha_n| + 2M \frac{1}{2^n}$$

for $n \in \mathbb{N}$, we have $\limsup\limits_{n \to \infty}(\|y_{n+1} - y_n\| - \|x_{n+1} - x_n\|) \le 0$. So, by Lemma 6.33, we have $\lim\limits_{n \to \infty} \|x_n - y_n\| = 0$. Fix $t \in \mathbb{R}$ with $0 < t < \alpha$. Then by Lemma 17.1, there exists a subsequence $\{\alpha_{n_k}\}$ of $\{\alpha_n\}$ converging to t. Since C is compact, there exists a subsequence $\{x_{n_{k_j}}\}$ of $\{x_{n_k}\}$ converging to some point $z_t \in C$. We have

$$\Big\| \Big(1 - \sum_{l=1}^{\infty} t^l\Big) T_1 z_t + \sum_{l=2}^{\infty} t^{l-1} T_l z_t - z_t \Big\|$$

$$\le \Big\| \Big(1 - \sum_{l=1}^{\infty} t^l\Big) T_1 z_t + \sum_{l=2}^{\infty} t^{l-1} T_l z_t - y_n \Big\| + \|y_n - x_n\|$$

$$+ \|x_n - z_t\|$$

$$= \Big\| \Big(1 - \sum_{l=1}^{\infty} t^l\Big) T_1 z_t + \sum_{l=2}^{\infty} t^{l-1} T_l z_t$$

$$- \Big(1 - \sum_{l=1}^{n-1} \alpha_n^l\Big) T_1 x_n - \sum_{l=2}^{n} \alpha_n^{l-1} T_l z_t \Big\| + \|y_n - x_n\|$$

$$+ \|x_n - z_t\|$$

$$\le \Big(1 - \sum_{l=1}^{\infty} t^l\Big) \|T_1 z_t - T_1 x_n\| + \sum_{l=2}^{\infty} t^{l-1} \|T_l z_t - T_l x_n\|$$

$$+ \left| \sum_{l=1}^{n-1} t^l - \sum_{l=1}^{n-1} \alpha_n^l \right| \| - T_1 x_n \| + \sum_{l=2}^{n} \left| t^{l-1} - \alpha_n^{l-1} \right| \| T_l x_n \|$$

$$+ \sum_{l=n}^{\infty} t^l \| - T_1 x_n \| + \sum_{l=n+1}^{\infty} t^{l-1} \| T_l x_n \| + \| y_n - x_n \|$$

$$+ \| x_n - z_t \|$$

$$\leq \left(1 - \sum_{l=1}^{\infty} t^l \right) \| z_t - x_n \| + \sum_{l=2}^{\infty} t^{l-1} \| z_t - x_n \|$$

$$+ \left| \sum_{l=1}^{n-1} t^l - \sum_{l=1}^{n-1} \alpha_n^l \right| M + \sum_{l=2}^{n} |t^{l-1} - \alpha_n^{l-1}| M$$

$$+ \sum_{l=n}^{\infty} t^l M + \sum_{l=n+1}^{\infty} t^{l-1} M + \| y_n - x_n \| + \| x_n - z_t \|$$

$$\leq 2 \| z_t - x_n \| + 8 |t - \alpha_n| M + \frac{2t^n}{1-t} M + \| y_n - x_n \|$$

for $n \in \mathbb{N}$, and hence

$$\left\| \left(1 - \sum_{l=1}^{\infty} t^l \right) T_1 z_t + \sum_{l=2}^{\infty} t^{l-1} T_l z_t - z_t \right\|$$

$$\leq \limsup_{j \to \infty} \left(2 \| z_t - x_{n_{k_j}} \| + 8 |t - \alpha_{n_{k_j}}| M \right.$$

$$\left. + \frac{2t^{n_{k_j}}}{1-t} M + \| y_{n_{k_j}} - x_{n_{k_j}} \| \right) = 0.$$

Therefore we have $\left(1 - \sum_{l=1}^{\infty} t^l \right) T_1 z_t + \sum_{l=2}^{\infty} t^{l-1} T_l z_t = z_t$ for all $t \in \mathbb{R}$ with $0 < t < \alpha$. Since C is compact, there is a real sequence $\{t_n\}$ in $(0, \alpha)$ such that $\lim_{n \to \infty} t_n = 0$, and $\{z_{t_n}\}$ converges strongly to some point $w \in C$. By Lemma 17.9, we obtain that such w is a common fixed point of $\{T_n : n \in \mathbb{N}\}$. We note that w is a cluster point of $\{x_n\}$ because so are $\{z_{t_n}\}$ for all $n \in \mathbb{N}$. Hence, $\liminf_{n \to \infty} \| x_n - w \| = 0$. We also have $\{\| x_n - w \|\}$ is non-increasing by Lemma 17.3. Thus, $\lim_{n \to \infty} \| x_n - w \| = 0$. This completes the proof. $\qquad \square$

Theorem 17.12. *Let C be a compact convex subset of a Banach space E. Let $\{T_n : n \in \mathbb{N}\}$ be an infinite family of commuting nonexpansive mappings on C. Fix $\lambda \in (0, 1)$. Let $\{\alpha_n\}$ be a sequence in $[0, \frac{1}{2}]$ satisfying*

$$\liminf_{n \to \infty} \alpha_n = 0, \ \limsup_{n \to \infty} \alpha_n > 0, \ \text{and} \ \lim_{n \to \infty} (\alpha_{n+1} - \alpha_n) = 0.$$

Define a sequence $\{x_n\}$ in C by $x_1 \in C$ and

$$x_{n+1} = \lambda\Big(1 - \sum_{k=1}^{\infty} \alpha_n^k\Big)T_1 x_n + \lambda \sum_{k=2}^{\infty} \alpha_n^{k-1} T_k x_n + (1-\lambda)x_n$$

for $n \in \mathbb{N}$. Then, $\{x_n\}$ converges strongly to a common fixed point of $\{T_n : n \in \mathbb{N}\}$.

As direct consequences, we obtain the following.

Theorem 17.13. Let C be a compact convex subset of a Banach space E. Let S and T be nonexpansive mappings on C with $ST = TS$. Let $\{\alpha_n\}$ be a sequence in $[0,1]$ satisfying $\liminf\limits_{n\to\infty} \alpha_n = 0$, $\limsup\limits_{n\to\infty} \alpha_n > 0$, and $\lim\limits_{n\to\infty}(\alpha_{n+1} - \alpha_n) = 0$. Define a sequence $\{x_n\}$ in C by $x_1 \in C$ and

$$x_{n+1} = \frac{1-\alpha_n}{2}Sx_n + \frac{\alpha_n}{2}Tx_n + \frac{1}{2}x_n$$

for $n \in \mathbb{N}$. Then, $\{x_n\}$ converges strongly to a common fixed point of S and T.

Theorem 17.14. Let C be a compact convex subset of a Banach space E. Let $l \in \mathbb{N}$ with $l \geq 2$ and let $\{T_1, T_2, ..., T_l\}$ be a finite family of commuting nonexpansive mappings on C. Let $\{\alpha_n\}$ be a sequence in $[0, \frac{1}{2}]$ satisfying

$$\liminf\limits_{n\to\infty} \alpha_n = 0, \quad \limsup\limits_{n\to\infty} \alpha_n > 0, \quad \text{and} \quad \lim\limits_{n\to\infty}(\alpha_{n+1} - \alpha_n) = 0.$$

Define a sequence $\{x_n\}$ in C by x_1 and

$$x_{n+1} = \frac{1}{2}\Big(1 - \sum_{k=1}^{l-1} \alpha_n^k\Big)T_1 x_n + \frac{1}{2}\sum_{k=2}^{l} \alpha_n^{k-1} T_k x_n + \frac{1}{2}x_n$$

for $n \in \mathbb{N}$. Then, $\{x_n\}$ converges strongly to a common fixed point of $\{T_1, T_2, ..., T_l\}$.

17.5 Historical Remarks

All the results of this chapter are due to Suzuki [463].

Chapter 18
Finite Families of Lipschitz Pseudo-contractive and Accretive Mappings

18.1 Introduction

In this chapter, we study an iteration process for approximating a common fixed point (assuming existence) for a family of Lipschitz pseudo-contractive mappings in arbitrary real Banach spaces.

18.2 Convergence Theorems

Let K be a nonempty closed convex subset of a real normed linear space. Let $T_1, T_2, ..., T_m : K \to K$ be m Lipschitz maps. We define the iterative sequence $\{x_n\}_{n=1}^{\infty}$ by

$$
\begin{cases}
u, \quad x_1 \in K \\
x_{n+1} = (1 - \lambda_n)x_n + \lambda_n T_1 x_n - \lambda_n \theta_n (x_n - u), \; if \; m = 1, n \geq 1. \\
\\
u, \quad x_1 \in K \\
x_{n+1} = (1 - \lambda_n)x_n + \lambda_n T_1 y_{1n} - \lambda_n \theta_n (x_n - u) \\
y_{1n} = (1 - \lambda_n)x_n + \lambda_n T_2 y_{2n} \\
\vdots \\
y_{(m-2)n} = (1 - \lambda_n)x_n + \lambda_n T_{m-1} y_{(m-1)n} \\
y_{(m-1)n} = (1 - \lambda_n)x_n + \lambda_n T_m x_n, \quad m \geq 2, \quad n \geq 1.
\end{cases}
\tag{18.1}
$$

We prove the following theorems.

Theorem 18.1. *Let K be a nonempty closed convex subset of a real Banach space E. Let $T_1, T_2, ..., T_m : K \to K$ be m Lipschitz mappings such that $F := \bigcap_{i=1}^{m} F(T_i) = \bigcap_{i=1}^{m} \{x \in K : T_i x = x\} \neq \emptyset$. Let $\{\lambda_n\}$ and $\{\theta_n\}$ be sequences in $(0,1)$ such that $\lambda_n = o(\theta_n)$, $\sum \lambda_n \theta_n = \infty$ and $\lambda_n(1+\theta_n) < 1$. Let $\{x_n\}_{n=1}^{\infty}$ be* *iteratively generated by (18.1). Suppose that T_1 is pseudo-contractive. Then, the sequence $\{x_n\}_{n=1}^{\infty}$ is bounded.*

Proof. For $m = 1$, the proof follows as in the proof of Theorem 11.5, by replacing x_1 by u in the iteration process used there. Now, for $m \geq 2$, let $p \in F$ be fixed. If $x_n = p \ \forall \, n \geq 1$, then we are done. So, let $n_0 \in \mathbb{N}$ be the first integer such that $x_{n_0} \neq p$. Clearly, there exist $N^* \geq n_0$, $r > 0$ and $M > 0$ such that $x_{N^*} \in B_r(p) := \{x \in K : \|x - p\| \leq r\}$, $u \in B_{\frac{r}{2}}(p)$ and $\forall \, n \geq N^*$, $\frac{1-\lambda_n^m L^m}{1-\lambda_n L} \leq M$, and $\frac{\lambda_n}{\theta_n} \leq \frac{1}{2[2LM+5+(2LM+\frac{7}{2})L]}$. It suffices to show by induction that $\{x_n\}$ belongs to $B := \overline{B_r(p)}$ for all integers $n \geq N^*$. By construction, we have that $x_{N^*} \in B$. Hence, we may assume $x_n \in B$ for any $n \geq N^*$ and prove that $x_{n+1} \in B$. Suppose x_{n+1} is not in B. Then $\|x_{n+1} - p\| > r$. From (18.1) we have the following estimates:

$$
\begin{aligned}
\|y_{1n} - p\| &= \|(1 - \lambda_n)(x_n - p) + \lambda_n(T_2 y_{2n} - p)\| \\
&\leq (1 - \lambda_n)\|x_n - p\| + \lambda_n L\|y_{2n} - p\| \\
&\leq \|x_n - p\| + \lambda_n L\big[(1 - \lambda_n)\|x_n - p\| + \lambda_n L\|T_3 y_{3n} - p\|\big] \\
&\leq \|x_n - p\| + \lambda_n L\|x_n - p\| + (\lambda_n L)^2\|y_{3n} - p\|
\end{aligned}
$$

$$
\begin{aligned}
&\leq \|x_n - p\| + \lambda_n L\|x_n - p\| + (\lambda_n L)^2\big[\|x_n - p\| + \lambda_n L\|y_{4n} - p\|\big] \\
&= \|x_n - p\| + \lambda_n L\|x_n - p\| + (\lambda_n L)^2\|x_n - p\| + (\lambda_n L)^3\|y_{4n} - p\| \\
&\ \ \vdots \\
&\leq \big(1 + \lambda_n L + (\lambda_n L)^2 + (\lambda_n L)^3 + ... + (\lambda_n L)^{m-1}\big)\|x_n - p\| \\
&\leq \Big(\frac{1 - (\lambda_n L)^m}{1 - \lambda_n L}\Big)\|x_n - p\| \\
&\leq M\|x_n - p\|.
\end{aligned}
$$

Similarly, we obtain that

$$
\|y_{2n} - p\| \leq M\|x_n - p\|.
$$

Furthermore,

$$
\begin{aligned}
\|T_1 x_{n+1} - T_1 y_{1n}\| &\leq L\|x_{n+1} - y_{1n}\| = L\|x_{n+1} - x_n + \lambda_n(x_n - T_2 y_{2n})\| \\
&= L\|\lambda_n[T_1 y_{1n} - x_n - \theta_n(x_n - u)] + \lambda_n(x_n - T_2 y_{2n})\| \\
&\leq \lambda_n L\big[\|T_1 y_{1n} - x_n\| + \theta_n\|x_n - u\| + \|x_n - T_2 y_{2n}\|\big] \\
&\leq \lambda_n L\big[L\|y_{1n} - p\| + \|p - x_n\| + \theta_n\|x_n - p\| \\
&\quad + \theta_n\|p - u\| + \|x_n - p\| + L\|p - y_{2n}\|\big] \\
&\leq \lambda_n L\big[(2LM + 3)\|x_n - p\| + \|u - p\|\big].
\end{aligned}
$$

Again, from (18.1) and using inequality (4.4), we obtain that

$$
\begin{aligned}
\|x_{n+1} - p\|^2 &= \|x_n - p - \lambda_n[(x_n - T_1 y_{1n}) + \theta_n(x_n - u)]\|^2 \\
&\leq \|x_n - p\|^2 - 2\lambda_n\langle x_n - T_1 y_{1n} + \theta_n(x_n - u), j(x_{n+1} - p)\rangle \\
&\leq \|x_n - p\|^2 - 2\lambda_n\theta_n\|x_{n+1} - p\|^2 \\
&\quad + 2\lambda_n\theta_n\langle (x_{n+1} - x_n) + (u - p), j(x_{n+1} - p)\rangle \\
&\quad - 2\lambda_n\langle x_n - T_1 y_{1n}, j(x_{n+1} - p)\rangle \\
&= \|x_n - p\|^2 - 2\lambda_n\theta_n\|x_{n+1} - p\|^2 \\
&\quad + 2\lambda_n\theta_n\langle (x_{n+1} - x_n) + (u - p), j(x_{n+1} - p)\rangle \\
&\quad - 2\lambda_n\langle (x_n - x_{n+1}) + (x_{n+1} - T_1 x_{n+1}) \\
&\quad + (T_1 x_{n+1} - T_1 y_{1n}), j(x_{n+1} - p)\rangle \\
&\leq \|x_n - p\|^2 - 2\lambda_n\theta_n\|x_{n+1} - p\|^2 \\
&\quad + 2\lambda_n\theta_n\big[\|x_{n+1} - x_n\| + \|u - p\|\big]\|x_{n+1} - p\| \\
&\quad + 2\lambda_n\big[\|x_{n+1} - x_n\| + \|T_1 x_{n+1} - T_1 y_{1n}\|\big]\|x_{n+1} - p\| \\
&\quad - 2\lambda_n\langle x_{n+1} - T_1 x_{n+1}, j(x_{n+1} - p)\rangle.
\end{aligned}
$$

But $\langle x_{n+1} - T_1 x_{n+1}, j(x_{n+1} - p)\rangle \geq 0$ (since T_1 is pseudocontractive). Thus,

$$
\begin{aligned}
\|x_{n+1} - p\|^2 &\leq \|x_n - p\|^2 - 2\lambda_n\theta_n\|x_{n+1} - p\|^2 \\
&\quad + \big[4\lambda_n\|x_{n+1} - x_n\| + 2\lambda_n\theta_n\|u - p\| \\
&\quad + \|T_1 x_{n+1} - T_1 y_{1n}\|\big]\|x_{n+1} - p\| \\
&\leq \|x_n - p\|^2 - 2\lambda_n\theta_n\|x_{n+1} - p\|^2 \\
&\quad + \Big[4\lambda_n\|\lambda_n(T_1 y_{1n} - x_n + \theta_n(x_n - u))\| + 2\lambda_n\theta_n\|u - p\| \\
&\quad + 2\lambda_n\Big(\lambda_n L\{(2LM + 3)\|x_n - p\| + \|u - p\|\}\Big)\Big]\|x_{n+1} - p\| \\
&\leq \|x_n - p\|^2 - 2\lambda_n\theta_n\|x_{n+1} - p\|^2 \\
&\quad + 4\lambda_n^2\Big[LM\|x_n - p\| + 2\|x_n - p\| + \|u - p\|\Big]\|x_{n+1} - p\| \\
&\quad + 2\lambda_n\theta_n\|u - p\|\|x_{n+1} - p\| \\
&\quad + 2\lambda_n^2 L\Big((2LM + 3)\|x_n - p\| + \|u - p\|\Big)\|x_{n+1} - p\|.
\end{aligned}
$$

Now, since $\|x_{n+1} - p\| > r$, we have that

$$
2\lambda_n\theta_n\|x_{n+1} - p\| \leq 4\lambda_n^2\big[LMr + 2r + \tfrac{r}{2}\big] + 2\lambda_n\theta_n\tfrac{r}{2} + 2\lambda_n^2 L\big[(2LM + 3)r + \tfrac{r}{2}\big].
$$

Thus,

$$\|x_{n+1} - p\| \le \frac{\lambda_n}{\theta_n}\left[2LM + 5 + \left(2LM + \frac{7}{2}\right)L\right]r + \frac{r}{2}.$$

which implies $\|x_{n+1} - p\| \le r$, a contradiction. So, $x_n \in B \ \forall \, n \ge N^*$. Hence, $\{x_n\}_{n=1}^{\infty}$ is bounded. This completes the proof. □

Remark 18.2. Since $\{x_n\}_{n=1}^{\infty}$ is bounded, there exists $\delta > 0$ sufficiently large such that $x_n \in \overline{B} := B_\delta(p) \ \forall \, n \in \mathbb{N}$. Furthermore, the set $K \cap \overline{B}$ is a bounded closed and convex nonempty subset of E. If we define a map $\varphi : E \to \mathbb{R}$ by

$$\varphi(y) = \mu_n\|x_n - y\|^2,$$

(where μ is a Banach limit) then φ is continuous, convex and $\varphi(y) \to +\infty$ as $\|y\| \to +\infty$. Thus, if E is a reflexive Banach space, then there exists $z^* \in K \cap \overline{B}$ such that

$$\varphi(z^*) = \min_{y \in K \cap \overline{B}} \varphi(y).$$

So, the set

$$K_{min} = \left\{x \in K \cap \overline{B} : \varphi(x) = \min_{y \in K \cap \overline{B}} \varphi(y)\right\} \ne \emptyset.$$

Remark 18.3. For the remainder of this chapter, $\{\lambda_n\}_{n=1}^{\infty}$ and $\{\theta_n\}_{n=1}^{\infty}$ will be as in Theorem 18.1.

Theorem 18.4. *Let K be a nonempty closed convex subset of a real Banach space E. Let $T_1, T_2,..., T_m : K \to K$ be m Lipschitz mappings such that $F := \bigcap_{i=1}^{m} F(T_i) = \bigcap_{i=1}^{m} \{x \in K : T_i x = x\} \ne \emptyset$. Let T_1 be pseudo-contractive. Let $\{x_n\}_{n=1}^{\infty}$ be iteratively generated by (18.1). Suppose that $K_{min} \cap F \ne \emptyset$. Then, the sequence $\{x_n\}_{n=1}^{\infty}$ converges strongly to a common fixed point of $\{T_i\}_{i=1}^{m}$.*

Proof. Let $\alpha \in (0,1)$ and $z \in K_{min} \cap F$. Then by convexity of $K \cap \overline{B}$ we have that $(1-\alpha)z + \alpha u \in K \cap \overline{B}$. Thus, $\varphi(z) \le \varphi((1-\alpha)z + \alpha u)$. Again, using inequality (4.4), we obtain that

$$\|x_n - z - t(u-z)\|^2 \le \|x_n - z\|^2 - 2\alpha\langle u - z, j(x_n - z - \alpha(u-z))\rangle \ \forall \, n \in \mathbb{N}.$$

Taking Banach limit over $n \in \mathbb{N}$ gives

$$\mu_n\|x_n - z - t(u-z)\|^2 \le \mu_n\|x_n - z\|^2 - 2\alpha\mu_n\langle u - z, j(x_n - z - \alpha(u-z))\rangle.$$

That is,

$$\varphi((1-\alpha)z + \alpha u) \le \varphi(z) - 2\alpha\mu_n\langle u - z, j(x_n - z - \alpha(u-z))\rangle.$$

This implies, $\mu_n\langle u - z, j(x_n - z - \alpha(u - z))\rangle \leq 0 \; \forall \; n \in \mathbb{N}$. Furthermore, the fact that the normalized duality mapping is norm-to-$weak^*$ uniformly continuous on bounded subsets of E gives, as $\alpha \to 0$, that

$$\langle u - z, j(x_n - z)\rangle - \langle u - z, j(x_n - z - \alpha(u - z))\rangle \to 0.$$

Thus, for all $\epsilon > 0$, there exists $\delta_\epsilon > 0$ such that for all $\alpha \in]0, \delta_\epsilon[$ and for all $n \in \mathbb{N}$,

$$\langle u - z, j(x_n - z)\rangle - \langle u - z, j(x_n - z - \alpha(u - z))\rangle < \epsilon.$$

Thus,

$$\mu_n\langle u - z, j(x_n - z)\rangle - \mu_n\langle u - z, j(x_n - z - \alpha(u - z))\rangle \leq \epsilon,$$

which implies (since $\epsilon > 0$ is arbitrary) that

$$\mu_n\langle u - z, j(x_n - z)\rangle \leq 0.$$

Moreover, since both $\{x_n\}_{n \geq 1}$ and $\{T_1 y_{1n}\}_{n \geq 1}$ are bounded, we have that

$$\|x_{n+1} - x_n\| \leq \lambda_n(\|x_n\| + \|T_1 y_{1n}\| + \theta_n\|x_n - u\|) \leq \lambda_n M_0,$$

for some constant $M_0 \geq 0$. Thus, $\lim_{n \to \infty} \|x_{n+1} - x_n\| = 0$, since $\lambda_n \to 0$ as $n \to \infty$. Again, using the fact that the normalized duality mapping j is norm-to-weak* uniformly continuous on bounded subsets of E, we have that

$$\lim_{n \to \infty} \left(\langle u - z, j(x_{n+1} - z)\rangle - \langle u - z, j(x_n - z)\rangle\right) = 0.$$

The sequence $\{\langle u - z, j(x_n - z)\rangle\}_{n \geq 1}$, therefore, satisfies the conditions of Lemma 7.7. Hence,

$$\limsup_{n \to \infty} \langle u - z, j(x_n - z)\rangle \leq 0.$$

Define

$$\sigma_n = \max\{\langle u - z, j(x_n - z)\rangle, 0\}.$$

Then $\lim_{n \to \infty} \sigma_n = 0$ and

$$\langle u - z, j(x_n - z)\rangle \leq \sigma_n \; \forall \; n \in \mathbb{N}. \tag{18.2}$$

Again, using inequality (4.4), recursion formulas (18.1) and (18.2), we obtain

$$\|x_{n+1} - z\|^2 = \|x_n - z - \lambda_n[x_n - T_1 y_{1n} + \theta_n(x_n - u)]\|^2$$
$$\leq \|x_n - z\|^2 - 2\lambda_n\langle x_n - T_1 y_{1n} + \theta_n(x_n - u), j(x_{n+1} - z)\rangle$$

$$\leq \|x_n - z\|^2 - 2\lambda_n\theta_n\|x_{n+1} - z\|^2$$

$$+2\lambda_n\theta_n\langle(x_{n+1} - x_n) + u - z, j(x_{n+1} - z)\rangle$$

$$-2\lambda_n\langle(x_n - x_{n+1}) + (x_{n+1} - T_1x_{n+1})$$

$$+(T_1x_{n+1} - T_1y_{1n}), j(x_{n+1} - z)\rangle$$

$$\leq \|x_n - z\|^2 - 2\lambda_n\theta_n\|x_{n+1} - z\|^2 + (\lambda_n{}^2\theta_n + \lambda_n^2)M_1$$

$$+2\lambda_n\theta_n\langle u - z, j(x_{n+1} - z)\rangle$$

$$\leq \|x_n - z\|^2 - 2\lambda_n\theta_n\|x_{n+1} - z\|^2 + \lambda_n{}^2(\theta_n + 1)M_1$$

$$+2\lambda_n\theta_n\sigma_{n+1}$$

$$= \|x_n - z\|^2 - 2\lambda_n\theta_n\|x_{n+1} - z\|^2 + \delta_n,$$

where $\delta_n := \lambda_n{}^2(\theta_n+1)M_1+2\lambda_n\theta_n\sigma_{n+1} = o(\lambda_n\theta_n)$ for some constant $M_1 > 0$. Hence, by Lemma 8.17, we have that the sequence $\{x_n\}_{n=1}^\infty$ converges strongly to $z \in F$. This completes the proof. $\qquad\square$

Corollary 18.5 *Let K be a nonempty closed convex subset of a real Banach space E. Let T_1, T_2, ..., $T_m : K \to K$ be m Lipschitz pseudo-contractive mappings such that $F := \bigcap_{i=1}^m F(T_i) \neq \emptyset$. Let the sequence $\{x_n\}_{n=1}^\infty$ be iteratively generated by (18.1). Then, the sequence $\{x_n\}_{n=1}^\infty$ converges strongly to a common fixed point of T_i, $i = 1, 2, ..., m$, provided $K_{min} \cap F \neq \emptyset$.*

Proof. Boundedness of $\{x_n\}_{n=1}^\infty$ follows from Theorem 18.1. The rest follows as in Theorem 18.4.

We showed in Chapter 10 that the class of mappings which are strictly pseudo-contractive in the sense of Browder and Petryshyn is a proper subclass of the class of Lipschitz pseudo-contractions. Thus, we have the following corollary. $\qquad\square$

Corollary 18.6 *Let K be a nonempty closed convex subset of a real Banach space E. Let T_1, T_2, ..., $T_m : K \to K$ be m mappings which are strictly pseudo-contractive in the sense of Browder and Petryshyn such that $F := \bigcap_{i=1}^m F(T_i) \neq \emptyset$. Let the sequence $\{x_n\}_{n=1}^\infty$ be iteratively generated by (18.1). Then, the sequence $\{x_n\}_{n=1}^\infty$ converges strongly to a common fixed point of T_i, $i = 1, 2, ..., m$, provided $K_{min} \cap F \neq \emptyset$.*

We now consider a situation where the condition $K_{min} \cap F \neq \emptyset$ is easily satisfied.

Corollary 18.7 *Let E be a real strictly convex Banach space with a uniformly Gâteaux differentiable norm possessing uniform normal structure. Let K be a nonempty closed convex subset of E and let T_1, T_2,...,$T_m : K \to K$ be m nonexpansive mappings such that $F := \bigcap_{i=1}^m F(T_i) = \bigcap_{i=1}^m \{x \in K : T_ix =$*

$x\} \neq \emptyset$. For $\gamma_i \in (0,1)$, define $S_i = (1 - \gamma_i)I + \gamma_i T_i$, $i = 1, 2, ..., m$ and put $G := S_m S_{m-1}...S_1$. Let the sequence $\{x_n\}_{n=1}^{\infty}$ be iteratively generated by (18.1). Then, $\{x_n\}_{n=1}^{\infty}$ converges strongly to a common fixed point of T_1, $T_2,...,T_m$.

Proof. First observe that $G := S_m S_{m-1}...S_1$ is nonexpansive. Thus, the set $K_{min} = \{x \in K \cap \overline{B} : \varphi(x) = \min_{y \in K \cap \overline{B}} \varphi(y)\}$ is a nonempty closed convex subset of K that has property (P) (shown in Lim and Xu [307] since every nonexpansive mapping is asymptotically nonexpansive). It is known that $\bigcap_{i=1}^{r} F(S_i) = \bigcap_{i=1}^{r} F(T_i)$ and $\bigcap_{i=1}^{r} F(S_i) = F(S_m S_{m-1}...S_1) = F(G)$ (see, e.g., Chidume *et al.* [184], Takahashi *et al.* [481]). Hence, $K_{min} \cap F(G) \neq \emptyset$ and $K_{min} \cap F(G) = K_{min} \cap F$. The rest follows as in the proof of Theorem 18.1 and Theorem 18.4. This completes the proof. $\square$

18.3 Finite Families of Lipschitz Accretive Operators

Let $A_1, A_2, ..., A_m : E \to E$ be m Lipschitz *accretive* mappings. We introduce the following iteration process for the approximation of common zero (assuming existence) of this family of mappings:

$$\begin{cases} u, \ x_1 \in E, \\ x_{n+1} = (1 - \lambda_n)x_n + \lambda_n x_n - \lambda_n A_1 x_n \\ \quad -\lambda_n \theta_n(x_n - u), \ if \ m = 1, n \geq 1. \\ x_{n+1} = (1 - \lambda_n)x_n + \lambda_n y_{1n} - \lambda_n A_1 y_{1n} - \lambda_n \theta_n(x_n - u) \\ y_{1n} = (1 - \lambda_n)x_n + \lambda_n y_{2n} - \lambda_n A_2 y_{2n} \\ \vdots \\ y_{(m-2)n} = (1 - \lambda_n)x_n + \lambda_n y_{(m-1)n} - \lambda_n A_{m-1} y_{(m-1)n} \\ y_{(m-1)n} = x_n - \lambda_n A_m x_n, \ m \geq 2, \ n \geq 1, \end{cases} \quad (18.3)$$

where $\{\lambda_n\}_{n=1}^{\infty}$ and $\{\theta_n\}_{n=1}^{\infty}$ are as above.

Theorem 18.8. *Let E be a real Banach space with uniformly Gâteaux differentiable norm, $A_1, A_2, ..., A_m : E \to E$ be m Lipschitz accretive operators such that $N := \bigcap_{i=1}^{m} N(A_i) := \bigcap_{i=1}^{m} \{x \in E : A_i x = 0\} \neq \emptyset$. Let $\{x_n\}_{n \geq 1}$ be iteratively generated by (18.3). Then, $\{x_n\}_{n \geq 1}$ is bounded. Moreover, if E is a reflexive Banach space with uniformly Gâteaux differentiable norm and $K_{min} \cap N \neq \emptyset$ (where K_{min} is constructed as in Remark 18.2), then, $\{x_n\}_{n \geq 1}$ converges strongly to a common zero of $A_1, A_2, ..., A_m$.*

Proof. Recall that A_i is accretive if and only if $T_i := I - A_i$ is pseudocontractive, $i = 1, 2, ..., m$. Then, if we replace A_i by $I - T_i$, $i = 1, 2, ..., m$

in (18.3), then (18.3) reduces to (18.1) and boundedness of $\{x_n\}$ therefore follows as in the proof of Theorem 18.1. Furthermore, $p^* \in K_{min} \cap N$ implies $A_i p^* = 0$, $i = 1, 2, ..., m$. This implies $T_i p^* = p^*$. Thus, $K_{min} \cap \bigcap_{i=1}^{m} F(T_i) \neq \emptyset$, $i = 1, 2, ..., m$. The rest follows as in the proof of Theorem 18.4. This completes the proof.   □

18.4 Some Applications

Following the methods of proof used in the last section, the following theorems are obtained.

Theorem 18.9. *Let K be a nonempty closed and convex subset of a real reflexive Banach space with uniformly Gâteaux differentiable norm. Let $T_1, T_2, ..., T_m : K \rightarrow K$ be m Lipschitz hemi-contractive mappings such that $F^* := \bigcap_{i=1}^{m} F(T_i) \neq \emptyset$. Let $\{x_n\}_{n \geq 1}$ be iteratively generated by (18.1). Suppose that $K_{min} \cap F^* \neq \emptyset$. Then $\{x_n\}_{n \geq 1}$ converges strongly to some $p^* \in F^*$.*

Proof. Boundedness of $\{x_n\}_{n \geq 1}$ follows as in the proof of Theorem 18.1. The rest then follows as in the proof of Theorem 18.4.   □

Theorem 18.10. *Let E be a real reflexive Banach space with uniformly Gâteaux differentiable norm. Let $A_1, A_2, ..., A_m : E \rightarrow E$ be m Lipschitz quasi-accretive operators such that $N^* := \bigcap_{i=1}^{m} N(A_i) \neq \emptyset$. Let $\{x_n\}_{n \geq 1}$ be iteratively generated by (18.3). Suppose that $K_{min} \cap N^* \neq \emptyset$. Then, $\{x_n\}_{n \geq 1}$ converges strongly to some $x^* \in N^*$.*

Proof. The mapping A_i is quasi-accretive if and only if $T_i := (I - A_i)$ is hemi-contractive $i = 1, 2, ..., m$. Thus, replacing A_i by $I - T_i$ in (18.3), the result follows as in Theorem 18.8.   □

Remark 18.11. Prototypes for the iteration parameters of this chapter are, for example, $\lambda_n = \frac{1}{(n+1)^a}$, and $\theta_n = \frac{1}{(n+1)^b}$, where $0 < a + b < 1$ and $a > b$.

18.5 Historical Remarks

Related results for strong convergence for a common zero of a finite family of m-accretive mappings and α-inverse strongly accretive mappings can be found in Zegeye and Shahzad ([541], [545]). All the results of this chapter are due to Ofoedu and Shehu [364].

Chapter 19
Generalized Lipschitz Pseudo-contractive and Accretive Mappings

19.1 Introduction

In this chapter, we construct an iterative sequence for the approximation of common fixed points of finite families of *generalized Lipschitz* pseudo-contractive and generalized Lipschitz accretive operators (assuming existence). These classes of mappings have been defined in Chapter 12. Furthermore, the iteration scheme introduced here and the method of proof are of independent interest.

19.2 Generalized Lipschitz Pseudo-contractive Mappings

Let K be a nonempty closed convex subset of a real normed space E. Let $T_1, T_2, ..., T_m : K \to K$ be m generalized Lipschitz mappings . We define the iterative sequence $\{x_n\}$ by

$$
\begin{cases}
u, \ x_1 \in K, \\
x_{n+1} \quad = (1 - \alpha_n \lambda_n)x_n + \alpha_n \lambda_n T_1 y_{1n} - \lambda_n \theta_n (x_n - u) \\
y_{1n} \quad = (1 - \alpha_n)x_n + \alpha_n T_2 y_{2n} \\
\vdots \\
y_{(m-2)n} \quad = (1 - \alpha_n)x_n + \alpha_n T_{m-1} y_{(m-1)n} \\
y_{(m-1)n} \quad = (1 - \alpha_n)x_n + \alpha_n T_m x_n, \ m \geq 2, \ n \geq 1,
\end{cases}
\tag{19.1}
$$

where $\{\alpha_n\}_{n=1}^{\infty}$, $\{\lambda_n\}_{n=1}^{\infty}$ and $\{\theta_n\}_{n=1}^{\infty}$ are sequence in $(0, 1)$ and u, x_1 are fixed vectors in K.

C. Chidume, *Geometric Properties of Banach Spaces and Nonlinear Iterations*,
Lecture Notes in Mathematics 1965,
© Springer-Verlag London Limited 2009

We prove the following theorems.

Theorem 19.1. *Let K be a nonempty closed convex subset of a real Banach space. Let $T_1, T_2, ..., T_m : K \to K$ be m generalized Lipschitz mappings such that $F := \cap_{i=1}^{m} F(T_i) = \cap_{i=1}^{m} \{x \in K : T_i x = x\} \neq \emptyset$. Let $\{\alpha_n\}_{n=1}^{\infty}$, $\{\lambda_n\}_{n=1}^{\infty}$ and $\{\theta_n\}_{n=1}^{\infty}$ be sequences in $(0, 1)$ such that $\alpha_n = o(\theta_n)$; $\lim_{n \to \infty} \lambda_n = 0$ and $\lambda_n(\alpha_n + \theta_n) < 1$. Let $\{x_n\}_{n \geq 1}$ be iteratively generated by (19.1). Suppose that T_1 is pseudo-contractive. Then, the sequence $\{x_n\}_{n=1}^{\infty}$ is bounded.*

Proof. Fix $p \in F$. If $x_n = p$ for all integers $n \geq 1$, then we are done. So, let $n_0 \in \mathbb{N}$ be the first integer such that $x_{n_0} \neq p$. Clearly, there exist $N^* \geq n_0$, $r > 0$ and $M > 0$ such that $x_{N^*} \in B_r(p) := \{x \in K : \|x-p\| \leq r\}$, $u \in B_{\frac{r}{2}}(p)$ and for all $n \geq N^*$, $\alpha_n < \min\{\frac{1}{L}, \theta_n\}$, $\lambda_n < \frac{r}{8[L(1+Mr+M)+\frac{5}{2}r]}$, $\frac{\alpha_n}{\theta_n} < \frac{r}{8[1+2L(1+Mr+M)+\frac{7}{2}r]}$ and $\frac{1-\alpha_n^m L^m}{1-\alpha_n L} \leq M$. It suffices to show by induction that $x_n \in B := B_r(p)$ for all $n \geq N^*$. By construction, $x_{N^*} \in B$. Suppose that $x_n \in B$ for $n > N^*$, we prove that $x_{n+1} \in B$. Suppose for contradiction that x_{n+1} is not in B. Then $\|x_{n+1} - p\| > r$. From (19.1), we have the following estimates:

$$
\begin{aligned}
\|y_{1n} - p\| &= \|(1 - \alpha_n)(x_n - p) + \alpha_n(T_2 y_{2n} - p)\| \\
&\leq (1 - \alpha_n)\|x_n - p\| + \alpha_n L(1 + \|y_{2n} - p\|) \\
&\leq \|x_n - p\| + \alpha_n L\|(1 - \alpha_n)(x_n - p) + \alpha_n(T_3 y_{3n} - p)\| + \alpha_n L \\
&\leq \|x_n - p\| + \alpha_n L\Big[\|x_n - p\| + \alpha_n L(1 + \|y_{3n} - p\|)\Big] + \alpha_n L \\
&= \|x_n - p\| + \alpha_n L\|x_n - p\| + \alpha_n^2 L^2\|y_{3n} - p\| + \alpha_n L + \alpha_n^2 L^2 \\
&\leq \|x_n - p\| + \alpha_n L\|x_n - p\| + \alpha_n^2 L^2\Big[\|x_n - p\| + \alpha_n L(1 + \|y_{4n} - p\|)\Big] \\
&\quad + \alpha_n L + \alpha_n^2 L^2 \\
&= \|x_n - p\| + \alpha_n L\|x_n - p\| + \alpha_n^2 L^2\|x_n - p\| + \alpha_n^3 L^3\|y_{4n} - p\| \\
&\quad + \alpha_n L + \alpha_n^2 L^2 + \alpha_n^3 L^3 \\
&\ \ \vdots \\
&\leq \Big(1 + \alpha_n L + \alpha_n^2 L^2 + \alpha_n^3 L^3 + ... + \alpha_n^{m-1} L^{m-1}\Big)\|x_n - p\| \\
&\quad + \Big(\alpha_n L + \alpha_n^2 L^2 + \alpha_n^3 L^3 + ... + \alpha_n^{m-1} L^{m-1}\Big) \\
&\leq \Big(1 + \alpha_n L + \alpha_n^2 L^2 + \alpha_n^3 L^3 + ... + \alpha_n^{m-1} L^{m-1}\Big)\|x_n - p\| \\
&\quad + \Big(1 + \alpha_n L + \alpha_n^2 L^2 + \alpha_n^3 L^3 + ... + \alpha_n^{m-1} L^{m-1}\Big) \\
&\leq \Big(\frac{1 - \alpha_n^m L^m}{1 - \alpha_n L}\Big)\|x_n - p\| + \Big(\frac{1 - \alpha_n^m L^m}{1 - \alpha_n L}\Big) \\
&\leq M\|x_n - p\| + M.
\end{aligned}
$$

Again, from (19.1) and using inequality (4.4), we get,

$$\|x_{n+1} - p\|^2 \le \|x_n - p\|^2 - 2\lambda_n\theta_n\|x_{n+1} - p\|^2$$
$$+ 2\lambda_n\theta_n\langle(x_{n+1} - x_n) + (u - p), j(x_{n+1} - p)\rangle$$
$$+ 2\lambda_n\alpha_n\langle(x_{n+1} - x_n) + (p - T_1x_{n+1}) - (T_1y_{1n} - p), j(x_{n+1} - p)\rangle$$

$$\le \|x_n - p\|^2 - 2\lambda_n\theta_n\|x_{n+1} - p\|^2$$
$$+ 2\lambda_n(\theta_n + \alpha_n)\|x_{n+1} - x_n\|.\|x_{n+1} - p\|$$
$$+ 2\lambda_n\theta_n\|u - p\|\|x_{n+1} - p\| + 2\lambda_n\alpha_nL(1 + \|x_{n+1} - p\|)\|.\|x_{n+1} - p\|$$
$$+ 2\lambda_n\alpha_nL(1 + \|y_{12} - p\|)\|x_{n+1} - p\|$$
$$\le \|x_n - p\|^2 - 2\lambda_n\theta_n\|x_{n+1} - p\|^2 + 2\lambda_n^2(\theta_n + \alpha_n)\Big[\alpha_n\|x_n - p\|$$
$$+ \alpha_nL(1 + M\|x_n - p\| + M) + \theta_n\|x_n - p\| + \theta_n\|u - p\|\Big]\|x_{n+1} - p\|$$
$$+ 2\lambda_n\theta_n\|u - p\|\|x_{n+1} - p\| + 2\lambda_n\alpha_nL\Big[1 + \alpha_n\lambda_n\|x_n - p\|$$
$$+ \alpha_n\lambda_nL(1 + M\|x_n - p\| + M) + \lambda_n\theta_n\|x_n - p\|$$
$$+ \lambda_n\theta_n\|u - p\| + \|x_n - p\|\Big]\|x_{n+1} - p\|$$
$$+ 2\lambda_n\alpha_nL(1 + M\|x_n - p\| + M)\|x_{n+1} - p\|.$$

Since $\|x_{n+1} - p\| > r$, we have

$$2\lambda_n\theta_n\|x_{n+1} - p\|^2 \le 4\lambda_n^2\theta_n\Big[L(1 + Mr + M) + r + r + \frac{r}{2}\Big] + 2\lambda_n\theta_n\frac{r}{2}$$
$$+ 2\alpha_n\lambda_nL\Big[1 + L(1 + Mr + M) + r + r + r + \frac{r}{2}\Big]$$
$$+ 2\lambda_n\alpha_nL\Big[1 + Mr + M\Big],$$

which implies, $\|x_{n+1} - p\| \le \frac{r}{4} + \frac{r}{2} + \frac{r}{8} < r$, a contradiction. So, $x_n \in B \,\forall\, n \ge N^*$. Hence, $\{x_n\}_{n\ge1}$ is bounded. This completes the proof. □

Remark 19.2. It is now easy to see that we can find $R > 0$ sufficiently large such that $x_n \in B^* = \overline{B_R(p)} \,\forall\, n \in \mathbb{N}$. Besides, the set $K \cap B^*$ is a bounded closed and convex nonempty subset of E. If we define a map $\varphi : E \to \mathbb{R}$ by $\varphi(x) = \mu_n\|x_{n+1} - x\|^2$, where μ is a Banach limit, then, φ is continuous, convex and coercive (i.e., $\lim_{\|x\|\to+\infty} \varphi(x) = +\infty$). Thus, if E is a reflexive Banach space, then there exists $x^* \in K \cap B^*$ such that $\varphi(x^*) = \min_{x\in K\cap B^*} \varphi(x)$. Thus, the set $K^* := \{y \in K \cap B^* : \varphi(y) = \min_{x\in K\cap B^*} \varphi(x)\} \ne \emptyset$.

Remark 19.3. For the remainder of this chapter, $\{\alpha_n\}_{n\geq 1}$, $\{\lambda_n\}_{n\geq 1}$ and $\{\theta_n\}_{n\geq 1}$ will be as in Theorem 19.1 and in addition, $\sum_{n=1}^{\infty} \lambda_n \theta_n = \infty$ will be assumed.

We prove the following theorem.

Theorem 19.4. *Let K be a nonempty closed and convex subset of a real reflexive Banach space E with uniformly Gâteaux differentiable norm. Let $T_1, T_2, ..., T_m : K \to K$ be m generalized Lipschitz mappings such that T_1 is pseudo-contractive and $F := \cap_{i=1}^{m} F(T_i) \neq \emptyset$. Let the sequence $\{x_n\}_{n\geq 1}$ be iteratively generated by (19.1). Suppose that $K^* \cap F \neq \emptyset$. Then, $\{x_n\}_{n\geq 1}$ converges strongly to a common fixed point of $T_1, T_2, ..., T_m$.*

Proof. Let $x^* \in K^* \cap F$ and $t \in (0,1)$. Then, by the convexity of $K \cap B^*$, we have that $(1-t)x^* + tu \in K \cap B^*$. It then follows that $\varphi(x^*) \leq \varphi((1-t)x^* + tu)$. Using inequality (4.4), taking Banach limits over $n \geq 1$, and using the fact that the normalized duality mapping is norm-to-weak* uniformly continuous on bounded subsets of E, (following the method of proof of Theorem 12.3) we obtain that $limsup_{n\to\infty} \langle u - x^*, j(x_n - x^*) \rangle \leq 0$. Define $\varepsilon_n := \max \{\langle u - x^*, j(x_{n+1} - x^*) \rangle, 0\}$. Then, $\lim \varepsilon_n = 0$, and $\langle u - x^*, j(x_{n+1} - x^*) \rangle \leq \varepsilon_n$. Again, using inequality (4.4), the recursion formula (19.1) and these estimates, we obtain that

$$\begin{aligned}
\|x_{n+1} - x^*\|^2 &= \|x_n - x^* - \lambda_n[\alpha_n x_n - \alpha_n T_1 y_{1n} + \theta_n(x_n - u)]\|^2 \\
&\leq \|x_n - x^*\|^2 - 2\lambda_n \theta_n \|x_{n+1} - x^*\|^2 + (\lambda_n^2 \theta_n \\
&\quad + \lambda_n \alpha_n)C + 2\lambda_n \theta_n \varepsilon_n \\
&\leq \|x_n - x^*\|^2 - 2\lambda_n \theta_n \|x_{n+1} - x^*\|^2 + \delta_n
\end{aligned}$$

where $\delta_n := (\lambda_n^2 \theta_n + \lambda_n \alpha_n)C + 2\lambda_n \theta_n \varepsilon_n = o(\lambda_n \theta_n)$ for some constant $C > 0$. Hence, by Lemma 8.17, we have that the sequence $\{x_n\}_{n\geq 1}$ converges strongly to $x^* \in F$. This completes the proof. $\qquad\square$

We obtain the following as an easy consequence of Theorem 19.4

Corollary 19.5 *Let K be a nonempty closed and convex subset of a real reflexive Banach space E with uniformly Gâteaux differentiable norm. Let $T_1, T_2, ..., T_m : K \to K$ be m generalized Lipschitz pseudo-contractive mappings such that $F := \cap_{i=1}^{m} F(T_i) \neq \emptyset$. Let the sequence $\{x_n\}_{n\geq 1}$ be iteratively generated by (19.1). Suppose that $K^* \cap F \neq \emptyset$. Then, $\{x_n\}_{n\geq 1}$ converges strongly to a common fixed point of $T_1, T_2, ..., T_m$.*

Proof. Boundedness of the sequence $\{x_n\}$ follows as in the proof of Theorem 19.1 and the rest follows as in the proof of Theorem 19.4. $\qquad\square$

19.3 Generalized Lipschitz Accretive Operators

Let $A_1, A_2, ..., A_m : E \to E$ be m generalized Lipschitz *accretive* mappings. We introduce the following iteration process for the approximation of a common zero (assuming existence) of this family of mappings:

$$
\begin{cases}
u, \ x_1 \in K, \\
x_{n+1} = (1 - \alpha_n\lambda_n)x_n + \alpha_n\lambda_n y_{1n} - \alpha_n\lambda_n A_1 y_{1n} - \lambda_n\theta_n(x_n - u) \\
y_{1n} = (1 - \alpha_n)x_n + \alpha_n y_{2n} - \alpha_n A_2 y_{2n} \\
\vdots \\
y_{(m-2)n} = (1 - \alpha_n)x_n + \alpha_n y_{(m-1)n} - \alpha_n A_{m-1} y_{(m-1)n} \\
y_{(m-1)n} = x_n - \alpha_n A_m x_n, \ m \geq 2, \ n \geq 1,
\end{cases}
\tag{19.2}
$$

where $\{\alpha_n\}_{n=1}^{\infty}$, $\{\lambda_n\}_{n=1}^{\infty}$ and $\{\theta_n\}_{n=1}^{\infty}$ are as above.

Theorem 19.6. *Let E be a real Banach space with uniformly Gâteaux differentiable norm, $A_1, A_2, ..., A_m : E \to E$ be m generalized Lipschitz accretive operators such that $N := \cap_{i=1}^{m} N(A_i) := \cap_{i=1}^{m}\{x \in E : A_i x = 0\} \neq \emptyset$. Let $\{x_n\}_{n\geq 1}$ be iteratively generated by (19.2). Then, $\{x_n\}_{n\geq 1}$ is bounded. Moreover, if E is a reflexive Banach space with uniformly Gâteaux differentiable norm and $K^* \cap N \neq \emptyset$ (where K^* is constructed as in Remark 19.2), then $\{x_n\}_{n\geq 1}$ converges strongly to a common zero of $A_1, A_2, ..., A_m$.*

Proof. Recall that A_i is accretive if and only if $T_i := I - A_i$ is pseudo-contractive, $i = 1, 2, ..., m$. Then if we replace A_i by $I - T_i$, $i = 1, 2, ..., m$ in (19.2), then (19.2) reduces to (19.1) and boundedness of $\{x_n\}$ therefore follows as in the proof of Theorem 19.1. Furthermore, $p^* \in K^* \cap N$ implies $A_i p^* = 0$, $i = 1, 2, ..., m$. This implies $T_i p^* = p^*$. Thus, $K^* \cap \left(\cap_{i=1}^{m} F(T_i)\right) \neq \emptyset$, $i = 1, 2, ..., m$. The rest follows as in the proof of Theorem 19.4. This completes the proof. $\qquad\square$

19.4 Some Applications

Following the method of proofs used in the last section, the following theorems are easily obtained.

Theorem 19.7. *Let K be a nonempty closed and convex subset of a real reflexive Banach space with uniformly Gâteaux differentiable norm. Let $T_1, T_2, ..., T_m : K \to K$ be m generalized Lipschitz hemi-contractive mappings such that $F^* := \cap_{i=1}^{m} F(T_i) \neq \emptyset$. Let $\{x_n\}_{n\geq 1}$ be iteratively generated by (19.1). Suppose that $K^* \cap F^* \neq \emptyset$. Then, $\{x_n\}_{n\geq 1}$ converges strongly to some $p^* \in F^*$.*

Proof. Boundedness of $\{x_n\}_{n\geq 1}$ follows as in the proof of Theorem 19.1. The rest then follows as in the proof of Theorem 19.4. $\qquad\square$

Theorem 19.8. *Let E be a real reflexive Banach space with uniformly Gâteaux differentiable norm. Let $A_1, A_2, ..., A_m : E \to E$ be m generalized Lipschitz quasi-accretive operators such that $N^* := \cap_{i=1}^m N(A_i) \neq \emptyset$. Let $\{x_n\}_{n\geq 1}$ be iteratively generated by (19.2). Suppose that $K^* \cap N^* \neq \emptyset$. Then, $\{x_n\}_{n\geq 1}$ converges strongly to some $x^* \in N^*$.*

Proof. The mapping A_i is quasi-accretive if and only if $T_i := (I - A_i)$ is hemi-contractive, $i = 1, 2, ..., m$. Thus, replacing A_i by $I - T_i$ in (19.2), the result follows as in Theorem 19.4.   □

Remark 19.9. Prototypes for the iteration parameters for the theorems of this chapter are, for example, $\lambda_n = \frac{1}{(n+1)^a}$, $\alpha_n = \frac{1}{(n+1)}$ and $\theta_n = \frac{1}{(n+1)^b}$, where $a + b < 1$.

19.5 Historical Remarks

All the theorems of this chapter are due to Chidume and Ofoedu [152, 153].

Chapter 20
Finite Families of Non-self Asymptotically Nonexpansive Mappings

20.1 Introduction

In Chapter 15, we studied the Halpern-type recursion formula

$$x_{n+1} = \lambda_{n+1} u + (1 - \lambda_{n+1}) S_{n+1} x_n,$$

for the approximation of a common fixed point for *a finite family* of nonexpansive mappings. In Chapter 16, we studied the following Halpern-type recursion formula

$$x_{n+1} = \alpha_n u + \sum_{i \geq 1} \sigma_{i,n} S_i x_n, \ n \geq 1,$$

to approximate a common fixed point for *a countable family* of nonexpansive mappings. In Chapter 17, we studied the recursion formula

$$x_{n+1} = \lambda \left(1 - \sum_{k=1}^{\infty} \alpha_n^k\right) T_1 x_n + \lambda \sum_{k=1}^{\infty} \alpha_n^{k-1} T_k x_n + (1 - \lambda) x_n, n \in \mathbb{N},$$

for the approximation of a common fixed point for *a commuting family* of nonexpansive mappings defined on a compact convex subset of a real Banach space.

There is a fourth iteration formula which has been studied for the same problem of approximating common fixed points of families of nonexpansive mappings. We explore this in this chapter. We shall study this fourth iteration method for approximating common fixed points of a family of *non-self asymptotically nonexpansive* mappings. This problem has been studied by various authors (e.g., Bruck *et al.* [60], Chidume *et al.* ([143], [164], [164]), Chidume and Ofoedu [153], Oka [362], Shahzad and Udomene [445], Shioji and Takahashi [451], Sun [458], Chang *et al.* [84] and a host of other

C. Chidume, *Geometric Properties of Banach Spaces and Nonlinear Iterations*,
Lecture Notes in Mathematics 1965,
© Springer-Verlag London Limited 2009

authors). As an immediate consequence, we shall obtain convergence theorems for common fixed points for *nonexpansive self and non-self mappings of K*.

The concept of *non-self asymptotically nonexpansive* mappings was introduced by Chidume *et al.* [150] as an important generalization of asymptotically nonexpansive self-mappings.

Definition 20.1. (Chidume *et al.* [150]) Let K be a nonempty subset of a real normed space E. Let $P : E \to K$ be a nonexpansive retraction of E onto K. A non-self mapping $T : K \to E$ is called *asymptotically nonexpansive* if there exists a sequence $\{k_n\} \subset [1, \infty)$ with $k_n \to 1$ as $n \to \infty$ such that for every $n \in \mathbb{N}$,

$$\|T(PT)^{n-1}x - T(PT)^{n-1}y\| \leq k_n \|x - y\| \text{ for every } x, y \in K.$$

Using an Ishikawa-like scheme, Takahashi and Tamura [477] proved strong and weak convergence of a sequence defined by $x_{n+1} = \alpha_n S[\beta_n T x_n + (1 - \beta_n)x_n] + (1 - \alpha_n)x_n$ to a common fixed point of *a pair* of nonexpansive mappings T and S. Wang [498] used a similar scheme and the definition of Chidume *et al.* [150] to prove strong and weak convergence theorems for *a pair* of *non-self asymptotically nonexpansive mappings*. More precisely he proved the followings theorems in which $F(T) := \{x \in K : Tx = x\}$.

Theorem 20.2. *(Wang, [498]) Let K be a nonempty closed convex subset of uniformly convex Banach space E. Suppose $T_1, T_2 : K \to E$ are two non-self asymptotically nonexpansive mappings with sequences $\{k_n\}$ and $\{l_n\} \in [1, \infty)$ such that $\sum_{n=1}^{\infty} (k_n - 1) < \infty$, $\sum_{n=1}^{\infty} (l_n - 1) < \infty, k_n \to 1, l_n \to 1$ as $n \to \infty$, respectively. Let $\{x_n\}$ be generated by*

$$\begin{cases} x_1 \in K, \\ x_{n+1} = P((1 - \alpha_n)x_n + \alpha_n T_1(PT_1)^{n-1}y_n), \\ y_n = P((1 - \beta_n)x_n + \beta_n T_2(PT_2)^{n-1}x_n), \quad n \geq 1, \end{cases} \tag{20.1}$$

where $\{\alpha_n\}$ and $\{\beta_n\}$ are sequences in $[\epsilon, 1 - \epsilon]$ for some $\epsilon \in (0, 1)$. If one of T_1 and T_2 is completely continuous, and $F(T_1) \cap F(T_2) \neq \emptyset$, then $\{x_n\}$ converges strongly to a common fixed point of T_1 and T_2.

Theorem 20.3. *(Wang, [498]) Let $K, T_1, T_2, \{k_n\}, \{l_n\}$, and $\{x_n\}$ be as in Theorem 20.2. If one of T_1 and T_2 is semi-compact, and $F(T_1) \cap F(T_2) \neq \emptyset$ then, $\{x_n\}$ converges strongly to a common fixed point of T_1 and T_2.*

Theorem 20.4. *(Wang, [498]) Let $K, T_1, T_2, \{k_n\}, \{l_n\}$, and $\{x_n\}$ be as in Theorem 20.2. If E satisfies Opial's condition, and $F(T_1) \cap F(T_2) \neq \emptyset$, then $\{x_n\}$ converges weakly to a common fixed point of T_1 and T_2.*

In this chapter, we study *an iteration process* introduced by Chidume and Ali [122] for approximating common fixed points for *finite families of non-self asymptotically nonexpansive mappings*. For these families of operators, strong convergence theorems are proved in uniformly convex Banach spaces and weak convergence theorems are proved in real uniformly convex Banach spaces that satisfy Opial's condition, or have Fréchet differentiable norms, or whose dual spaces have the *Kadec-Klee property*.

20.2 Preliminaries

A mapping $T : K \to K$ is said to be *semi-compact* if, for any bounded sequence $\{x_n\}$ in K such that $\|x_n - Tx_n\| \to 0$ as $n \to \infty$, there exists a subsequence, say $\{x_{n_j}\}$, of $\{x_n\}$ such that $\{x_{n_j}\}$ converges strongly to some x^* in K. Recall that T is said to be *completely continuous* if for every bounded sequence $\{x_n\}$, there exists a subsequence, say $\{x_{n_j}\}$ of $\{x_n\}$ such that the sequence $\{Tx_{n_j}\}$ converges strongly to some element of the range of T.

A Banach space E is said to have the *Kadec-Klee property* if, for every sequence $\{x_n\}$ in E, $x_n \rightharpoonup x$ and $\|x_n\| \to \|x\|$ imply $\|x_n - x\| \to 0$.

In what follows we shall use the following results.

Lemma 20.5. *(Schu, [439]) Let E be a real uniformly convex Banach space and $0 \le p \le t_n \le q < 1$ for all positive integers $n \ge 1$. Suppose that $\{x_n\}$ and $\{y_n\}$ are two sequences of E such that*

$$\limsup_{n \to \infty} \|x_n\| \le r, \ \limsup_{n \to \infty} \|y_n\| \le r \ and \ \lim_{n \to \infty} \|t_n x_n + (1 - t_n)y_n\| = r$$

hold for some $r \ge 0$, then $\lim_{n \to \infty} \|x_n - y_n\| = 0$.

Lemma 20.6. *(Chidume et al., [150]) Let E be a real uniformly convex Banach space, K a nonempty closed subset of E, and let $T : K \to E$ be an asymptotically nonexpansive mapping with a sequence $\{k_n\} \subset [1, \infty)$ and $k_n \to 1$ as $n \to \infty$, then (I-T) is demiclosed at zero.*

Lemma 20.7. *(Kaczor, [267]) Let E be a real uniformly convex Banach space whose dual E^* satisfies the Kadec-Klee property. Let $\{x_n\}$ be a bounded sequence in E and $x^*, q^* \in \omega_\omega(\{x_n\})$ (where $\omega_\omega(\{x_n\})$ denote the weak limit set of $\{x_n\}$). Suppose $\lim_{n \to \infty} \|tx_n + (1-t)x^* - q^*\|$ exists for all $t \in [0, 1]$. Then, $x^* = q^*$.*

Theorem 20.8. *(Falset et al., [222]) Let E be a uniformly convex Banach space and K be a convex subset of E. Then, there exists a strictly increasing continuous convex function $\phi : \mathbb{R} \to \mathbb{R}$ with $\phi(0) = 0$ such that for each Lipschitz mapping $S : K \to K$ with Lipschitz constant L, we have,*

$$\|\alpha Sx + (1 - \alpha)Sy - S(\alpha x + (1 - \alpha)y)\| \le L\phi^{-1}\left(\|x - y\| - \frac{1}{L}\|Sx - Sy\|\right)$$

for all $x, y \in K$ and $0 < \alpha < 1$.

20.3 Strong Convergence Theorems

In this section we state and prove the main results of this chapter. In the sequel, we designate the set $\{1, 2, \ldots, m\}$ by I and we always assume $\bigcap_{i=1}^{m} F(T_i) \neq \emptyset$. We now introduce the following iteration scheme.

$$
\begin{cases}
x_1 \in K, \\
x_{n+1} = P\Big[(1 - \alpha_n)x_n + \alpha_n T_1 (PT_1)^{n-1} y_{n+m-2}\Big] \\
y_{n+m-2} = P\Big[(1 - \alpha_n)x_n + \alpha_n T_2 (PT_2)^{n-1} y_{n+m-3}\Big] \\
\quad \vdots \\
y_n = P\Big[(1 - \alpha_n)x_n + \alpha_n T_m (PT_m)^{n-1} x_n\Big], \qquad n \geq 1.
\end{cases}
\tag{20.2}
$$

We illustrate the iteration scheme (20.2) for the case of three maps T_1, T_2, T_3. In this case, the scheme (20.2) becomes;

$$
\begin{cases}
x_1 \in K \\
x_{n+1} = P\Big[(1 - \alpha_n)x_n + \alpha_n T_1 (PT_1)^{n-1} y_{n+1}\Big] & (i) \\
y_{n+1} = P\Big[(1 - \alpha_n)x_n + \alpha_n T_2 (PT_2)^{n-1} y_n\Big] & (ii) \\
y_n = P\Big[(1 - \alpha_n)x_n + \alpha_n T_3 (PT_3)^{n-1} x_n\Big]. & (iii)
\end{cases}
$$

Starting with arbitrary $x_1 \in K$ (i.e., $n = 1$) we compute y_1 from (iii), using y_1 we compute y_2 from (ii), we then use y_2 to compute x_2 from (i). Then, using x_2 we come back to (iii) to compute y_2, then with y_2 in (ii) we get y_3 and with y_3 in (i) we compute x_3, and, continuing in this way, we generate the sequence $\{x_n\}$.

Now, let $x^* \in \bigcap_{i=1}^{m} F(T_i)$. We start by presenting a proposition in which boundedness of the sequence $\{x_n\}$ defined by (20.2) and the existence of the limit, $\lim_{n \to \infty} \|x_n - x^*\|$, are proved for *four* non-self asymptotically nonexpansive maps T_1, T_2, T_3 *and* T_4 of K into E. The purpose of this is to illustrate the pattern of the proof of the general case.

Proposition 20.9. *Let E be a real normed linear space and K be a nonempty subset of E which is also a nonexpansive retract with retraction P. Let T_1, T_2, T_3 and T_4 be asymptotically nonexpansive mappings of K into E with sequences $\{k_{in}\}_{n=1}^{\infty}$, respectively, $i = 1, 2, 3, 4$ satisfying $k_{in} \to 1$ as $n \to \infty$ and $\sum_{n=1}^{\infty} (k_{in} - 1) < \infty$. Let $\{\alpha_n\}_{n=1}^{\infty}$ be a sequence in $[\epsilon, 1 - \epsilon]$, $\epsilon \in (0, 1)$. Let $\{x_n\}$ be a sequence defined iteratively by*

$$\begin{cases} x_1 \in K \\ x_{n+1} = P\Big[(1-\alpha_n)x_n + \alpha_n T_1(PT_1)^{n-1}y_{n+2}\Big] \\ y_{n+2} = P\Big[(1-\alpha_n)x_n + \alpha_n T_2(PT_2)^{n-1}y_{n+1}\Big] \\ y_{n+1} = P\Big[(1-\alpha_n)x_n + \alpha_n T_3(PT_3)^{n-1}y_n\Big] \\ y_n = P\Big[(1-\alpha_n)x_n + \alpha_n T_4(PT_4)^{n-1}x_n\Big], \qquad n \geq 1. \end{cases}$$

Then, $\{x_n\}$ is bounded and $\lim\limits_{n\to\infty} ||x_n - x^*||$ exists.

Proof. Set $k_{in} = 1 + u_{in}$ so that $\sum\limits_{n=1}^{\infty} u_{in} < \infty$ for each $i \in \{1,...,4\}$. Let $w_n := \sum\limits_{i=1}^{4} \mu_{in}$ so that $\mu_{in} \leq w_n$ for $i = 1,2,3,4$ and let $x^* \in \bigcap_{i=1}^{4} F(T_i)$. Then, we have the following estimates,

$$||x_{n+1} - x^*|| \leq (1-\alpha_n)||x_n - x^*|| + \alpha_n(1+u_{1n})\Big[(1-\alpha_n)||x_n - x^*||$$
$$+ \alpha_n(1+u_{2n})\Big((1-\alpha_n)||x_n - x^*|| + \alpha_n(1+u_{3n})[(1-\alpha_n)||x_n - x^*||$$
$$+ \alpha_n(1+u_{4n})||x_n - x^*||]\Big)\Big]$$
$$\leq ||x_n - x^*||\Big[1 - \alpha_n + \alpha_n(1+u_{1n})(1-\alpha_n) + \alpha_n^2(1-\alpha_n)(1+u_{1n})(1+u_{2n})$$
$$+ \alpha_n^3(1-\alpha_n)(1+u_{1n})(1+u_{2n})(1+u_{3n})$$
$$+ \alpha_n^4(1+u_{1n})(1+u_{2n})(1+u_{3n})(1+u_{4n})\Big]$$

$$= ||x_n - x^*||\Big[1 - \alpha_n + \alpha_n(1+u_{1n} - \alpha_n(1+u_{1n})) + \alpha_n^2(1+u_{1n})$$
$$+ \alpha_n^2 u_{2n}(1+u_{1n}) - \alpha_n^3(1+u_{1n})(1+u_{2n})$$
$$+ \alpha_n^3(1+u_{1n})(1+u_{2n})$$
$$+ \alpha_n^3 u_{3n}(1+u_{1n})(1+u_{2n}) - \alpha_n^4(1+u_{1n})(1+u_{2n})(1+u_{3n})$$
$$+ \alpha_n^4(1+u_{1n})(1+u_{2n})(1+u_{3n})$$
$$+ \alpha_n^4 u_{4n}(1+u_{1n})(1+u_{2n})(1+u_{3n})\Big]$$

$$= ||x_n - x^*||\Big[1 + \alpha_n u_{1n} + \alpha_n^2 u_{2n}(1+u_{1n}) + \alpha_n^3 u_{3n}(1+u_{1n})(1+u_{2n})$$
$$+ \alpha_n^4 u_{4n}(1+u_{1n})(1+u_{2n})(1+u_{3n})\Big]$$
$$\leq ||x_n - x^*||\Big[1 + u_{1n} + u_{2n}(1+u_{1n}) + u_{3n}(1+u_{1n})(1+u_{2n})$$
$$+ u_{4n}(1+u_{1n})(1+u_{2n})(1+u_{3n})\Big]$$

$$= \|x_n - x^*\| \Big[1 + u_{1n} + u_{2n} + u_{3n} + u_{4n} + u_{1n}u_{2n} + u_{1n}u_{3n} + u_{1n}u_{4n}$$

$$+ u_{2n}u_{3n} + u_{2n}u_{4n} + u_{3n}u_{4n} + u_{1n}u_{2n}u_{3n} + u_{1n}u_{2n}u_{4n} + u_{1n}u_{3n}u_{4n}$$

$$+ u_{2n}u_{3n}u_{4n} + u_{1n}u_{2n}u_{3n}u_{4n} \Big]$$

$$\leq \|x_n - x^*\| \Big[1 + 4w_n + 6w_n^2 + 4w_n^3 + w_n^4 \Big]$$

$$= \|x_n - x^*\| \Big[1 + \binom{4}{1} w_n + \binom{4}{2} w_n^2 + \binom{4}{3} w_n^3 + \binom{4}{4} w_n^4 \Big]$$

$$\leq \|x_n - x^*\| (1 + 15 w_n) \leq \|x_n - x^*\| e^{15 w_n}$$

$$\leq \|x_1 - x^*\| e^{15 \sum\limits_{n=1}^{\infty} w_n} < \infty.$$

Hence, $\{x_n\}$ is bounded and so there exists a positive integer M such that $\|x_{n+1} - x^*\| \leq \|x_n - x^*\| + 15 M w_n$. By Lemma 6.32, $\lim\limits_{n\to\infty} \|x_n - x^*\|$ exists. This completes the proof of Proposition 20.9. □

Now, for finitely many maps, we prove the following lemma.

Lemma 20.10. *Let E be a real normed linear space and K be a nonempty subset of E which is also a nonexpansive retract with retraction P. Let $T_1, T_2, ..., T_m$ be asymptotically nonexpansive mappings of K into E with sequences $\{k_{in}\}_{n=1}^{\infty}$ satisfying $k_{in} \to 1$ as $n \to \infty$ and $\sum\limits_{n=1}^{\infty} (k_{in} - 1) < \infty$, $i = 1, 2, ..., m$. Let $\{\alpha_n\}_{n=1}^{\infty}$ be a sequences in $[\epsilon, 1-\epsilon]$, $\epsilon \in (0, 1)$. Let $\{x_n\}$ be a sequence defined iteratively by (20.2). Then, $\{x_n\}$ is bounded and $\lim \|x_n - x^*\|$ exists.*

Proof. Set $k_{in} = 1 + u_{in}$ so that $\sum\limits_{n=1}^{\infty} u_{in} < \infty$ for each $i \in I$. Let $w_n := \sum\limits_{i=1}^{m} u_{in}$. Let $x^* \in \bigcap_{i=1}^{m} F(T_i)$. Then we have, for some positive integer h, $2 \leq h < m$,

$$\|x_{n+1} - x^*\| = \|(1 - \alpha_n)x_n + \alpha_n T_1 (PT_1)^{n-1} y_{n+m-2} - x^*\|$$

$$\leq (1 - \alpha_n)\|x_n - x^*\| + \alpha_n (1 + u_{1n})\|y_{n+m-2} - x^*\|$$

$$\leq (1 - \alpha_n)\|x_n - x^*\| + \alpha_n (1 + u_{1n}) \Big[(1 - \alpha_n)\|x_n - x^*\|$$

$$+ \alpha_n (1 + u_{2n})\|y_{n+m-3} - x^*\| \Big]$$

$$\leq (1 - \alpha_n)\|x_n - x^*\| + \alpha_n (1 - \alpha_n)(1 + u_{1n})\|x_n - x^*\|$$

$$+ ... + \alpha_n^{h-1}(1 - \alpha_n)(1 + u_{1n})(1 + u_{2n})...(1 + u_{h-1n})\|x_n - x^*\|$$

$$+ ... + \alpha_n^m (1 + u_{1n})(1 + u_{2n})...(1 + u_{mn})\|x_n - x^*\|$$

$$= \|x_n - x^*\| \Big[1 + \alpha_n u_{1n} + \alpha_n^2 u_{2n} + \ldots + \alpha_n^m u_{mn}$$

$$+ \alpha_n^2 u_{1n} u_{2n} + \alpha_n^3 u_{1n} u_{3n} + \ldots + \alpha_n^m u_{1n} u_{2n} \ldots u_{mn} \Big]$$

$$\leq \|x_n - x^*\| \Big[1 + u_{1n} + u_{2n}(1 + u_{1n}) + u_{3n}(1 + u_{1n})(1 + u_{2n}) + \ldots$$

$$+ u_{mn}(1 + u_{1n})(1 + u_{2n}) \ldots (1 + u_{m-1n}) \Big]$$

$$= \|x_n - x^*\| \Big[1 + u_{1n} + u_{2n} + \ldots + u_{mn} + u_{1n} u_{2n} + u_{1n} u_{3n} + \ldots + u_{1n} u_{2n} u_{3n}$$

$$+ \ldots + u_{1n} u_{2n} u_{3n} \ldots u_{mn} \Big]$$

$$\leq \|x_n - x^*\| \Big[1 + \binom{m}{1} w_n + \binom{m}{2} w_n^2 + \ldots + \binom{m}{m} w_n^m \Big]$$

$$\leq \|x_n - x^*\| (1 + \delta_m w_n) \leq \|x_n - x^*\| e^{\delta_m w_n}$$

$$\leq \|x_1 - x^*\| e^{\delta_m \sum_{n=1}^{\infty} w_n} < \infty,$$

where δ_m is a positive real number defined by $\delta_m := \Big[\binom{m}{1} + \binom{m}{2} + \ldots + \binom{m}{m} \Big]$.

This implies that $\{x_n\}$ is bounded and so there exists a positive integer M such that $\|x_{n+1} - x^*\| \leq \|x_n - x^*\| + \delta_m M w_n$. By Lemma 6.32, $\lim_{n \to \infty} \|x_n - x^*\|$ exists. This completes the proof of Lemma 20.10. □

We also prove the following lemma which will be the main tool in the sequel.

Lemma 20.11. *Let E be a real uniformly convex Banach space and K be a closed convex nonempty subset of E which is also a nonexpansive retract with retraction P. Let $T_1, T_2, \ldots, T_m : K \to E$ be asymptotically nonexpansive mappings with sequences $\{k_{in}\}_{n=1}^{\infty}$ satisfying $k_{in} \to 1$ as $n \to \infty$ and $\sum_{n=1}^{\infty} (k_{in} - 1) < \infty, i = 1, 2, \ldots, m$. Let $\{\alpha_n\}_{n=1}^{\infty}$ be a sequence in $[\epsilon, 1 - \epsilon], \epsilon \in (0, 1)$. Let $\{x_n\}$ be a sequence defined iteratively by (20.2). Then,*

$$\lim_{n \to \infty} \|x_n - T_1 x_n\| = \lim_{n \to \infty} \|x_n - T_2 x_n\| = \ldots = \lim_{n \to \infty} \|x_n - T_m x_n\| = 0.$$

Proof. By Lemma 20.10 , $\lim_{n \to \infty} \|x_n - x^*\|$ exists. Let

$$\lim_{n \to \infty} \|x_n - x^*\| = l. \tag{20.3}$$

Observe that for any positive integer $h, 2 \leq h < m$ we have, using the notations of Lemma 20.10, the following estimates.

$$\|y_{n+m-h} - x^*\| \leq \|x_n - x^*\| \Big[1 + u_{hn} + u_{h+1n}(1 + u_{hn})$$

$$+ u_{h+2n}(1 + u_{hn})(1 + u_{h+1n})$$

$$+ \ldots + u_{mn}(1 + u_{hn})(1 + u_{h+1n}) \ldots (1 + u_{m-1n}) \Big] \tag{20.4}$$

and

$$\|T_h(PT_h)^{n-1}y_{n+m-h-1} - x^*\| \le (1 + u_{hn})\|y_{n+m-h-1} - x^*\|. \quad (20.5)$$

From (20.3) and (20.4) we have

$$\limsup_{n \to \infty} \|y_{n+m-h} - x^*\| \le l \qquad (20.6)$$

and from (20.5),

$$\limsup_{n \to \infty} \|T_h(PT_h)^{n-1}y_{n+m-h-1} - x^*\| \le l. \qquad (20.7)$$

Since

$$l = \lim_{n \to \infty} \|x_n - x^*\| = \lim_{n \to \infty} \|x_{n+1} - x^*\|$$

$$= \lim_{n \to \infty} \|(1 - \alpha_n)(x_n - x^*) + \alpha_n(T_1(PT_1)^{n-1}y_{n+m-2} - x^*)\|, \quad (20.8)$$

by using (20.3), (20.7) and Lemma 20.5 we have

$$\lim_{n \to \infty} \|x_n - T_1(PT_1)^{n-1}y_{n+m-2}\| = 0. \qquad (20.9)$$

Using (20.7) we also have

$$\limsup_{n \to \infty} \|T_1(PT_1)^{n-1}y_{n+m-2} - x^*\| \le l. \qquad (20.10)$$

Using (20.9) and the following estimate

$$\|x_n - x^*\| \le \|x_n - T_1(PT_1)^{n-1}y_{n+m-2}\| + \|T_1(PT_1)^{n-1}y_{n+m-2} - x^*\|$$

$$\le \|x_n - T_1(PT_1)^{n-1}y_{n+m-2}\| + (1 + u_{1n})\|y_{n+m-2} - x^*\|,$$

we have

$$l \le \liminf_{n \to \infty} \|y_{n+m-2} - x^*\|. \qquad (20.11)$$

From (20.6) and (20.11) we get

$$\lim_{n \to \infty} \|y_{n+m-2} - x^*\| = l. \qquad (20.12)$$

Thus, $\lim_{n \to \infty} \|(1 - \alpha_n)(x_n - x^*) + \alpha_n(T_2(PT_2)^{n-1}y_{n+m-3} - x^*)\| = l$, and similarly by (20.3), (20.7) and Lemma 20.5 we get $\lim_{n \to \infty} \|x_n - T_2(PT_2)^{n-1} y_{n+m-3}\| = 0$. Continuing, we observe that for $2 \le h < m$,

$$\lim_{n \to \infty} \|x_n - T_h(PT_h)^{n-1}y_{n+m-h-1}\| = 0 \qquad (20.13)$$

and

$$\lim_{n\to\infty} ||y_{n+m-h} - x^*|| = l. \tag{20.14}$$

Now,

$$
\begin{aligned}
||x_n - T_h(PT_h)^{n-1}x_n|| &\leq ||x_n - T_h(PT_h)^{n-1}y_{n+m-h-1}|| \\
&\quad + ||T_h(PT_h)^{n-1}y_{n+m-h-1} - T_h(PT_h)^{n-1}x_n|| \\
&\leq ||x_n - T_h(PT_h)^{n-1}y_{n+m-h-1}|| \\
&\quad + (1 + u_{hn})||y_{n+m-h-1} - x_n|| \tag{20.15} \\
&= ||x_n - T_h(PT_h)^{n-1}y_{n+m-h-1}|| + (1 + u_{hn})||(1 - \alpha_n)x_n \\
&\quad + \alpha_n T_{h+1}(PT_{h+1})^{n-1}y_{n+m-h-2} - x_n|| \\
&\leq ||x_n - T_h(PT_h)^{n-1}y_{n+m-h-1}|| \\
&\quad + \alpha_n(1 + u_{hn})||x_n - (T_{h+1}PT_{h+1})^{n-1}y_{n+m-h-2}||,
\end{aligned}
$$

and so

$$\lim_{n\to\infty} ||x_n - T_h(PT_h)^{n-1}x_n|| = 0. \tag{20.16}$$

For each $i \in I$, T_i is asymptotically nonexpansive, and so is Lipschitzian with Lipschitz constants L_i. If $L = \max_{1 \leq i \leq m}\{L_i\}$, then, noting that $x_n = Px_n \ \forall n \geq 1$, we have the following estimate.

$$
\begin{aligned}
||x_n - T_h x_n|| &\leq ||x_n - T_h(PT_h)^{n-1}x_n|| \\
&\quad + ||T_h(PT_h)^{n-1}x_n - T_h(PT_h)^{n-1}y_{n+m-h-1}|| \\
&\quad + ||T_h(PT_h)^{n-1}y_{n+m-h-1} - T_h x_n|| \\
&\leq ||x_n - T_h(PT_h)^{n-1}x_n|| + (1 + u_{hn})||x_n - y_{n+m-h-1}|| \\
&\quad + L||T_h(PT_h)^{n-2}y_{n+m-h-1} - x_n,||
\end{aligned}
$$

and from (20.15) and (20.16) we have

$$\lim_{n\to\infty} ||x_n - T_h x_n|| = 0. \tag{20.17}$$

Note that

$$
\begin{aligned}
||y_n - x^*|| &= ||(1 - \alpha_n)(x_n - x^*) + \alpha_n(T_m(PT_m)^{n-1}x_n - x^*)|| \\
&\leq [1 - \alpha_n + \alpha_n + \alpha_n u_{mn}]||x_n - x^*||
\end{aligned}
$$

which implies

$$\limsup_{n\to\infty} ||y_n - x^*|| \leq l. \tag{20.18}$$

On the other hand,

$$||x_n - x^*|| \leq ||x_n - T_{m-1}(PT_{m-1})^{n-1}y_n|| + ||T_{m-1}(PT_{m-1})^{n-1}y_n - x^*||$$
$$\leq ||x_n - T_{m-1}(PT_{m-1})^{n-1}y_n|| + (1 + u_{(m-1)n})||y_n - x^*||.$$

$$(20.19)$$

From this and (20.13), we get

$$\liminf_{n\to\infty} ||y_n - x^*|| \geq l. \tag{20.20}$$

From (20.18) and (20.20), we have $\lim_{n\to\infty} ||y_n - x^*|| = l$. Hence,

$$\lim_{n\to\infty} ||(1 - \alpha_n)(x_n - x^*) + \alpha_n(T_m(PT_m)^{n-1}x_n - x^*)|| = l.$$

By (20.3), Lemma 20.5 and the following estimate,

$$\limsup_{n\to\infty} ||T_m(PT_m)^{n-1}x_n - x^*|| \leq \limsup_{n\to\infty}(1 + u_{mn})||x_n - x^*|| = l,$$

we have that $\lim_{n\to\infty} ||x_n - T_m(PT_m)^{n-1}x_n|| = 0$. Moreover,

$$||x_n - T_m x_n|| \leq ||x_n - T_m(PT_m)^{n-1}x_n|| + ||T_m(PT_m)^{n-1}x_n - T_m x_n||$$
$$\leq ||x_n - T_m(PT_m)^{n-1}x_n|| + L||T_m(PT_m)^{n-2}x_n - x_n||,$$

so that, $\lim_{n\to\infty} ||x_n - T_m x_n|| = 0$ and this completes the proof. $\quad\square$

We now prove the following convergence theorem.

Theorem 20.12. *Let E be a real uniformly convex Banach space and K be a closed convex nonempty subset of E which is also a nonexpansive retract with retraction P. Let $T_1, T_2, \ldots, T_m : K \to E$ be asymptotically nonexpansive mappings of K into E with sequences $\{k_{in}\}_{n=1}^{\infty}$ and $\{\alpha_n\}_{n=1}^{\infty}$ be as in Lemma 20.11. If one of $\{T_i\}_{i=1}^{m}$ is either completely continuous or semi-compact, then, the sequence $\{x_n\}$ defined by (20.2) converges strongly to a common fixed point of $\{T_i\}_{i=1}^{m}$.*

Proof. If one of T_i's is semi-compact, say $T_s, s \in \{1, 2, \ldots, m\}$, from the fact that $\lim_{n\to\infty} ||x_n - T_s x_n|| = 0$ and $\{x_n\}$ is bounded, there exists a subsequence, say $\{x_{n_j}\}$, of $\{x_n\}$ that converges strongly to some x^* in K. By Lemma 20.6, $T_s x^* = x^*$. From $\lim_{n\to\infty} ||x_n - T_i x_n|| = 0$ and the continuity of T_i, $i = 1, 2, \ldots, m$, and using $x_{n_j} \to x^*$ as $j \to \infty$, we obtain that $x^* \in \cap_{i=1}^{m} Fix(T_i)$. By Lemma 20.10, $\lim_{n\to\infty} ||x_n - x^*||$ exists and so $\{x_n\}$ converges strongly to x^*.

If, on the other hand, one of T_i's is completely continuous, say T_s, then $\{T_s x_n\}$ is bounded (since $\{x_n\}$ is). Hence, there exists a subsequence $\{T_s x_{n_j}\}$ of $\{T_s x_n\}$ converging strongly to some x^*. By Lemma 20.11, $\lim_{n\to\infty} ||x_{n_j} -$

$T_s x_{n_j}\| = 0$, and by the continuity of T_s, we have $\lim_{j \to \infty} \|x_{n_j} - x^*\| = 0$. Using Lemma 20.6 again, we have $x^* \in \cap_{i=1}^m F(T_i)$. Thus, since $\lim_{n \to \infty} \|x_n - x^*\|$ exists by Lemma 20.10, we have $\lim_{n \to \infty} \|x_n - x^*\| = 0$ and the proof is complete. $\square$

The following corollary follows from Theorem 20.12.

Corollary 20.13 *Let K be a nonempty closed convex subset of a real uniformly convex Banach space E. Let $T_1, T_2, \ldots, T_m : K \to K$ be asymptotically nonexpansive mappings with sequences $\{k_{in}\}_{n=1}^\infty$ and $\{\alpha_n\}_{n=1}^\infty$ be as in Lemma 20.11. If one of $\{T_i\}_{i=1}^m$ is either completely continuous or semicompact, then the sequence $\{x_n\}$ defined by*

$$
\begin{cases}
x_1 \in K, \\
x_{n+1} = (1 - \alpha_n)x_n + \alpha_n(T_1)^n y_{n+m-2} \\
y_{n+m-2} = (1 - \alpha_n)x_n + \alpha_n(T_2)^n y_{n+m-3} \\
\vdots \\
y_n = (1 - \alpha_n)x_n + \alpha_n(T_m)^n x_n, \qquad n \geq 1, m \geq 2,
\end{cases}
\tag{20.21}
$$

converges strongly to a common fixed point of $\{T_i\}_{i=1}^m$.

20.4 Weak Convergence Theorems

We now prove weak convergence theorems.

Theorem 20.14. *Let E be a real uniformly convex Banach space and K be a closed convex nonempty subset of E which is also a nonexpansive retract with retraction P. Let $T_1, T_2, \ldots, T_m : K \to E$ be asymptotically nonexpansive mappings of K into E with sequences $\{k_{in}\}_{n=1}^\infty$ and $\{\alpha_n\}_{n=1}^\infty$ be as in Lemma 20.11. If E satisfies Opial's condition or has a Fréchet differentiable norm, then, the sequence $\{x_n\}$ defined by (20.2) converges weakly to a common fixed point of $\{T_i\}_{i=1}^m$.*

Proof. If E satisfies Opial's condition the proof follows as in the proof of Theorem 3.2 of Tan and Xu [488]. If E has Fréchet differentiable norm, the proof follows as in the proof of Theorem 3.10 of Chidume *et al.* [150], using Lemma 20.7 instead of Lemma 3.9 of Chidume *et al.* [150]. $\square$

The following corollary follows from Theorem 20.14.

Corollary 20.15 *Let K be a nonempty closed convex subset of a real uniformly convex Banach space E. Let $T_1, T_2, \ldots, T_m : K \to K$ be asymptotically nonexpansive mappings with sequences $\{k_{in}\}_{n=1}^\infty$ and $\{\alpha_n\}_{n=1}^\infty$ be as in Lemma 20.11. If E satisfies Opial's condition or has a Fréchet differentiable norm, then, the sequence $\{x_n\}$ defined by (20.21) converges weakly to a common fixed point of $\{T_i\}_{i=1}^m$.*

For our next theorem, we shall need the following Lemma.

Lemma 20.16. *Let E be a real uniformly convex Banach space, and K be a closed convex subset of E. Let $T_1, T_2, \ldots, T_m : K \to K$ be nonexpansive mappings. Let the sequence $\{\alpha_n\}_{n=1}^{\infty}$ be as in Lemma 20.11. For $t \in [0, 1]$ and $u, v \in \bigcap_{i=1}^{m} F(T_i)$, if $\{x_n\}$ is a sequence defined by*

$$\begin{cases} x_1 \in K, \\ x_{n+1} = (1 - \alpha_n)x_n + \alpha_n T_1 y_{n+m-2} \\ y_{n+m-2} = (1 - \alpha_n)x_n + \alpha_n T_2 y_{n+m-3} \\ \vdots \\ y_n = (1 - \alpha_n)x_n + \alpha_n T_m x_n, \qquad n \geq 1, m \geq 2, \end{cases} \qquad (20.22)$$

then, $\lim_{n\to\infty} \|tx_n + (1-t)u - v\|$ *exists.*

Proof. Since $\{x_n\}$ is bounded, there exists a positive real number r such that $\{x_n\} \subseteq D \equiv B_r(0) \cap K$, so that D is a closed convex nonempty subset of K. Let

$$g_n(t) := \|tx_n + (1-t)u - v\|. \qquad (20.23)$$

Then, $\lim_{n\to\infty} g_n(1) = \lim_{n\to\infty} \|x_n - v\|$ and $\lim_{n\to\infty} g_n(0) = \|u - v\|$ exist. Now let $t \in (0, 1)$ and define a map $Q_n : D \to D$ by

$$\begin{cases} Q_n x = (1 - \alpha_n)x + \alpha_n T_1 x^{m-2} \\ x^{m-2} = (1 - \alpha_n)x + \alpha_n T_2 x^{m-3} \\ \vdots \\ x^0 = (1 - \alpha_n)x + \alpha_n T_m x, \qquad n \geq 1, m \geq 2. \end{cases}$$

Denoting x^{m-h} by y_{n+m-h}, $2 \leq h \leq m$, we note that $Q_n x_n = x_{n+1}$. Observe also that $\bigcap_{n=1}^{m} F(T_i) \subseteq F(Q_n)$. We show Q_n is nonexpansive. For $x, z \in D$,

$$\begin{aligned} \|Q_n x - Q_n z\| &= \|(1 - \alpha_n)x + \alpha_n T_1 x^{m-2} - [(1 - \alpha_n)z + \alpha_n T_1 z^{m-2}]\| \\ &\leq (1 - \alpha_n)\|x - z\| + \alpha_n \|x^{m-2} - z^{m-2}\| \\ &\leq \|x - z\|\Big[1 - \alpha_n + \alpha_n(1 - \alpha_n) + \alpha_n^2(1 - \alpha_n) + \ldots \\ &\quad + \alpha_n^{m-1}(1 - \alpha_n) + \alpha_n^m\Big] \\ &= \|x - z\|. \end{aligned}$$

If we define $S_{n,d} := Q_{n+d-1} Q_{n+d-2} \ldots Q_n$, $d \geq 1$ an integer and

$$b_{n,d} := \|S_{n,d}(tx_n + (1-t)u) - (tS_{n,d}x_n + (1-t)u)\|,$$

we then have, $\|S_{n,d}x - S_{n,d}y\| \leq \|x - y\|$, $S_{n,d}x_n = x_{n+d}$ and $S_{n,d}x^* = x^*$ $\forall x^* \in \bigcap_{i=1}^{m} F(T_i)$. By Theorem 20.8 and the fact that $\lim\limits_{n\to\infty} \|x_n - x^*\|$ exists, we get $\lim\limits_{d\to\infty} b_{n,d} = 0$. Note also that,

$$\begin{aligned}
g_{n+d}(t) &= \|tx_{n+d} + (1-t)u - v\| \\
&= \|tS_{n,d}x_n + (1-t)u - v\| \\
&\leq \|(tS_{n,d}x_n + (1-t)u) - S_{n,d}(tx_n + (1-t)u)\| \\
&\quad + \|S_{n,d}(tx_n + (1-t)u) - v\| \\
&\leq b_{n,d} + \|S_{n,d}(tx_n + (1-t)u) - S_{n,d}v\| \\
&\leq b_{n,d} + g_n(t),
\end{aligned}$$

which implies $\limsup\limits_{n\to\infty} g_n(t) \leq \liminf\limits_{n\to\infty} g_n(t)$ and the proof is complete. $\qquad\square$

Theorem 20.17. *Let E be a real uniformly convex Banach space whose dual E^* satisfies the Kadec-Klee property. Let K be a nonempty closed convex subset of E. Let $T_1, T_2, \ldots, T_m : K \to K$ be nonexpansive mappings. Let the sequence $\{\alpha_n\}_{n=1}^{\infty}$ be as in Lemma 20.11 and $\{x_n\}$ be defined iteratively by (20.22). Then, $\{x_n\}$ converges weakly to some common fixed point of $\{T_i\}_{i=1}^{m}$.*

Proof. Since $\{x_n\}$ is bounded, by the reflexivity of E, there exists a subsequence $\{x_{n_j}\}$ of $\{x_n\}$ that converges weakly to some x^* in K. By Lemma 20.6 and Lemma 20.11, we have $T_i x^* = x^*$ for each $i \in I$, and this implies $x^* \in \bigcap_{i=1}^{m} F(T_i)$. Suppose we have another subsequence say, $\{x_{n_k}\}$ of $\{x_n\}$ converging weakly to say q^*, then $x^*, q^* \in \omega_w(\{x_n\}) \cap \bigcap_{i=1}^{m} F(T_i)$ (where $\omega_w(\{x_n\})$ denotes the weak limit set of $\{x_n\}$). By Lemma 20.16, $\lim\limits_{n\to\infty} \|tx_n + (1-t)x^* - q^*\|$ exists for all $t \in [0, 1]$ and by Lemma 20.7, $x^* = q^*$ and the proof is complete. $\qquad\square$

20.5 The Case for Nonexpansive Mappings

For completeness and easy reference, we conclude with the following theorems for finite families of *nonexpansive* mappings.

Theorem 20.18. *Let E be a real uniformly convex Banach space and K be a closed convex nonempty subset of E which is also a nonexpansive retract with retraction P. Let $T_1, T_2, \ldots, T_m : K \to E$ be nonexpansive mappings of K into E. Let the sequence $\{\alpha_n\}_{n=1}^{\infty}$ be as in lemma 20.11. If one of $\{T_i\}_{i=1}^{m}$ is either completely continuous or semi-compact, then the sequence $\{x_n\}$ defined by (20.2) converges strongly to a common fixed point of $\{T_i\}_{i=1}^{m}$.*

Theorem 20.19. *Let E be a real uniformly convex Banach space and K be a closed convex nonempty subset of E which is also a nonexpansive retract with* *retraction P. Let $T_1, T_2, \ldots, T_m : K \to E$ be nonexpansive mappings of K into E. Let the sequence $\{\alpha_n\}_{n=1}^{\infty}$ be as in Lemma 20.11. If E satisfies Opial's condition or has a Frèchet differentiable norm, then the sequence $\{x_n\}$ defined by (20.2) converges weakly to a common fixed point of $\{T_i\}_{i=1}^{m}$.*

20.6 Historical Remarks

More on metric projection operators and their applications in iterative methods for non-self mappings in Banach spaces and on approximation of a common fixed point for families of asymptotically nonexpansive mappings can be found in Al'ber ([3], [4]), Al'ber *et al.* ([5], [6], [12], [13], [11]), Al'ber and Ryazantseva [14], Chidume and Ali [123], Chidume *et al.* [141], Chidume and Li [142], Jung and Kim [266], Tan and Zhou [483], Takahashi and Kim [476], Xu and Yin [520], Al'ber and Guerre-delabriere [7], Al'ber and Reich [10], Li [297], [298]), Falset *et al.* [222]. All the theorems of this chapter are due to Chidume and Ali [122].

Chapter 21
Families of Total Asymptotically Nonexpansive Maps

21.1 Introduction

Let K be a nonempty subset of a real normed linear space, E. Recall that a mapping $T : K \to K$ is said to be *asymptotically nonexpansive in the intermediate sense* (see e.g., Bruck [60]) if it is continuous and the following inequality holds:

$$\limsup_{n\to\infty} \sup_{x,y\in K} (\|T^n x - T^n y\| - \|x - y\|) \leq 0. \tag{21.1}$$

If $F(T) := \{x \in K : Tx = x\} \neq \emptyset$ and (21.1) holds for all $x \in K$, $y \in F(T)$, then T is called *asymptotically quasi-nonexpansive in the intermediate sense*. Observe that if we define

$$a_n := \sup_{x,y\in K} (\|T^n x - T^n y\| - \|x - y\|), \; and \; \sigma_n = max\{0, a_n\}, \tag{21.2}$$

then $\sigma_n \to 0$ as $n \to \infty$ and (21.1) reduces to

$$\|T^n x - T^n y\| \leq \|x - y\| + \sigma_n, \; for \; all \; x,y \in K, \; n \geq 1. \tag{21.3}$$

Recently, Alber *et al.* [6] introduced the class of *total asymptotically nonexpansive mappings*.

Definition 21.1. A mapping $T : K \to K$ is said to be *total asymptotically nonexpansive* if there exist nonnegative real sequences $\{\mu_n\}$ and $\{l_n\}$, $n \geq 1$ with μ_n, $l_n \to 0$ as $n \to \infty$ and a strictly increasing continuous function $\phi : \mathbb{R}^+ \to \mathbb{R}^+$ with $\phi(0) = 0$ such that for all $x, y \in K$,

$$\|T^n x - T^n y\| \leq \|x - y\| + \mu_n \phi(\|x - y\|) + l_n, \; n \geq 1. \tag{21.4}$$

Remark 21.2. If $\phi(\lambda) = \lambda$, then (21.4) reduces to

$$\|T^n x - T^n y\| \leq (1 + \mu_n)\|x - y\| + l_n, \; n \geq 1.$$

C. Chidume, *Geometric Properties of Banach Spaces and Nonlinear Iterations*,
Lecture Notes in Mathematics 1965,
© Springer-Verlag London Limited 2009

In addition, if $l_n = 0$ for all $n \geq 1$, then total asymptotically nonexpansive mappings coincide with asymptotically nonexpansive mappings. If $\mu_n = 0$ and $l_n = 0$ for all $n \geq 1$, we obtain from (21.4) the class of mappings that includes the class of nonexpansive mappings. If $\mu_n = 0$ and $l_n = \sigma_n = max\{0, a_n\}$, where $a_n := \sup_{x,y \in K} (\|T^n x - T^n y\| - \|x - y\|)$ for all $n \geq 1$, then (21.4) reduces to (21.1) which has been studied as mappings asymptotically nonexpansive in the intermediate sense.

The idea of Definition 21.1 is to unify various definitions of classes of mappings associated with the class of asymptotically nonexpansive mappings and which are extensions of nonexpansive mappings; and to prove general convergence theorems applicable to all these classes.

Shahzad and Udomene [444] established necessary and sufficient conditions for convergence of an Ishikawa-type iteration sequence to a common fixed point of two asymptotically quasi-nonexpansive mappings in arbitrary real Banach spaces. They also established a sufficient condition for the convergence of the Ishikawa-type iteration sequences to a common fixed point of two uniformly continuous asymptotically quasi-nonexpansive mappings in real uniformly convex Banach spaces.

Also, Quan *et al.* [396] studied necessary and sufficient conditions for the so called *finite-step iterative sequences with mean errors* for a finite family of *asymptotically quasi-nonexpansive-type* mappings in Banach spaces to converge to a common fixed point of members of the family; where a mapping T is said to be of the asymptotically quasi-nonexpansive-type if T is continuous and

$$\limsup_{n \to \infty} \left\{ \sup_{x \in E, p \in F(T)} \left[\|T^n x - p\|^2 - \|x - p\|^2 \right] \right\} \leq 0.$$

Remark 21.3. Observe that

$$\limsup_{n \to \infty} \left\{ \sup_{x \in E, p \in F(T)} \left[\|T^n x - p\|^2 - \|x - p\|^2 \right] \right\} \leq 0$$

implies

$$\limsup_{n \to \infty} \left\{ \sup_{x \in E, p \in F(T)} \left(\|T^n x - p\| - \|x - p\| \right) \left(\|T^n x - p\| + \|x - p\| \right) \right\} \leq 0$$

which implies $\limsup_{n \to \infty} \left\{ \sup_{x \in E, p \in F(T)} \left[\|T^n x - p\| - \|x - p\| \right] \right\} \leq 0$, so that asymptotically quasi-nonexpansive-type mappings studied by Jing Quan *et al.* [396] reduce to mappings which are asymptotically quasi-nonexpansive in the intermediate sense.

In this chapter, we study an iterative sequence for the approximation of common fixed points of finite families of *total asymptotically nonexpansive mappings*; and give necessary and sufficient conditions for the convergence of the scheme to common fixed points of the mappings in arbitrary real Banach spaces. Furthermore, a sufficient condition for convergence of the iteration process to a common fixed point of these mappings is established in real uniformly convex Banach spaces. In fact, the theorems of this chapter hold for the slightly more general class of *total asymptotically quasi-nonexpansive mappings*.

21.2 Convergence Theorems

Let K be a nonempty closed convex subset of a real normed space E. Let $T_1, T_2, ..., T_m : K \to K$ be m total asymptotically nonexpansive mappings. We define the iterative sequence $\{x_n\}$ by

$$\begin{cases} x_1 \in K, \\ x_{n+1} = (1 - \alpha_n)x_n + \alpha_n T_1^n x_n, \ if \ m = 1, \ n \geq 1, \\ x_1 \in K, \\ x_{n+1} = (1 - \alpha_n)x_n + \alpha_n T_1^n y_{1n} \\ y_{1n} = (1 - \alpha_n)x_n + \alpha_n T_2^n y_{2n} \\ \vdots \\ y_{(m-2)n} = (1 - \alpha_n)x_n + \alpha_n T_{m-1}^n y_{(m-1)n} \\ y_{(m-1)n} = (1 - \alpha_n)x_n + \alpha_n T_m^n x_n, \ if \ m \geq 2, \ n \geq 1, \end{cases} \tag{21.5}$$

where $\{\alpha_n\}_{n=1}^\infty$ is a sequence in $[0, 1]$ bounded away from 0 and 1.

We prove the following theorem.

Theorem 21.4. *Let E be a real Banach space, K be a nonempty closed convex subset of E and $T_i : K \to K$, $i = 1, 2, ..., m$ be m total asymptotically nonexpansive mappings with sequences $\{\mu_{in}\}$, $\{l_{in}\}$ $n \geq 1$, $i = 1, 2, ..., m$ such that $F := \cap_{i=1}^m F(T_i) \neq \emptyset$. Let $\{x_n\}$ be given by (21.5). Suppose $\sum\limits_{n=1}^\infty \mu_{in} < \infty$, $\sum\limits_{n=1}^\infty l_{in} < \infty$, $i = 1, 2, ..., m$ and suppose that there exist M_i, $M_i^* > 0$ such that $\phi_i(\lambda_i) \leq M_i^* \lambda_i$ for all $\lambda_i \geq M_i$, $i = 1, 2, ..., m$. Then, the sequence $\{x_n\}$ is bounded and $\lim\limits_{n \to \infty} \|x_n - p\|$ exists, $p \in F$.*

Proof. Let $p \in F$. Then for $m = 1$, we have from (21.5) that $x_1 \in K$, $x_{n+1} = (1 - \alpha_n)x_n + \alpha_n T_1^n x_n$. Using this we obtain

$$\|x_{n+1} - p\| = \|(1 - \alpha_n)(x_n - p) + \alpha_n(T_1^n x_n - p)\|$$
$$\leq (1 - \alpha_n)\|x_n - p\| + \alpha_n \left[\|x_n - p\| + \mu_{1n}\phi_1\left(\|x_n - p\| \right) + l_{1n} \right].$$

Since ϕ_1 is an increasing function, it follows that $\phi_1(\lambda_1) \leq \phi_1(M_1)$ whenever $\lambda_1 \leq M_1$ and (by hypothesis) $\phi_1(\lambda_1) \leq M_1^* \lambda_1$ if $\lambda_1 \geq M_1$. In either case, we have $\phi_1(\|x_n - p\|) \leq \phi_1(M_1) + M_1^* \|x_n - p\|$ for some constants $M_1 > 0$,

$M_1^* > 0$. Thus,

$$\|x_{n+1} - p\| \le \|x_n - p\| + \alpha_n \mu_{1n} \phi_1(M_1) + \alpha_n \mu_{1n} M_1^* \|x_n - p\| + \alpha_n l_{1n},$$

from which we obtain $\|x_{n+1} - p\| \le (1 + \mu_{1n} Q_1) \|x_n - p\| + (\mu_{1n} + l_{1n}) Q_1$ for some constant $Q_1 > 0$. Next, for $m = 2$, we obtain from (21.5) that

$$\begin{cases} x_1 \in K, \\ x_{n+1} = (1 - \alpha_n) x_n + \alpha_n T_1^n y_{1n} \\ y_{1n} = (1 - \alpha_n) x_n + \alpha_n T_2^n x_n, \end{cases}$$

and from this we have,

$$\begin{aligned} \|x_{n+1} - p\| &= \|(1 - \alpha_n)(x_n - p) + \alpha_n (T_1^n y_{1n} - p)\| \\ &\le (1 - \alpha_n) \|x_n - p\| + \alpha_n \Big[\|y_{1n} - p\| + \mu_{1n} \phi_1 \Big(\|y_{1n} - p\| \Big) + l_{1n} \Big], \end{aligned}$$

$$\begin{aligned} \|y_{1n} - p\| &\le \|(1 - \alpha_n)(x_n - p) + \alpha_n (T_2^n x_n - p)\| \\ &\le (1 - \alpha_n) \|x_n - p\| + \alpha_n \Big[\|x_n - p\| + \mu_{2n} \phi_2 \Big(\|x_n - p\| \Big) + l_{2n} \Big]. \end{aligned}$$

Again, since ϕ_i is an increasing function for $i = 1, 2$, it follows that $\phi_i(\lambda_i) \le \phi_i(M_i) + M_i^* \lambda_i$ for some constants $M_i > 0$, $M_i^* > 0$, $i = 1, 2$. Hence,

$$\|x_{n+1} - p\| \le (1 - \alpha_n) \|x_n - p\| + \alpha_n \|y_{1n} - p\| + \alpha_n \mu_{1n} \phi_1(M_1)$$

$$+ \alpha_n \mu_{1n} M_1^* \|y_{1n} - p\| + \alpha_n l_{1n} \tag{21.6}$$

and

$$\|y_{1n} - p\| \le \|x_n - p\| + \alpha_n \mu_{2n} \phi_2(M_2)$$

$$+ \alpha_n \mu_{2n} M_2^* \|x_n - p\| + \alpha_n l_{2n}. \tag{21.7}$$

Substituting (21.7) in (21.6) gives

$$\|x_{n+1} - p\| \le (1 - \alpha_n)\|x_n - p\| + \alpha_n \Big[\|x_n - p\| + \alpha_n \mu_{2n}\phi_2(M_2)$$

$$+\alpha_n\mu_{2n}M_2^*\|x_n - p\| + \alpha_n l_{2n}\Big] + \alpha_n\mu_{1n}\phi_1(M_1)$$

$$+\alpha_n\mu_{1n}M_1^*\Big[\|x_n - p\| + \alpha_n\mu_{2n}\phi_2(M_2)$$

$$+\alpha_n\mu_{2n}M_2^*\|x_n - p\| + \alpha_n l_{2n}\Big] + \alpha_n l_{1n}$$

$$\le \|x_n - p\| + \alpha_n\mu_{2n}\phi_2(M_2) + \alpha_n\mu_{2n}M_2^*\|x_n - p\| + \alpha_n l_{2n}$$

$$+\alpha_n\mu_{1n}\phi_1(M_1) + \alpha_n\mu_{1n}M_1^*\|x_n - p\| + \alpha_n\mu_{1n}\mu_{2n}M_1^*\phi_2(M_2)$$

$$+\alpha_n\mu_{1n}\mu_{2n}M_1^*M_2^*\|x_n - p\| + \alpha_n\mu_{1n}l_{2n}M_1^* + \alpha_n l_{1n}.$$

Thus we have

$$\|x_{n+1} - p\| \le \Big(1 + (\mu_{1n} + \mu_{2n})Q_2\Big)\|x_n - p\| + (\mu_{1n} + \mu_{2n} + l_{1n} + l_{2n})Q_2$$

for some constant $Q_2 > 0$. Following the computations as above, we obtain that

$$\|x_{n+1} - p\| \le \Big(1 + Q\sum_{j=1}^{m}\mu_{jn}\Big)\|x_n - p\| + Q\sum_{j=1}^{m}(\mu_{jn} + l_{jn})$$

for some constant $Q > 0$. Hence,

$$\|x_{n+1} - p\| \le (1 + \delta_n)\|x_n - p\| + \gamma_n, \ n \ge 1, \tag{21.8}$$

where $\delta_n := Q\sum_{j=1}^{m}\mu_{jn}$ and $\gamma_n := Q\sum_{j=1}^{m}(\mu_{jn} + l_{jn})$. Observe that $\sum_{n=1}^{\infty}\delta_n < \infty$ and $\sum_{n=1}^{\infty}\gamma_n < \infty$. So, from (21.8) and Lemma 10.5, we obtain that the sequence $\{x_n\}$ is bounded and that $\lim_{n\to\infty}\|x_n - p\|$ exists. This completes the proof. $\square$

21.2.1 Necessary and Sufficient Conditions for Convergence in Real Banach Spaces

Theorem 21.5. *Let E be a real Banach space, K be a nonempty closed convex subset of E and $T_i : K \to K$, $i = 1, 2, ..., m$ be m continuous total asymptotically nonexpansive mappings with sequences $\{\mu_{in}\}$, $\{l_{in}\}$ $n \ge 1$, $i = 1, 2, ..., m$ such that $F := \cap_{i=1}^{m}F(T_i) \ne \emptyset$. Let $\{x_n\}$ be given* *by (21.5). Suppose* $\sum_{n=1}^{\infty} \mu_{in} < \infty$, $\sum_{n=1}^{\infty} l_{in} < \infty$ $i = 1, 2, ..., m$ *and suppose that there exist* M_i, $M_i^* > 0$ *such that* $\phi_i(\lambda_i) \leq M_i^* \lambda_i$ *for all* $\lambda_i \geq M_i$, $i = 1, 2, ..., m$. *Then, the sequence* $\{x_n\}$ *converges strongly to a common fixed point of* T_i, $i = 1, 2, ..., m$ *if and only if* $\liminf_{n\to\infty} d\left(x_n, F\right) = 0$, *where* $d\left(x_n, F\right) = \inf_{y \in F} \|x_n - y\|$, $n \geq 1$.

Proof. **Necessity:** Since (21.8) holds for all $p \in F$, we obtain from it that

$$d\left(x_{n+1}, F\right) \leq (1 + \delta_n) d\left(x_n, F\right) + \gamma_n, \quad n \geq 1. \tag{21.9}$$

Lemma 10.5 then implies that $\lim_{n\to\infty} d\left(x_n, F\right)$ exists. But, $\liminf_{n\to\infty} d\left(x_n, F\right) = 0$. Hence, $\lim_{n\to\infty} d\left(x_n, F\right) = 0$.

Sufficiency: We first show that $\{x_n\}$ is a Cauchy sequence in E. For all integers $k \geq 1$, we obtain from inequality (21.8) that

$$\|x_{n+k} - p\| \leq \prod_{j=n}^{n+k-1} \left(1 + \delta_j\right) \|x_n - p\| + \left(\sum_{j=n}^{n+k-1} \gamma_j\right) \prod_{j=n}^{n+k-1} \left(1 + \delta_j\right)$$

$$\leq exp\left(\sum_{j=n}^{n+k-1} \delta_j\right) \|x_n - p\| + \left(\sum_{j=n}^{n+k-1} \gamma_j\right) exp\left(\sum_{j=n}^{n+k-1} \delta_j\right),$$

so that for all integers $k \geq 1$ and all $p \in F$,

$$\|x_{n+k} - x_n\| \leq \|x_{n+k} - p\| + \|x_n - p\|$$

$$\leq \left[1 + exp\left(\sum_{j=n}^{n+k-1} \delta_j\right)\right] \|x_n - p\| + \left(\sum_{j=n}^{n+k-1} \gamma_j\right) exp\left(\sum_{j=n}^{n+k-1} \delta_j\right).$$

We therefore have that

$$\|x_{n+k} - x_n\| \leq D\|x_n - p\| + D\left(\sum_{j=n}^{\infty} \gamma_j\right), \tag{21.10}$$

for some constant $D > 0$. This yields

$$\|x_{n+k} - x_n\| \leq Dd\left(x_n, F\right) + D\left(\sum_{j=n}^{\infty} \gamma_j\right). \tag{21.11}$$

Now, since $\lim\limits_{n\to\infty} d\left(x_n, F\right) = 0$ and $\sum\limits_{j=1}^{\infty} \gamma_j < \infty$, given $\epsilon > 0$, there exists

an integer $N_1 > 0$ such that for all $n \geq N_1$, $d\left(x_n, F\right) < \dfrac{\epsilon}{2(D+1)}$ and

$\sum\limits_{j=n}^{\infty} \gamma_j < \dfrac{\epsilon}{2(D+1)}$. So, for all integers $n \geq N_1$, $k \geq 1$, we obtain from

(21.11) that $\|x_{n+k} - x_n\| < \epsilon$. Hence, $\{x_n\}$ is a Cauchy sequence in E; and
since E is complete, there exists $l^* \in E$ such that $x_n \to l^*$ as $n \to \infty$.
We now show that l^* is a common fixed point of T_i, $i = 1, 2, ..., m$, that
is, we show that $l^* \in F$. Suppose for contradiction that $l^* \in F^c$ (where F^c
denotes the complement of F). Since F is a closed subset of E (recall each
T_i, $i = 1, 2, ..., m$ is continuous), we have that $d\left(l^*, F\right) > 0$. But, for all
$p \in F$, we have $\|l^* - p\| \leq \|l^* - x_n\| + \|x_n - p\|$ which implies

$$d\left(l^*, F\right) \leq \|x_n - l^*\| + d\left(x_n, F\right),$$

so that as $n \to \infty$, we obtain $d\left(l^*, F\right) = 0$ which contradicts $d\left(l^*, F\right) > 0$.
Thus, l^* is a common fixed point of T_i, $i = 1, 2, ..., m$. This completes the
proof. $\qquad\square$

Remark 21.6. If $T_1, T_2, ..., T_m$ are asymptotically nonexpansive mappings,
then $l_{in} = 0$ for all $n \geq 1$, $i = 1, 2, ..., m$ and $\phi_i(\lambda_i) = \lambda_i$ so that the assumption that there exist M_i, $M_i^* > 0$ such that $\phi_i(\lambda_i) \leq M_i^* \lambda_i$ for all
$\lambda_i \geq M_i$, $i = 1, 2, ..., m$ in the above theorems will no longer be needed.
Thus, we have the following corollary.

Corollary 21.7 *Let E be a real Banach space, K be a nonempty closed
convex subset of E and $T_i : K \to K$, $i = 1, 2, ..., m$ be m asymptotically
nonexpansive mappings with sequences $\{\mu_{in}\}$, $n \geq 1$, $i = 1, 2, ..., m$ such
that $F = \cap_{i=1}^{m} F(T_i) \neq \emptyset$. Let $\{x_n\}$ be given by (21.5). Suppose $\sum\limits_{n=1}^{\infty} \mu_{in} <
\infty$, $i = 1, 2, ..., m$. Then, the sequence $\{x_n\}$ is bounded and $\lim\limits_{n\to\infty} \|x_n - p\|$
exists, $p \in F$. Moreover, $\{x_n\}$ converges strongly to a common fixed point of
T_i, $i = 1, 2, ..., m$ if and only if $\liminf\limits_{n\to\infty} d\left(x_n, F\right) = 0$.*

21.2.2 *Convergence Theorem in Real Uniformly Convex Banach Spaces*

Theorem 21.8. *Let E be a real uniformly convex Banach space, K be a
nonempty closed convex subset of E and $T_i : K \to K$, $i = 1, 2, ..., m$ be m uniformly continuous total asymptotically nonexpansive mappings with sequences*

$\{\mu_{in}\}$, $\{l_{in}\} \subset [0, \infty)$ such that $\sum_{n=1}^{\infty} \mu_{in} < \infty$, $\sum_{n=1}^{\infty} l_{in} < \infty$, $i = 1, 2, ..., m$ and $F := \cap_{i=1}^{m} F(T_i) \neq \emptyset$. Let $\{\alpha_n\} \subset [\varepsilon, 1 - \varepsilon]$ for some $\varepsilon \in (0, 1)$. From arbitrary $x_1 \in E$, define the sequence $\{x_n\}$ by (21.5). Suppose that there exist M_i, $M_i^* > 0$ such that $\phi_i(\lambda_i) \leq M_i^* \lambda_i$ whenever $\lambda_i \geq M_i$, $i = 1, 2, ..., m$. Then, $\lim_{n \to \infty} \|x_n - T_i^n x_n\| = 0$, $i = 1, 2, ..., m$.

Proof. Let $p \in F$. Then by Theorem 21.4, $\lim_{n \to \infty} \|x_n - p\|$ exists. Let $\lim_{n \to \infty} \|x_n - p\| = r$. If $r = 0$, then by continuity of T_i, $i = 1, 2, ..., m$, we are done. Now suppose $r > 0$. We show that $\lim_{n \to \infty} \|x_n - T_i^n x_n\| = 0$, $i = 1, 2, ..., m$. Now, for $m = 1$, we get from (21.5) that $x_1 \in K$, $x_{n+1} = (1 - \alpha_n)x_n + \alpha_n T_1^n x_n$. Using this and inequality (4.32), we have for some constant $R_1 > 0$ that

$$\|x_{n+1} - p\|^2 \leq (1 - \alpha_n)\|x_n - p\|^2 + \alpha_n \Big[\|x_n - p\| + \mu_{1n}\phi_1(\|x_n - p\|) + l_{1n}\Big]^2$$
$$- w_2(\alpha_n)g(\|T_1^n x_n - x_n\|)$$
$$\leq (1 - \alpha_n)\|x_n - p\|^2 + \alpha_n \Big[\|x_n - p\| + \mu_{1n}\phi_1(M_1)$$
$$+ \mu_{1n}M_1^*\|x_n - p\| + l_{1n}\Big]^2 - w_2(\alpha_n)g(\|T_1^n x_n - x_n\|)$$
$$\leq \|x_n - p\|^2 + (\mu_{1n} + l_{1n})R_1 - \varepsilon^2 g(\|T_1^n x_n - x_n\|).$$

Thus, $\varepsilon^2 g(\|T_1^n x_n - x_n\|) \leq \|x_n - p\|^2 - \|x_{n+1} - p\|^2 + (\mu_{1n} + l_{1n})R_1$ which implies

$$\varepsilon^2 \sum_{n=1}^{\infty} g(\|T_1^n x_n - x_n\|) \leq \|x_1 - p\|^2 + R_1 \sum_{n=1}^{\infty} (\mu_{1n} + l_{1n}) < \infty.$$

Hence, $\varepsilon^2 \sum_{n=1}^{\infty} g(\|T_1^n x_n - x_n\|) < \infty$. So, $\lim_{n \to \infty} g(\|T_1^n x_n - x_n\|) = 0$; and properties of g imply $\lim_{n \to \infty} \|T_1^n x_n - x_n\| = 0$. For $m = 2$, (21.5) becomes

$$\begin{cases} x_1 \in K, \\ x_{n+1} = (1 - \alpha_n)x_n + \alpha_n T_1^n y_{1n} \\ y_{1n} = (1 - \alpha_n)x_n + \alpha_n T_2^n x_n, \end{cases} \qquad (21.12)$$

so we have from this that

$$\|y_{1n} - p\| \leq (1 - \alpha_n)\|x_n - p\| + \alpha_n \mu_{2n}\phi_2(M_2)$$
$$+ \alpha_n(1 + \mu_{2n}M_2^*)\|x_n - p\| + \alpha_n l_{2n}$$
$$\leq \Big(1 + \mu_{2n}R_2\Big)\|x_n - p\| + (\mu_{2n} + + l_{2n})R_2, \qquad (21.13)$$

for some constant $R_2 > 0$. Using (21.12), (21.13) and inequality (4.32), we obtain, for some constant $R_3 > 0$, that

$$\|x_{n+1} - p\|^2 \leq (1 - \alpha_n)\|x_n - p\|^2$$
$$+\alpha_n\|T_1^n y_{1n} - p\|^2 - w_2(\alpha_n)g(\|T_1^n y_{1n} - x_n\|)$$
$$\leq (1 - \alpha_n)\|x_n - p\|^2 + \alpha_n\Big[\|y_{1n} - p\| + \mu_{1n}\phi_1(\|y_{1n} - p\|) + l_{1n}\Big]^2$$
$$-w_2(\alpha_n)g(\|T_1^n y_{1n} - x_n\|)$$
$$\leq \|x_n - p\|^2 + (\mu_{2n} + l_{2n} + \mu_{1n} + l_{1n})R_3 - \varepsilon^2 g(\|T_1^n y_n - x_n\|).$$

Thus, $\varepsilon^2 g(\|T_1^n y_{1n} - x_n\|) \leq \|x_n - p\|^2 - \|x_{n+1} - p\|^2 + (\mu_{2n} + l_{2n} + \mu_{1n} + l_{1n})R_3$ which implies that

$$\varepsilon^2 \sum_{n=1}^{\infty} g(\|T_1^n y_{1n} - x_n\|) \leq \|x_1 - p\|^2 + R_3 \sum_{n=1}^{\infty} (\mu_{2n} + l_{2n} + \mu_{1n} + l_{1n}) < \infty.$$

So, $\lim_{n\to\infty} g(\|T_1^n y_{1n} - x_n\|) = 0$; and properties of g yield $\lim_{n\to\infty} \|T_1^n y_{1n} - x_n\| = 0$. Next,

$$\|x_n - p\| \leq \|x_n - T_1^n y_{1n}\| + \|T_1^n y_{1n} - p\|$$
$$\leq \|x_n - T_1^n y_{1n}\| + \|y_{1n} - p\| + \mu_{1n}M + l_{1n},$$

for some constant $M > 0$. Hence, we deduce from this that $r \leq \liminf_{n\to\infty} \|y_{1n} - p\|$. Also from (21.13), we obtain that $\|y_{1n} - p\| \leq \Big(1 + \mu_{2n}R_2\Big)\|x_n - p\| + (\mu_{2n} + l_{2n})R_2$. This gives $\limsup_{n\to\infty} \|y_{1n} - p\| \leq r$. Thus,

$$\lim_{n\to\infty} \|y_{1n} - p\| = r; \ i.e., \ \lim_{n\to\infty} \|(1 - \alpha_n)(x_n - p) + \alpha_n(T_2^n x_n - p)\| = r.$$

Again, we have that

$$\|T_2^n x_n - p\| \leq \|x_n - p\| + \mu_{2n}\phi_2(\|x_n - p\|) + l_{2n}$$
$$\leq \|x_n - p\| + \mu_{2n}\phi_2(M_2) + \mu_{2n}M_2^*\|x_n - p\| + l_{2n}.$$

This implies that $\limsup_{n\to\infty} \|T_2^n x_n - p\| \leq r$. Hence, by inequality (4.32), we have that $\lim_{n\to\infty} \|T_2^n x_n - x_n\| = 0$. Also, $\|T_1^n x_n - x_n\| \leq \|T_1^n x_n - T_1^n y_{1n}\| + \|T_1^n y_{1n} - x_n\|$. Uniform continuity of T_1 and the fact that $\|x_n - y_{1n}\| \to 0$ as $n \to \infty$ give $\lim_{n\to\infty} \|T_1^n x_n - x_n\| = 0$. Hence, $\lim_{n\to\infty} \|T_1^n x_n - x_n\| = \lim_{n\to\infty} \|T_2^n x_n - x_n\| = 0$. Continuing, we get that $\lim_{n\to\infty} \|T_i^n x_n - x_n\| = 0$, $i = 1, 2, ..., m$. This completes the proof. □

Theorem 21.9. *Let E be a real uniformly convex Banach space, K be a nonempty closed convex subset of E and $T_i : K \to K$, $i = 1, 2, ..., m$ be m uniformly continuous total asymptotically nonexpansive mappings with sequences $\{\mu_{in}\}$, $\{l_{in}\} \subset [0, \infty)$ such that $\sum_{n=1}^{\infty} \mu_{in} < \infty$, $\sum_{n=1}^{\infty} l_{in} < \infty$, $i = 1, 2, ..., m$ and $F := \cap_{i=1}^{m} F(T_i) \neq \emptyset$. Let $\{\alpha_{in}\} \subset [\varepsilon, 1 - \varepsilon]$ for some $\varepsilon \in (0, 1)$. From arbitrary $x_1 \in E$, define the sequence $\{x_n\}$ by (21.5). Suppose that there exist M_i, $M_i^* > 0$ such that $\phi_i(\lambda_i) \leq M_i^* \lambda_i$ whenever $\lambda_i \geq M_i$, $i = 1, 2, ..., m$; and that one of $T_1, T_2, ..., T_m$ is compact, then $\{x_n\}$ converges strongly to some $p \in F$.*

Proof. We obtain from Theorem 21.8 that

$$\lim_{n \to \infty} \|T_i^n x_n - x_n\| = 0, \ i = 1, 2, ..., m. \tag{21.14}$$

We also have that

$$\lim_{n \to \infty} \|T_1^n y_{1n} - x_n\| = 0. \tag{21.15}$$

Without loss of generality, let T_1 be compact. Since T_1 is continuous and compact, it is completely continuous. Thus, there exists a subsequence $\{T_1^{n_k} x_{n_k}\}$ of $\{T_1^n x_n\}$ such that $T_1^{n_k} x_{n_k} \to x^*$ as $k \to \infty$ for some $x^* \in E$. Thus $T_1^{n_k+1} x_{n_k} \to T_1 x^*$ as $k \to \infty$ and from (21.14), we have that $\lim_{k \to \infty} x_{n_k} = x^*$. Also from (21.14), $T_2^{n_k} x_{n_k} \to x^*$, $T_3^{n_k} x_{n_k} \to x^*$, ..., $T_m^{n_k} x_{n_k} \to x^*$ as $k \to \infty$. Thus, $T_2^{n_k+1} x_{n_k} \to T_2 x^*$, $T_3^{n_k+1} x_{n_k} \to T_3 x^*$, ..., $T_m^{n_k+1} x_{n_k} \to T_m x^*$ as $k \to \infty$. Now, since $\|x_{n_k+1} - x_{n_k}\| \leq \|x_{n_k} - T_1^{n_k} y_{1n_k}\|$, it follows (using (21.15)) that $x_{n_k+1} \to x^*$ as $k \to \infty$. Next, we show that $x^* \in F$. Observe that

$$\|x^* - T_1 x^*\| \leq \|x^* - x_{n_k+1}\| + \|x_{n_k+1} - T_1^{n_k+1} x_{n_k+1}\|$$
$$+ \|T_1^{n_k+1} x_{n_k+1} - T_1^{n_k+1} x_{n_k}\| + \|T_1^{n_k+1} x_{n_k} - T_1 x^*\|.$$

Taking limit as $k \to \infty$ and using the fact that T_1 is uniformly continuous we have that $x^* = T_1 x^*$ and so $x^* \in F(T_1)$. Also,

$$\|x^* - T_2 x^*\| \leq \|x^* - x_{n_k+1}\| + \|x_{n_k+1} - T_2^{n_k+1} x_{n_k+1}\|$$
$$+ \|T_2^{n_k+1} x_{n_k+1} - T_2^{n_k+1} x_{n_k}\| + \|T_2^{n_k+1} x_{n_k} - T_2 x^*\|.$$

Taking limit as $k \to \infty$ and using the fact that T_2 is uniformly continuous we obtain that $x^* = T_2 x^*$; that is, $x^* \in F(T_2)$. Again,

$$\|x^* - T_3 x^*\| \leq \|x^* - x_{n_k+1}\| + \|x_{n_k+1} - T_3^{n_k+1} x_{n_k+1}\|$$
$$+ \|T_3^{n_k+1} x_{n_k+1} - T_3^{n_k+1} x_{n_k}\| + \|T_3^{n_k+1} x_{n_k} - T_3 x^*\|.$$

As $k \to \infty$, we have that $x^* \in F(T_3)$. Eventually, we have that, $x^* \in F$. But by Theorem 21.4, $\lim_{n\to\infty} \|x_n - p\|$ exists, $p \in F$. Hence, $\{x_n\}$ converges strongly to $x^* \in F$. This completes the proof. $\qquad\qquad\qquad\qquad\qquad\qquad\square$

In view of Remark 21.6, the following corollary is now obvious.

Corollary 21.10 *Let E be a real uniformly convex Banach space, K be a nonempty closed convex subset of E and $T_i : K \to K$, $i = 1, 2, ..., m$ be m uniformly continuous asymptotically nonexpansive mappings with sequences $\{\mu_{in}\} \subset [0, \infty)$ such that $\sum_{n=1}^{\infty} \mu_{in} < \infty$, $i = 1, 2, ..., m$ and $F = \cap_{i=1}^{m} F(T_i) \neq \emptyset$. Let $\{a_{in}\} \subset [\varepsilon, 1 - \varepsilon]$ for some $\varepsilon \in (0, 1)$. From arbitrary $x_1 \in E$, define the sequence $\{x_n\}$ by (21.5). Then $\{x_n\}$ converges strongly to some $p \in F$.*

Remark 21.11. Observe that the theorems of this chapter remain true for mappings $T_1, T_2, ..., T_m$ satisfying (21.3) provided $\sum_{n=1}^{\infty} \sigma_n < \infty$. In this case, the requirement that there exist M_i, $M_i^* > 0$ such that $\phi_i(\lambda_i) \leq M_i^* \lambda_i$ for all $\lambda_i \geq M_i$, $i = 1, 2, ..., m$ is not needed.

Remark 21.12. A prototype for $\phi_i : [0, \infty) \to [0, \infty)$ satisfying the conditions of the theorems of this chapter is $\phi_i(\lambda_i) = \lambda_i^s$, $0 < s \leq 1$, $i = 1, 2, ..., m$.

Definition 21.13. A mapping $T : K \to K$ is said to be *total asymptotically quasi-nonexpansive* if $F(T) \neq \emptyset$ and there exist nonnegative real sequences $\{\mu_n\}$ and $\{l_n\}$, $n \geq 1$ with μ_n, $l_n \to 0$ as $n \to \infty$ and strictly increasing continuous function $\phi : \mathbb{R}^+ \to \mathbb{R}^+$ with $\phi(0) = 0$ such that for all $x \in E$, $x^* \in F(T)$,

$$\|T^n x - x^*\| \leq \|x - x^*\| + \mu_n \phi(\|x - x^*\|) + l_n, \ n \geq 1. \qquad (21.16)$$

Remark 21.14. If $\phi(\lambda) = \lambda$, then (21.16) reduces to

$$\|T^n x - x^*\| \leq (1 + \mu_n)\|x - x^*\| + l_n, \ n \geq 1.$$

In addition, if $l_n = 0$ for all $n \geq 1$, then total asymptotically quasi-nonexpansive mappings coincide with asymptotically quasi-nonexpansive mappings studied by various authors. If $\mu_n = 0$ and $l_n = 0$ for all $n \geq 1$, we obtain from (21.16) the well known class of quasi-nonexpansive mappings studied by various authors. Observe that the class of total asymptotically nonexpansive mappings with nonempty fixed point sets belongs to the class of total asymptotically quasi-nonexpansive mappings. Moreover, if $\mu_n = 0$ and $l_n = \sigma_n$, then (21.16) reduces to (21.3) with $y = x^* \in F(T)$, part of which is equivalent to the class of mappings studied by Jing Quan *et al.* [396]. It is trivial to observe that all the theorems of this chapter carry over to the class of total asymptotically quasi-nonexpansive mappings with little modifications.

21.3 The Case of Non-self Maps

Definition 21.15. Let K be a nonempty closed and convex subset of E. Let $P : E \to K$ be the nonexpansive retraction of E onto K. A non-self map $T : K \to E$ is said to be *total asymptotically nonexpansive* if there exist sequences $\{\mu_n\}_{n \geq 1}$, $\{l_n\}_{n \geq 1}$ in $[0, +\infty)$ with $\mu_n, l_n \to 0$ as $n \to \infty$ and a strictly increasing continuous function $\phi : [0, \infty) \to [0, +\infty)$ with $\phi(0) = 0$ such that for all $x, y \in K$,

$$\|T(PT)^{n-1}x - T(PT)^{n-1}y\| \leq \|x - y\| + \mu_n \phi(\|x - y\|) + l_n, \ n \geq 1.$$

Let $T_1, T_2, ..., T_m : K \to E$ be m total asymptotically nonexpansive *non-self* maps; assuming existence of common fixed points of these operators, the theorems and method of proof of this chapter easily carry over to this class of mappings using the iterative sequence $\{x_n\}$ defined by

$$\begin{cases} x_1 \in K \\ x_{x+1} = P\Big((1 - \alpha_n)x_n + \alpha_n T_1(PT_1)^{n-1}x_n\Big) \ if \ m = 1, \ n \geq 1 \\ x_1 \in K, \\ x_{n+1} = P\Big((1 - \alpha_n)x_n + \alpha_n T_1(PT_1)^{n-1}y_{1n}\Big) \\ y_{1n} = P\Big((1 - \alpha_n)x_n + \alpha_n T_2(PT_2)^{n-1}y_{2n}\Big) \\ \vdots \\ y_{(m-2)n} = P\Big((1 - \alpha_n)x_n + \alpha_n T_{m-1}(PT_{m-1})^{n-1}y_{(m-1)n}\Big) \\ y_{(m-1)n} = P\Big((1 - \alpha_n)x_n + \alpha_n T_m(PT_m)^{n-1}x_n\Big), \ if \ m \geq 2 \ n \geq 1, \end{cases}$$

instead of (21.5), provided the well definedness of P as a sunny nonexpansive retraction is guaranteed.

21.4 Historical Remarks

All the theorems of this chapter are due to Chidume and Ofoedu [153], and in particular, they unify, extend and generalize the corresponding results of Alber *et al.* [6], Shahzad and Udomene [444], Quan *et al.* [396], Zegeye and Shahzad [447] and a host of other results on the approximation of common fixed points of finite families of several classes of nonlinear mappings. For other related results, we refer the reader to Zegeye and Shahzad [546].

Chapter 22
Common Fixed Points
for One-parameter Nonexpansive
Semigroup

22.1 Introduction

Let K be a nonempty subset of a normed space E. A family of mappings $\{T(t) : t \geq 0\}$ is called a *one-parameter strongly continuous semigroup of nonexpansive mappings* (or, briefly, *a nonexpansive semigroup*) on K if the following conditions hold:

(i) $T(s + t) = T(s)oT(t)$ for all $s, t \geq 0$;
(ii) For each $x \in K$, the mapping $t \to T(t)x$ from $[0, \infty)$ into K is continuous;
(iii) For each $t \geq 0, T(t)$ is a nonexpansive mapping on K.

The existence of a common fixed point for $\{T(t) : t \geq 0\}$ when E is a uniformly convex Banach space and K is a nonempty closed convex and bounded subset of E is well known (see e.g., Browder [42], Bruck [57]).

Let $\cap_{t \geq 0} F(T(t))$ denote the set of common fixed points of the semigroup $\{T(t) : t \geq 0\}$. In Suzuki [470] proved the following interesting result:

$$\bigcap_{t \geq 0} F(T(t)) = F(T(1)) \cap F(T(\sqrt{2}));$$

i.e., that the set of common fixed points of the nonexpansive semigroup $\{T(t) : t \geq 0\}$ is simply the set of common fixed points of only two mappings $T(1)$ and $T(\sqrt{2})$. Recently, he improved on this result by proving that $\cap_{t \geq 0} F(T(t))$ is actually the set of all fixed points of a single nonexpansive map. He proved the following theorem.

Theorem 22.1. *(Suzuki, [465]) Let $\{T(t) : t \geq 0\}$ be a strongly continuous semigroup of nonexpansive mappings on a subset K of a Banach space E. Let α and β be positive real numbers satisfying $\alpha/\beta \in \mathbb{R}$ and $\alpha/\beta \notin \mathbb{Q}$. Then,*

$$\bigcap_{t \geq 0} F(T(t)) = F(\lambda T(\alpha) + (1 - \lambda)T(\beta))$$

C. Chidume, *Geometric Properties of Banach Spaces and Nonlinear Iterations*,
Lecture Notes in Mathematics 1965,
© Springer-Verlag London Limited 2009

holds for every $\lambda \in (0,1)$, where $\lambda T(\alpha) + (1-\lambda)T(\beta)$ is a mapping from K into E defined by

$$(\lambda T(\alpha) + (1-\lambda)T(\beta))x = \lambda T(\alpha)x + (1-\lambda)T(\beta)x, \ \ x \in K.$$

An immediate corollary of theorem 22.1 is the following.

Corollary 22.2 *Let $\{T(t) : t \geq 0\}$ be a strongly continuous semigroup of nonexpansive mappings on a subset K of a Banach space E. Let α and β be positive real numbers satisfying $\alpha/\beta \in \mathbb{R}$ and $\alpha/\beta \notin \mathbb{Q}$. Then,*

$$\bigcap_{t \geq 0} F(T(t)) = F(T(\alpha)) \cap F(T(\beta)).$$

Suzuki also proved the following theorem.

Theorem 22.3. *(Suzuki [465], theorem 2, p.1015). Let $\{T(t) : t \geq 0\}$ be a strongly continuous semigroup of nonexpansive mappings on a subset K of a Banach space E. Let α and β be different positive real numbers. Then,*

$$F(T(\alpha)) \cap F(T(\beta)) = F(\lambda T(\alpha) + (1-\lambda)T(\beta))$$

holds for every $\lambda \in (0,1)$.

22.2 Existence Theorems

The following is a corollary of a theorem of Bruck [42].

Theorem 22.4. *(Bruck [42]). Let $\{T(t) : t \geq 0\}$ be a strongly continuous semigroup of nonexpansive mappings on a closed convex bounded nonempty subset K of a Banach space E. Assume that the following hold:*

(i) K is either weakly compact, or bounded and separable;
(ii) Every nonexpansive mapping G on K has a fixed point in every nonempty closed convex G−invariant subset of K. Then, $\{T(t) : t \geq 0\}$ has a common fixed point.

Using the ideas of section 21.1, the following generalization of theorem 22.4 is obtained.

Theorem 22.5. *(Suzuki [465], theorem 4, p.1016). Let $\{T(t) : t \geq 0\}$ be a strongly continuous semigroup of nonexpansive mappings on a closed convex bounded nonempty subset K of a Banach space E. Assume that every nonexpansive mapping on K has a fixed point. Then, $\{T(t) : t \geq 0\}$ has a common fixed point.*

22.3 Convergence Theorems

The following convergence theorems have been proved.

Theorem 22.6. *(Suzuki [465], theorem 6). Let K be a compact convex subset of a Banach space E and $\{T(t) : t \geq 0\}$ be a strongly continuous semigroup of nonexpansive mappings on K. For $\kappa, \lambda > 0$ with $\kappa + \lambda < 1$. define a sequence $\{x_n\}$ in K by $x_1 \in K$, and*

$$x_{n+1} = (1 - \kappa - \lambda)x_n + \kappa T(\alpha)x_n + \lambda T(\beta)x_n, n \in \mathbb{N}.$$

Then, $\{x_n\}$ converges strongly to a common fixed point of $\{T(t) : t \geq 0\}$.

Theorem 22.7. *(Suzuki [465], theorem 7). Let K be a closed convex bounded nonempty subset of a Banach space E, and $\{T(t) : t \geq 0\}$ be a strongly continuous semigroup of nonexpansive mappings on K. Let α and β be positive real numbers satisfying $\alpha/\beta \in \mathbb{R}$ and $\alpha/\beta \notin \mathbb{Q}$, and fix $\lambda \in (0,1)$. Let $\{s_n\}$ and $\{t_n\}$ be sequences in $(0,1)$ and $[0,1]$, respectively, satisfying*

$$\lim_{n \to \infty} s_n = \lim_{n \to \infty} t_n = 0, \quad \sum_{n=1}^{\infty} t_n = \infty; \quad \text{and} \quad \sum_{n=1}^{\infty} |t_{n+1} - t_n| < \infty.$$

Fix $u \in K$, and define two sequences $\{x_n\}$ and $\{y_n\}$ in K by

$$x_n = (1 - s_n)\Big(\lambda T(\alpha) + (1 - \lambda)T(\beta)\Big)x_n + s_n u; \quad n \geq 1,$$

and $y_1 \in K$;

$$y_{n+1} = (1 - t_n)\Big(\lambda T(\alpha) + (1 - \lambda)T(\beta)\Big)y_n + t_n u; \quad n \geq 1.$$

Then, $\{x_n\}$ and $\{y_n\}$ converge strongly to a common fixed point of $\{T(t) : t \geq 0\}$.

22.4 Historical Remarks

For more results on approximation of fixed points of semi-groups, the reader may consult, for example, Zegeye and Shahzad [547]; Shimizu and Takahashi [449], [454], Suzuki [460], [465], [471], and the references contained therein.

Chapter 23
Single-valued Accretive Operators; Applications; Some Open Questions

23.1 Introduction

Set-valued accretive operators in Banach spaces have been extensively studied for several decades under various continuity assumptions. In the first part of this chapter we establish a recent incisive finding that every set-valued lower semi-continuous accretive mapping defined on a normed space is, indeed, single-valued on the interior of its domain. No reference to the well-known Michael's Selection Theorem is needed. In Section 23.3, this result is used to extend known theorems concerning the existence of zeros for such operators, as well as, showing existence of solutions for variational inclusions. In Section 23.4, we make some general comments on some fixed point theorems; the rest of the chapter is devoted to some examples of accretive operators; examples of nonexpansive retracts; open problems; and some suggestions for further reading.

23.2 Lower Semi-continuous Accretive Operators are Single-valued

Zeros of *set-valued* accretive operators (or, fixed points of *set-valued* pseudo-contractions) and solutions of *set-valued* inclusions have been studied extensively by various authors under varying continuity assumptions.

Let H be a real Hilbert space and $T : H \to C(H)$ (where $C(H)$ denotes the family of all nonempty subsets of H) be $H-$Lipschitz continuous and strongly monotone with respect to the first argument of a mapping $N(.,.) : H \times H \to H$. Liu and Li [317] basically proved the following interesting result : that $N(.,.)$ cannot be set-valued. In particular, they proved the following theorem.

C. Chidume, *Geometric Properties of Banach Spaces and Nonlinear Iterations,*
Lecture Notes in Mathematics 1965,
© Springer-Verlag London Limited 2009

Theorem LL (Liu and Li [317], Theorem 3.1) *Let the operator* $N(.,.)$ *be Lipschitz continuous with constant* $\beta > 0$ *with respect to the first argument. If* T *is* $H-Lipschitz$ *continuous with constant* $\mu > 0$ *and monotone with respect to the first argument of the operator* $N(.,.)$ *and, for each fixed* $k \in H, intD(N(T(.),k)) \neq \emptyset$, *then* $N(T(.),k)$ *cannot be set -valued in* $intD(N(T(.),k))$.

He [250] proved the following more general result:

Theorem H ([250]). *Let* E *be a real Banach space and* $A : D(A) \subset E \to E$ *be an accretive continuous mapping with* $intD(A) \neq \emptyset$. *Then* A *is single-valued.*

Chidume and Morales [147] extended theorem H to *lower semi-continuous* mappings. To present this result, we begin with the following preliminaries.

Let X be a real normed space and let $\phi : [0, \infty) \to [0, \infty)$ be a function with $\phi(0) = 0$ and $\liminf_{r \to r_0} \phi(r) > 0$ for every $r_0 > 0$. A mapping $A : D(A) \subset X \to 2^X$ is said to be ϕ-*expansive* if $||u - v|| \geq \phi(||x - y||)$ for $u \in A(x)$ and $v \in A(y)$, while A is said to be ϕ-*strongly accretive* if for every $x, y \in D(A)$ there exists $j(x - y) \in J(x - y)$ such that

$$\langle u - v, j(x - y) \rangle \geq \phi(||x - y||)||x - y|| \tag{23.1}$$

for $u \in A(x)$ and $v \in A(y)$, where $J : X \to 2^{X^*}$ is the normalized duality mapping.

Recall that for $\phi(r) = kr$, with $0 < k < 1$, the latter mapping A is called *strongly accretive* and if ϕ is chosen to be the zero function, A is called *accretive*. Consequently, every ϕ-strongly accretive mapping is ϕ-expansive and accretive; however, the converse of this fact does not hold true. On the other hand, if A is accretive, I is the identity mapping on X and the range of $A + \lambda I$ is precisely X for all $\lambda > 0$, then we say that A is $m-accretive$. Also, a mapping A is said to be *locally accretive* if for each $x \in D(A)$, there exists a neighbourhood $N(x)$ of x where A is globally accretive.

We shall prove in this section that *lower semi-continuous accretive operators defined on normed spaces are always single-valued*. In addition, we extend recent works on *generalized set-valued variational inclusions*, first studied by Noor *et al.* (see, for instance, [359], [358]) and later by Chang *et al.* [78], and a host of other authors.

In the sequel, we use $\mathcal{D}(A)$ and $\mathcal{R}(A)$ to denote the domain and the range of an operator A, respectively. For $u, v \in X$ we use $seg[u, v]$ to denote the segment $\{(1 - t)u + tv : t \in [0, 1]\}$.

Let X and Y be topological spaces. We say that the mapping $A : X \to 2^Y$ is *lower semi-continuous* if for every $x \in X$ and every neighbourhood $V(y)$ with $y \in A(x)$, there exists a neighbourhood $U(x)$ of x such that

$$A(u) \bigcap V(y) \neq \emptyset \quad \text{for all } u \in U(x).$$

The following result may be seen as a continuous selection theorem.

Theorem 23.1. *Let X be a real normed space and let $A : D(A) \subset X \to 2^X$ be a lower semi-continuous and locally accretive mapping with $int D(A) \neq \emptyset$. Then, A is a single-valued mapping on $int(D(A))$.*

Proof. Let $x_0 \in int(D(A))$. Then, there exists $r > 0$ such that the open ball $B(x_0; r) \subset D(A)$ with A accretive on $B(x_0; r)$. Suppose there exist $u_1, u_2 \in A(x_0)$ such that $u_1 \neq u_2$. Let $\epsilon = \|u_1 - u_2\|$. Since A is lower semi-continuous, there exists an open ball $B(x_0; r_0)$ with $r_0 < r$, such that

$$B(u_1; \epsilon/2) \cap A(x_1) \neq \emptyset \qquad \text{for all } x_1 \in B(x_0; r_0). \tag{23.2}$$

Select $\alpha \in (0, 1)$ such that $\alpha \|u_1 - u_2\| < r_0$. We now choose x_1 as

$$x_1 = x_0 + \alpha(u_2 - u_1), \tag{23.3}$$

and consequently this x_1 satisfies (23.2). This means, there exists $v \in A(x_1)$ so that $\|u_1 - v\| < \epsilon/2$, and, in particular, $\alpha \|u_1 - v\| < \epsilon/2$. On the other hand, since A is accretive, there exists $j(x_1 - x_0) \in J(x_1 - x_0)$ such that $\langle v - u_2, j(x_1 - x_0) \rangle \geq 0$. Additionally, we derive from (23.3) that

$$\langle x_1 - x_0, j(x_1 - x_0) \rangle = \langle \alpha(u_2 - u_1), j(x_1 - x_0) \rangle = \alpha \langle u_2 - u_1, j(x_1 - x_0) \rangle.$$

Therefore, we obtain that

$$\langle v - u_1, j(x_1 - x_0) \rangle = \langle v - u_2, j(x_1 - x_0) \rangle + \langle u_2 - u_1, j(x_1 - x_0) \rangle \geq \alpha \|u_2 - u_1\|^2.$$

Since $\|j(x_1 - x_0)\| = \alpha \|u_2 - u_1\|$, we get $\|u_2 - u_1\| \leq \|v - u_1\| < \epsilon/2$, which is a contradiction. $\square$

Corollary 23.2 *Let X be a real normed space and let $A : D(A) = X \to 2^X$ be a lower semi-continuous and accretive mapping. Then, A is single-valued on X.*

Corollary 23.2 extends Theorem 2.1 of He, [250] on two fundamental aspects. First, the corollary holds true for general normed spaces, which implies that the celebrated *Michael Selection Theorem* would not be applicable in this case, as it is in Theorem 2.1 of He [250]. Secondly, it holds for the larger class of accretive operators, which includes the ϕ-strongly accretive mappings.

We next prove an invariance of domain result for set-valued mappings, where continuity with respect to the *Hausdorff metric* is not needed. This result extends Theorem 3.1 of Kirk [285].

Corollary 23.3 *Let X be a real Banach space and let D be an open subset of X. Suppose $A : D \to 2^X$ is locally accretive and ϕ-expansive. If A is lower semi-continuous, then $A(D)$ is open in X.*

Proof. By Theorem 23.1, A is a single-valued mapping on D, and consequently, A is continuous on D. Since A is ϕ-expansive, then it is injective and maps closed sets onto closed sets. Therefore, by a theorem of Schöneberg [433], which is an extension of Theorem 3 of Deimling [198], $A(D)$ is open in X.   □

Another interesting consequence of Theorem 23.1 concerns the existence of a continuous selection for accretive operators defined on normed spaces. This result extends Lemma 2.4 of Chang [78].

Corollary 23.4 *Let X be a real normed space and let $A : D(A) = X \to 2^X$ be a lower semi-continuous and locally accretive mapping. Then, A admits a continuous $m-$accretive selection.*

Proof. We first observe that locally accretive mappings defined on the entire space X are globally accretive (see for instance, Kirk and Schöneberg [287]). Then, by Corollary 23.2, we derive that A is a single-valued continuous accretive operator defined on X. Consequently, by the result of Martin [323], we conclude that A is also m-accretive, which completes the proof.   □

We observe that the fact that a lower semi-continuous accretive operator is single-valued in the interior of its domain does not seem to hold on the closure of the domain. Consequently, the following theorem appears to be of interest, since it would extend several known results concerning the existence of zeros, as may be seen in [285], [337], and [338], among others.

Theorem 23.5. *Let X be a Banach space and let D be an open subset of X. Suppose $A : \overline{D} \to 2^X$ is a lower semi-continuous, accretive and ϕ-expansive mapping, which satisfies for some $z \in D$,*

$$t(x - z) \notin A(x) \text{ for all } x \in \partial D \text{ and } t < 0, \tag{23.4}$$

(with $\liminf_{r \to \infty} \phi(r) > 2|A(z)|$). Then, there exists a unique $x \in \overline{D}$ such that $0 \in A(x)$.

Proof. By replacing $A(x)$ with $A(x + z)$ and D by $D - z$, one may select $z = 0$ in (23.4). We first observe, since D is open that A is single-valued on D. The set E defined by

$$E = \{x \in D : A(x) = tx \text{ for some } t < 0\}$$

is bounded. To see this, let $x \in E$. Then $A(x) = tx$ for some $t < 0$. Since A is accretive, there exists $j(x) \in J(x)$ such that $\langle tx - A(0), j(x) \rangle \geq 0$. This implies $-t||x|| \leq ||A(0)||$, and thus, $||A(x)|| \leq ||A(0)||$. Hence, the ϕ-expansiveness of A implies that E is bounded. Therefore, there is no loss of generality in assuming D is bounded. Now, let $h_t : \overline{D} \to 2^X$ be defined by $h_t(x) = (1 - t)x + tA(x)$ for each $t \in [0, 1]$, and let

$$M = \{t \in [0,1] : h_t(x) = 0 \text{ for some } x \in D\}.$$

Clearly, $M \neq \emptyset$ since $0 \in M$. Our goal is to show that $1 \in M$. To see this, let $\{t_n\}$ be a sequence in M with $t_n \to t$ as $n \to \infty$. Then, for each n, there exists $x_n \in D$ so that $h_{t_n}(x_n) = 0$. This means, $(1 - t_n)x_n + t_n A(x_n) = 0$. If $t < 1$, we may choose $j(x_n - x_m) \in J(x_n - x_m)$ such that

$$
\begin{aligned}
0 &= \langle h_{t_n}(x_n) - h_{t_m}(x_m), j(x_n - x_m) \rangle \\
&= \langle (t_m - t_n)(x_n - A(x_n)) + t_m(A(x_n) - A(x_m)) \\
&\quad + (t_m - 1)(x_m - x_n), j(x_n - x_m) \rangle.
\end{aligned}
$$

This and the fact that A is accretive imply $(1 - t_m)\|x_m - x_n\| \leq |t_n - t_m|\|x_n - A(x_n)\|$. Since the sequence $\{x_n - Ax_n\}$ is bounded, we derive that $\{x_n\}$ is a Cauchy sequence in D. Now, if $t = 1$, we use the ϕ-expansiveness of A to obtain

$$
\begin{aligned}
\phi(\|x_n - x_m\|) &\leq \|A(x_n) - A(x_m)\| \\
&= \|(1 - t_n^{-1})x_n - (1 - t_m^{-1})x_m\|.
\end{aligned}
$$

Since the sequence $\{(1 - t_n^{-1})x_n\}$ is Cauchy, we derive once again that $\{x_n\}$ is a Cauchy sequence in D. Therefore $x_n \to x$ for some $x \in \overline{D}$. It remains to show that $x \in D$. Suppose $x \in \partial D$ and let $u \in A(x)$. Since A is lower semi-continuous at x, then for every $k \in \mathbb{N}$ there exists x_{n_k} such that

$$\|x_{n_k} - x\| < 1/k \quad \text{and} \quad \|Ax_{n_k} - u\| < 1/k.$$

This implies $0 = (1 - t_{n_k})x_{n_k} + t_{n_k} A x_{n_k} \to (1 - t)x + tu$ as $k \to \infty$. Hence $0 \in (1 - t)x + tA(x)$, which contradicts assumption (23.4). Therefore $x \in D$, and consequently, M is closed in $[0,1]$.

On the other hand, suppose M is not open. Then there exit $t \in M$ $(t < 1)$ and a sequence $\{t_n\}$ in $[0, 1)$ for which $t_n \notin M$ and $t_n \to t$. Then $h_t(x_0) = 0$ for some $x_0 \in D$, which means, $(1 - t)x_0 + tA(x_0) = 0$. Select an open ball B centered at x_0 and contained in D. Then

$$y_n = h_{t_n}(x_0) \in h_{t_n}(B) \quad \text{for each} \quad n \in \mathbb{N}, \tag{23.5}$$

while $0 \notin h_{t_n}(B)$. This implies there exists $u_n \in seg[0, y_n] \cap \partial h_{t_n}(B)$. Since h_{t_n} is continuous and strongly accretive, then for each $n \in \mathbb{N}$, we know that $h_{t_n}(B)$ is open (by Theorem 3 of Deimling [198]), while by (23.5), $h_{t_n}(\overline{B})$ is closed. Hence, we may derive that $\partial h_{t_n}(B) \subset h_{t_n}(\partial B)$, which yields the existence of a point $x_n \in \partial B$ so that $u_n = h_{t_n}(x_n)$. Since $y_n \to 0$, so does $\{u_n\}$. As before, we choose $j(x_n - x_m) \in J(x_n - x_m)$ such that

$$
\begin{aligned}
\langle u_n - u_m, j(x_n - x_m) \rangle = \langle (t_m - t_n)(x_n - A(x_n)) \\
+ t_m(A(x_n) - A(x_m)) + (t_m - 1)(x_m - x_n), j(x_n - x_m) \rangle,
\end{aligned}
$$

which implies,

$$(1 - t_m)||x_n - x_m|| \le ||u_n - u_m|| + |t_n - t_m|||x_n - A(x_n)||.$$

Hence $\{x_n\}$ is a Cauchy sequence which must converge to some $\overline{x} \in \partial B$. Then by a continuity argument we derive that $0 \in h_t(\overline{x})$, this is a contradiction. Therefore, M is open and the proof is complete.   □

As a consequence of Theorem 23.5, we obtain an extension of Theorem 3.4 of Kirk [285]. First of all, we derive a corollary that will be used in the proof of the next result. We will use $|A|$ to denote $inf\{||x|| : x \in A\}$.

Corollary 23.6 *Let X be a Banach space and let D be a bounded open subset of X. Suppose $A : \overline{D} \to 2^X$ is a lower semi-continuous and accretive mapping which satisfies for some $z \in D$*

$$|A(z)| < |A(x)| \text{for } x \in \partial D. \tag{23.6}$$

Then, $\inf\{|A(x)| : x \in D\} = 0$.

Proof. We first show that assumption (23.6) implies assumption (23.4). To this end, suppose there exists $x \in \partial D$ so that $t(x - z) \in A(x)$ for some $t < 0$. Then there exists $j(x - z) \in J(x - z)$ such that

$$\langle t(x - z) - v, j(x - z)\rangle \ge 0   \text{for } v \in A(z).$$

This implies that $-t \parallel x - z \parallel \le \parallel v \parallel$ for all $v \in A(z)$, and thus $|A(x)| \le |A(z)|$, which is a contradiction. Hence, (23.4) holds. We may assume without lost of generality, that $z = 0$ in (23.4). For $\lambda > 0$, the mapping $A_\lambda(x) = \lambda x + A(x)$ is strongly accretive and satisfies (23.4) as well. Hence, by Theorem 23.5, $A_\lambda(x) = 0$ for some $x \in D$. Select $\lambda_n \to 0^+$ as $n \to \infty$. Then there exists $x_n \in D$ such that $\lambda_n x_n + A(x_n) = 0$. Since D is bounded, $||A(x_n)|| \to 0$, which completes the proof.   □

Theorem 23.7. *Let D be a bounded open subset of a Banach space X, and let $A : \overline{D} \to 2^X$ be a lower semi-continuous accretive mapping. Suppose there exists $z \in D$ such that*

$$|A(z)| < |A(x)| \text{ for all } x \in \partial D. \tag{23.7}$$

Then, either A has a zero in D or there exists a single-valued nonexpansive mapping $g : \overline{D} \to D$ whose fixed points are zeros of A.

Proof. Suppose A has no zeros in D. Then by Corollary 23.6, we know $\inf\{|A(x)| : x \in D\} = 0$. This and the fact that A is single-valued on D, allow us to re-define z so that

$$||A(z)|| < inf\{|A(x)| : x \in \partial D\}.$$

Since D is bounded, we can choose $\rho > 0$ such that $D \subset B(z;\rho)$. Then, by following the proof of Theorem 1 of [338], we may derive that $g(y) = (rI + A)^{-1}(ry)$ is a nonexpansive mapping defined on $\overline{B}(z;\rho)$ and taking values in D for a suitable $r > 0$. In particular, g maps $\overline{D}$ into D, which completes the proof.   □

23.3 An Application to Variational Inequalities

The study of set-valued Variational Inclusion Problems was introduced by Noor *et al.* [359] in Hilbert spaces. Later, Chang *et al.* [78] extended these ideas to uniformly smooth Banach spaces. In this section, we extend these results to more general Banach spaces. In fact, we do not require that X be uniformly smooth, or that the operators S and T be Lipschitz with respect to the Hausdorff metric as is needed in Theorem 3.1 of Chang *et al.* [78].

Theorem 23.8. *Let X be a real Banach space, and let $S, T : X \to 2^X$ be lower semi-continuous set-valued mappings. Suppose $A : D(A) \subset X \to 2^X$ is m-accretive while $G : X \times X \to X$ is a continuous single-valued mapping such that*

> (i) $G(S(.), y)$ *is $\phi-$ strongly accretive for each $y \in X$;*
>
> (ii) $G(x, T(.))$ *is accretive for each $x \in X$.*

Then, for any $z \in X$, there exist $x \in D(A)$, $u \in S(x)$, and $v \in T(x)$ such that the triple (x, u, v) is a solution of the set-valued variational inclusion

$$z \in G(u, v) + \lambda A(x).$$

We observe that the accretivity conditions (i) and (ii), mentioned above, may be stated relative to the sets $S(X)$ and $T(X)$, respectively. However, for simplicity, we have chosen otherwise. For this apparently more general setting, the reader may consult, for example, Noor *et al.* [359], Chang *et al.* [78].

To prove Theorem 23.8, we need some basic results.

Lemma 23.9. *Let X, Y, Z be topological spaces, let $g : X \to 2^Y$ be a lower semi-continuous mapping, and let $F : \mathcal{R}(g) \subset Y \to Z$ be continuous mapping such that $\mathcal{D}(F) \subset \mathcal{R}(g)$. Then $F \circ g$ is lower semi-continuous.*

Proof. Let $x \in X$ and let $z \in F(g(x))$. Then there exists $u \in g(x)$ such that $z = F(u)$. Suppose V is an arbitrary neighborhood of z. Since F is continuous, there exists a neighborhood W of u such that $F(W) \subset V$. On the other hand, since g is lower semi-continuous at x, there exists a neighborhood U of x so that

$$g(v) \cap W(u) \neq \emptyset \quad \text{for all} \quad v \in U(x),$$

which implies

$$\emptyset \neq F[g(v) \cap W(u)] \subset F(g(v)) \cap V \quad \text{for all} \quad v \in U(x).$$

Therefore $F \circ g$ is lower semi-continuous.   □

As a consequence of Lemma 23.9, we derive the following proposition.

Proposition 23.10. *Let X be a Banach space, and let $S, T : X \to 2^X$ be set-valued mappings where T is lower semi-continuous. Suppose $G : X \times X \to X$ is a continuous single-valued mapping such that*

(i) $G(S(.), y)$ is $\phi-$ strongly accretive for each $y \in X$;

(ii) $G(x, T(.))$ is accretive for each $x \in X$.

Then the mapping $F : X \to 2^X$ defined by $F(x) = G(S(x), T(x))$ is ϕ-strongly accretive.

Proof. Let $u \in X$ be fixed, and let $G_u : X \to 2^X$ be defined by $G_u(x) = G(u, T(x))$. Then by Lemma 23.9, G_u is lower semi-continuous. Since, by assumption (ii), it is also accretive, Theorem 23.1 implies that G_u is a continuous single-valued accretive operator defined on X.

To see that F is ϕ-strongly accretive, consider arbitrary elements $x_1, x_2 \in X$ and let $y_i \in F(x_i)$, $i = 1, 2$. Then there exist $u_i \in S(x_i)$ and $v_i \in T(x_i)$ such that $y_i = G(u_i, v_i), i = 1, 2$. Since by assumption (i), $G(., v_1)$ is ϕ-strongly accretive, there exists $j_*(x_1 - x_2) \in J(x_1 - x_2)$ such that

$$\langle G(u_1, v_1) - G(u_2, v_1), j_*(x_1 - x_2) \rangle \geq \phi(\|x_1 - x_2\|)\|x_1 - x_2\|. \quad (23.8)$$

As we observed earlier, $G(u_2, .)$ is a continuous accretive operator defined on the entire space X, then by a result of Barbu [24], we know that

$$\langle G(u_2, v_1) - G(u_2, v_2), j(x_1 - x_2) \rangle \geq 0 \text{ for all } j(x_1 - x_2) \in J(x_1 - x_2). \quad (23.9)$$

Therefore, in particular, (23.8) and (23.9) hold for $j_*(x_1 - x_2) \in J(x_1 - x_2)$, and thus

$$\begin{aligned}
\langle y_1 - y_2, j_*(x_1 - x_2) \rangle &= \langle G(u_1, v_1) - G(u_2, v_2, j_*(x_1 - x_2) \rangle \\
&= \langle G(u_1, v_1) - G(u_2, v_1, j_*(x_1 - x_2) \rangle \\
&\quad + \langle G(u_2, v_1) - G(u_2, v_2, j_*(x_1 - x_2) \rangle \\
&\geq \phi(\|x_1 - x_2\|)\|x_1 - x_2\|,
\end{aligned}$$

which completes the proof.   □

Proof of Theorem 23.8. Due to Proposition 23.10, we know that the mapping $F : X \to 2^X$ defined by $F(x) = G(S(x), T(x))$ is ϕ-strongly accretive. In addition, since S is also lower semi-continuous, Lemma 23.9 implies that

$F = G \circ (S,T)$ is lower semi-continuous. Consequently, by Corollary 23.2, F is a single-valued continuous ϕ-strongly accretive operator defined on X. On the other hand, since A is m-accretive, Theorem 5.3 of [289] implies that $F + \lambda A$ is also m-accretive. Since, in addition, it is also ϕ-expansive, Theorem 8 of [227] implies that $F + \lambda A$ is surjective. Therefore, for each $z \in X$ and $\lambda > 0$, there exists a unique $x \in D(A)$ such that $z \in F(x) + \lambda A(x)$, which means, there exist $u \in S(x)$ and $v \in T(x)$ such that the variational inclusion $z \in G(u,v) + \lambda A(x)$ has a solution. □

23.4 General Comments on Some Fixed Point Theorems

The contents of this monograph have been chosen with the particular interest of the author in mind. Consequently, several important topics which fall within the scope of *geometric properties of Banach spaces and nonlinear iterations* have been omitted. In particular, we have confined our applications of the Banach space inequalities discussed in this monograph primarily to iterative methods for approximating solutions of *nonlinear equations of the accretive and pseudo-contractive types, using fixed point techniques.*

The study of nonlinear operators had its beginning about the start of the twentieth century with investigations into the existence properties of solutions to certain boundary value problems arising in differential equations. The earliest techniques, largely devised by E. Picard [384], involved the iteration of an integral operator to obtain solutions to such problems. In 1922, these techniques were given precise abstract formulation by S. Banach [23] and R. Cacciopoli [64] in what is now generally referred to as the Contraction Mapping Principle (Theorem 6.2). The classical importance of fixed point theory in functional analysis is due to its usefulness in differential equations. The existence or construction of a solution to a differential equation is often reduced to the existence or location of a fixed point for an operator defined on a subset of a space of functions. Theorem 6.2 (which is involved in many of the existence and uniqueness proofs of differential equations) is perhaps, the most useful fixed point theorem. Another fixed point theorem which has proved very useful in the proofs of many existence theorems of differential equations is the following.

Theorem (Schauder-Tychnov Theorem) *Let K be a compact convex subset of a locally convex topological linear space. If T is a continuous mapping of K into K, then T has a fixed point.*

However, since the compactness assumption of this theorem is often difficult to obtain in applications, considerable research has been done concerning possible weaker conditions for the domain which guarantee the existence of a fixed point. If domains which are only bounded, closed and convex are considered, Vidossich [496] showed that a nonexpansive mapping may fail to have a fixed point even under the additional assumption that the domain be compact in the weak* topology. Further research resulted in Kirk's theorem (Theorem 6.3). Let K be a weakly compact convex set in a Banach space X. Recall that K is said to have *the fixed point property for nonexpansive mappings* (i.e., K has the f.p.p.) if every nonexpansive $T : K \to K$ has a fixed point. We say that X has the f.p.p. if every $K \subseteq X$ has the f.p.p. One of the major unsolved problems of nonlinear functional analysis arising from Kirk's theorem which remained open for many years is the following:

Does every weakly compact convex subset K of a Banach space X have the fixed point property for nonexpansive mappings?

This is the same as asking whether the condition in Theorem 6.3 that K has normal structure can be dropped.

If X_r denotes the space l_2 renormed by $|||x|||_r := \max\{||x||_2, r||x||_\infty\}$ where $||.||_2$ denotes the l_2 norm and $||.||_\infty$ denotes the l_∞ norm, then X_r has normal structure only if $r < \sqrt{2}$. Nevertheless, Karlovitz [274] showed that the answer to the above problem was affirmative for the space $X_{\sqrt{2}}$. In fact, Karlovitz's proof works for all $r < 2$ (see e.g., Baillon and Schöneberg [22]). The first counter-example to the general problem was discovered by D. Alspach [2] who showed that $L_1[0,1]$ fails the f.p.p. Maurey [328], however, has proved that if $K \subseteq L_1[0,1]$ and is reflexive then K has the f.p.p.

Further fixed point theorems for nonexpansive mappings and some of their generalizations in various Banach spaces can be found in, for example, Elton *et al.* [219], Goebel and Kirk [231], [232], Goebel and Koter [235], [236], [237], Goebel and Reich [234], Holmes and Lau [253], Karlovitz [272], [273], [274], Kirk [282], [283], Kirk and Morales [286], Kirk and Schönberg [288], Krasnosel'skii [291], [292], and a host of other authors.

23.5 Examples of Accretive Operators

We mention some examples of accretive operators in $L_p(\Omega)$ spaces, $1 < p < \infty$.

1. Let β be a maximal monotone graph in $\mathbb{R}$, and let Ω be a bounded domain in $\mathbb{R}^n$ with smooth boundary $\partial\Omega$. With appropriate domains, the operator

$$(i) \quad A_1 u := -\triangle u + \beta(u)$$

with homogeneous Neumann boundary condition, and

$$(ii) \quad A_2 u := -\triangle u, \ -\frac{\partial u}{\partial n} \in \beta(u) \text{ on } \partial\Omega$$

are accretive on $L_p(\Omega)$.

The operator

$$(iii) \quad A_3u := -\sum \left(\frac{\partial}{\partial x_i}\right)\left(\left|\frac{\partial u}{\partial x_i}\right|^{r-1}\frac{\partial u}{\partial x_i}\right)$$

is accretive for $r \geq 1$.

2. In addition to L_p spaces, $(1 < p < \infty)$, other **uniformly smooth** and **smooth** spaces , for example **Orlic spaces** (see e.g., [312]) arise in applications. Accretive operators occur in such spaces too (see, e.g., Le, C.R. Acad. Sci. Paris **283** (1976), 469 - 472).

23.6 Examples of Nonexpansive Retracts

1. A closed linear subspace of L_p, $1 < p < \infty$ is a **nonexpansive retract** of the space if and only if it is isometric to another L_p space.
2. The sets
$$K_1 := \left\{f \in L_p(\Omega) : \|f(x)\| \leq 1, \ a.e. \ \in \Omega\right\}$$

and

$$K_2 := \text{ Positive cone in } L_p$$

are **nonexpansive retracts** in $L_p, 1 < p < \infty$.

3. The set
$$K_3 := \left\{f \in H_0^1(\Omega) : \|\triangledown f(x)\| \leq 1 \ a.e. \ in \ \Omega\right\}$$

is a **nonexpansive retract** of $L_p(\Omega)$, $p \geq 2$.

4. (Shioji and Takahashi, Theorem 1). Let K be a closed convex subset of a uniformly convex Banach space whose norm is uniformly Gâteaux differentiable and let T be an asymptotically nonexpansive mapping from K into itself such that the set $F(T)$ of fixed points of T is nonempty. Then, $F(T)$ is a **sunny nonexpansive retract** of K.

23.7 Some Questions of Interest

Question 1. For a sequence $\{\alpha_n\}$ of real numbers in $[0, 1]$ and an arbitrary $u \in K$, let the sequence $\{x_n\}$ in K be iteratively defined by $x_0 \in K$,

$$x_{n+1} := \alpha_n u + (1 - \alpha_n)Tx_n, n \geq 0, \tag{23.10}$$

where $T : K \to K$ is a nonexpansive map. Concerning this process, Reich [412] posed the following question.

Let E be a Banach space. Is there a sequence $\{\alpha_n\}$ such that whenever a weakly compact convex subset K of E has the fixed point property for nonexpansive mappings, then the sequence $\{x_n\}$ defined by (23.10) converges to a fixed point of T for arbitrary fixed $u \in K$ and all nonexpansive $T : K \to K$?

Question 2. Apart from the example given by Goebel and Kirk [231] of an asymptotically nonexpansive map which is not nonexpansive, there are very few nontrivial examples of operators belonging to this class. It is certainly of interest to find more nontrivial such examples.

Question 3. Chidume and Ofoedu [152] introduced the following recursion formula to approximate a fixed point of a *generalized* Lipschitz pseudo-contractive mapping:

$$x_1 \in K, \ x_{n+1} = (1 - \lambda_n \alpha_n)x_n + \lambda_n \alpha_n T x_n - \lambda_n \theta_n(x_n - x_1), \ n \geq 1,$$

where $\{\alpha_n\}, \{\lambda_n\}, \{\theta_n\}$ satisfy appropriate conditions.

Will the slightly simpler recursion formula studied by Chidume and Zegeye [174] defined by $x_1 \in K$,

$$x_{n+1} := (1 - \lambda_n)x_n + \lambda_n T x_n - \lambda_n \theta_n(x_n - x_1), n \in \mathbb{N},$$

under suitable conditions on the parameters θ_n, λ_n converge to a fixed point of a *generalized Lipschitz* pseudo-contractive mapping.?

Question 4. Although iterative methods for approximating fixed points of *asymptotically pseudo-contractive* mappings have been studied by some authors, there is *no known existence* theorem for this class of mappings. One is certainly desirable.

Question 5. Is the class of Lipschitz pseudo-contractive mappings a *proper* subclass of that of asymptotically pseudo-contractions (as defined by Schu [435], [437], [436])? If this is not the case, then it is, perhaps, desirable to give a new definition of asymptotically pseudo-contractive mappings which will include the class of Lipschitz pseudo-contractions as a *proper* subclass. Nontrivial examples of operators satisfying any given definition should be produced. Finally, and more importantly, for any definition given, an *existence theorem* will be desirable *before* approximation of a fixed point will be investigated.

Open question 6. Do Theorem 13.27 and Corollary 13.29 hold in L_p spaces for all p such that $1 < p \leq 2$?

Open question 7. Do Corollaries 13.25 and 13.30 hold in L_p spaces for all p such that $2 \leq p < \infty$?

23.8 Further Reading

Fixed point theorems for nonlinear *semigroups of nonexpansive mappings* have been proved by Bruck [58]. More information on this topic can also be found in Barbu [24], Crandall and Pazy [194], Lau [294], Lau and Takahashi [295], Nevanlinna and Reich [352], Oka [362], [365], and a host of other authors.

Further convergence theorems for fixed points of pseudo-contractive operators (or zeros of accretive operators) using the Mann or Ishikawa-type sequence can be found in, for example, Bethke [29], Dunn [215], Crandall and Pazy [194], Ghosh and Debnath [229], Kirk and Morales [286], Liang [302], Maruster [325], Minty [329], [330], Moore [333], [334], Moore and Nnoli [335], Morales [339], Morales and Chidume [340], Muller and Reinermann [342], Nadezhkina and Takahashi [344], [343], [345], [346], Nadler [347], Reinermann [414], Reich [407], [404], Tricomi [490], Vainberg [495], Zarantonello [539], [538], Zeng [550], [551] and in a host of more recent papers listed in the bibliography.

Numerous papers have also been published on approximation of solutions of variational inequalities using the Mann iteration process and some of the geometric inequalities presented in this monograph. For results in this direction, the reader may consult, for example, any of the following references: Chang [68], [69], [70], Chidume *et al.* [139], [140], [181], Hassouni and Moudafi [249], He [250], Liu [317], Noor [355], [356], Siddiqi and Ansari [455], [456], Udomene [494], Li [296], [297], LI and Whitaker [299], Li and Rhoades [300], Li and Park [301], Kazmi [277] and the references in them.

The Mann iteration process has also been applied in various other interesting situations not elaborated in this monograph. For instance, it has been applied in Chidume and Aneke [127], [128], to approximate solutions of $k-positive$ *definite operator equations* introduced by Petryshyn (see also, Chidume and Osilike [159], Osilike and Udomene [375]); and in Chidume and Lubuma [144], it is used to approximate *solutions of the Stokes system.*

For the study of geometric properties of Banach spaces, important references include: Al'ber and Notik [8], Birkhoff and Kellog [32], Bynuum [61], [62], Day [196], Lim [306], Lindenstrauss [309], [310], Lindenstrauss and Tzafriri [312], Nörlander [360], Petryshyn [381], Prus [386], Prus and Smarzewski [387], Prüss [388], and Xu and Roach [526], [528].

Finally, for further reading, we recommend particularly the following excellent textbooks for an in-depth study of geometric properties of Banach spaces, and much more: Aksoy and Khamsi [1], Barbu [24], Beauzamy [26], Berinde [28], Cudia [195], Ciorencscu [189], Deimling [199], Diestel [206], Goebel and Kirk [230], Goebel and Reich [234], Lindenstrauss and Tzafriri [312], Martin [322], Opial [367].

References

1. Aksoy, A. G. and Khamsi, M. A.; *Nonstandard methods in fixed point theory*, Springer, New York, 1990.
2. Alspach, D.; *A fixed point free nonexpansive map*, Proc. Amer. Math. Soc. **82** (1981) no. 3, 423–424.
3. Al'ber, Ya.; *Metric and generalized projection operators in Banach spaces: properties and applications.* In *Theory and Applications of Nonlinear Operators of Mootone and Accretive Type* (A. Kartsatos, editor), Marcel dekker, New York (1996), 15–50.
4. Al'ber, Ya., *A bound for the modulus of continuity for the metric projection in a uniformly convex and uniformly smooth Banach space*, J. Approx. Theory **85** (1996), no. 3, 237–249.
5. Al'ber, Ya., Chidume, C.E. and Zegeye, H.; *Regularization of nonlinear ill-posed equations with accretive operators on sets of Banach spaces*, Fixed Point Theory Appl. **(1)** (2005), no. 1, 11–33.
6. Al'ber, Ya., Chidume, C.E. and Zegeye, H.; *Approximating fixed points of total asymptotically nonexpansive mappings*, Fixed Point Theory Appl. **2006**, Art. ID 10673.
7. Al'ber, Ya. and Guerre-Delabriere, S.; *On the projecton methods for fixed point problems*, Analysis (Munich) **21** (2001), no. 1, 17–39.
8. Al'ber, Ya. and Notik, A. I.; *Geometric properties of Banach spaces and approximate method for solving nonlinear operator equations*, Soviet Math. Dokl. **29** (1984), 611–615.
9. Al'ber, Ya. and Notik, A.; *On some estimates for projection operator in Banach spaces*, Commun. Appl. Nonlinear Anal. **2** (1995), no. 1, 47–56.
10. Al'ber, Ya. and Reich, S.; *An iterative method for solving a class of nonlinear operator equations in Banach spaces*, PanAmer. Math. J. 4(1994), no. 2, 39–54.
11. Al'ber, Ya., Reich, S. and Shoikhet, D.; *Iterative approximations of null points of uniformly accretive operators with estimates of the convergence rate*, Commu. Appl. Anal. **6** (2002), no. 1, 89–104.
12. Al'ber, Ya., Reich, S. and Yao, J. C.; *Iterative methods for solving fixed-point problems with nonself-mappings in Banach spaces*, Proceedings of the International Conference on Fixed-Point Theory and its Applcations, 193–216, Hindawi Publ. Corp., Cairo, 2003.
13. Al'ber, Ya., Reich, S. and Yao, J. C.; *Iterative methods for solving fixed-point problems with nonself-mappings in Banach spaces*, Abstr. Appl. Anal. (2003), no. 4, 193–216.
14. Al'ber, Ya. and Ryazantseva, I.; *Nonlinear ill-posed problems of monotone type*, Springer, Dordrecht, (2006), xiv+410 pp.
15. Aoyama, K., Iiduka, H. and Takahashi, W.; *Weak convergence of an iterative sequence for accretive operators in Banach spaces*, Fixed Point Theory Appl. (2006), Art. ID 35390, 13 pp.

16. Aoyama, K, Kimura, Y., Takahashi, W. and Toyoda, M.; *Approximation of common fixed points of a countable family of nonexpansive mappings in a Banach space*, Nonlinear Anal. **67** (2007), 2350–2360.

17. Asplund, E.; *Positivity of duality mappings*, Bull. Amer. Math. Soc. **73** (1967).

18. Atsushiba, S., Shioji, N. and Takahashi, W.; *Approximating common fixed points by the Mann iteration procedure in Banach spaces,* J. Nonlinear Convex Anal. **1** (2002), no. 3, 351–361.

19. Atsushiba, S. and Takahashi, W.; *Weak and strong convergence theorems for nonexpansive semigroups in Banach spaces,* Fixed Point Theory Appl. **2005**, no. 3, 343–354.

20. Baillon, J. B.; *Un théoreme de type ergodic pour les contractions nonlinearies dans un espace de Hilbert*, C.R. Acad. Sci. Paris **280** (1975), 1511–1514.

21. Baillon, J. B. and Brézis, H.; *Une remarque sur le comportement asymtotique des semigroupes non lineaires,* Houston J. Math. **2** (1976), 5–7.

22. Baillon, J. B. and Schoneberg. R.; *Asymptotic normal astructure and fixed points of nonexpansive mappings,* Pacific J. Math. **53** (1974), 59–71.

23. Banach, S.; *Theorie des operations lineires,* Warsaw, 1932.

24. Barbu, V.; *Nonlinear Semigroups and Differential Equations in Banach Spaces,* Noordhoof, Leyden. The Netherlands, 1976.

25. Bauschke, H. H.; *The approximation of fixed points of compositions of nonexpansive mappings in Hilbert spaces,* J. Math. Anal. Appl. **202** (1996), no. 1, 150–159.

26. Beauzamy, B.; *Introduction to Banach spaces and their geometry*, North-Holland, Amsterdam, (1985).

27. Belluce, L.P. and Kirk, W.A., *Fixed-point theorem for families of contraction mappings,* Pacific J. Math., **18** (1966), 213–217.

28. Berinde, V., *Iterative approximation of fixed points* Lecture Notes in Mathematics 1912, Springer Berlin, 2007.

29. Bethke, M.; *Approximation von fixpunkten streng pseudo-kontractiver operatoren (Approximation of fixed points of strongly pseudo-contractve operators),* Math. Naturwiss Fak. **27** (1989), no. 2, 263–270.

30. Beurling, A. and Livingston, A. E.; *A theorem on duality mappings in Banach spaces,* Ark. Math. **4** (1962),405–411.

31. Bianchini, R. M. T.; *Su un problema di S. Reich riguardente la teoria dei punti fissi,* Boll. Un. Mat. Ital. **5** (1972), 103–108.

32. Birkhoff, D. G. and Kellog, O. D.; *Invariant points in function spaces,* Trans. Amer. Math. Soc. **23** (1922), 96–115.

33. Borwein, D. and Borwein, J. M.; *Fixed point iterations for real functions,* J. Math. Anal. Appl. **157** (1991), 112–126.

34. Bose, S. C.; *Weak convergence to a fixed point of an asymptotically nonexpansive map,* Proc. Amer. Math. Soc., **68** (1978), 305–308.

35. Brézis, H. and Browder, F. E.; *Some new results about Hammerstein equations,* Bull. Amer. Math. Soc. **80** (1974), 567–572.

36. Brézis, H. and Browder, F. E.; *Existence theorems for nonlinear integral equations of Hammerstein type,* Bull. Amer. Math. Soc. **81** (1975), 73–78.

37. Brézis, H. and Browder, F. E.; *Nonlinear integral equations and systems of Hammerstein type,* Bull. Amer. Math. Soc. **82** (1976), 115–147.

38. Brodskii, M. S. and Mil'man, D. P.; *On the center of convex sets,* Dokl. Akad. Nauk. SSR. **59** (1984), 837–840.

39. Browder, F. E.; *The solvability of nonlinear functional equations,* Duke Math. J. **30** (1963), 557–566.

40. Browder, F. E.; *Nonlinear elliptic boundary value problems,* Bull. Amer. Math. Soc. **69** (1963), 862–874.

41. Browder, F.E.; *Multi-valued and monotone nonlinear mappings and duality mapping in Banach spaces,* Trans. Amer. Math. Soc. **118** (1965), 338–361.

42. Browder, F. E.; *Nonexpansive nonlinear operators in Banach spaces*, Proc. Nat. Acad. Sci., U.S.A. **54** (1966), 1041–1044.

43. Browder, F. E.; *Nonlinear equations of evolution and nonlinear accretive operators in Banach spaces*, Bull. Amer. Math. Soc. **73** (1967), 470–475.

44. Browder, F. E.; *Convergence of approximants to fixed points of nonexpansive nonlinear mappings in Banach spaces*, Arch. Rat. Mech. Anal. **24** (1967), 82–90.

45. Browder, F. E.; *Nonlinear mappings of nonexpansive and accretive type in Banach spaces*, Bull. Amer. Math. Soc. **73** (1967), 875–882.

46. Browder, F. E.; *Nonlinear monotone and accretive operators in Banach space*, Proc. Nat. Acad. Sci. U.S.A. **61** (1968), 388–393.

47. Browder, F. E.; *Nonlinear operators and nonlinear equations of evolution in Banach spaces*, Proc. of Symposia in pure Math. Vol. XVIII, part 2, 1976.

48. Browder, F.E. and de Figueiredo, D.G.; *Monotone nonlinear operatrs*, Koukl. Nederl. Akad. Wetersch **69** (1966), 412–420.

49. Browder, F.E., de Figueiredo, D. G. and Gupta, C. P.; *Maximal monotone operators and nonlinear integral equations of Hammerstein type*, Bull. Amer. Math. Soc. **76** (1970), 700–705.

50. Browder, F.E. and Gupta, C. P.; *Monotone operators and nonlinear integral equations of Hammerstein type*, Bull. Amer. Math. Soc. **75** (1969), 1347–1353.

51. Browder, F.E. and Petryshyn, W. V.; *The solution by iteration of nonlinear functional equations in Banach spaces*, Bull. Amer. Math. Soc. **72** (1966), 571–575.

52. Browder, F.E. and Petryshyn, W. V.; *Construction of fixed points of nonlinear mappings in Hilbert spaces*, J. Math. Anal. Appl. **20** (1967), 197–228.

53. Bruck, R.E.; *The iterative solution of the equation $f \in x + Tx$ for a monotone operator T in Hilbert space*, Bull. Amer. Math. Soc. **79** (1973), 1258–1262.

54. Bruck, R.E., *Nonexpansive projections on subsets of Banach space,* Pacific J. Math., 47 (1973), 341–355.

55. Bruck, R. E., *Properties of fixed point sets of nonexpansive mappings in Banach spaces,* Trans. Amer. Math. Soc. **179** (1973), 251–262.

56. Bruck, R. E.; *A strongly convergent iterative method for the solution of $0 \in Ux$ for a maximal monotone operator U in Hilbert space*, J. Math. Anal. Appl. **48** (1974), 114–126.

57. Bruck, R. E.; *A common fixed point theorem for a commuting family of nonexpansive mappings*, Pacific J. Math. **53** (1974), 59–71.

58. Bruck, R. E.; *A simple proof of mean ergodic theorem for nonlinear contractions in Banach spaces*, Israel J. Math. **32** (1979), 279–282.

59. Bruck, R.E.; *Asymptotic behavior of nonexpansive mappings*, Contemporary Mathematics, **18**, Fixed Points and Nonexpansive Mappings, (R.C. Sine, Editor), AMS, Providence, RI, 1980.

60. Bruck, R. E., Kuczumow, T. and Reich, S.; *Convergence of iterates of asymptotically nonexpansive mappings in Banach spaces with the uniform Opial property*, Colloquium Mathematicum Vol. LXV, Fasc. 2 (1993), 169–179.

61. Bynum, W. L.; *Weak parallelogram laws for Banach spaces*, Canad. Math. Bull. **19**(3) (1976), 269–275.

62. Bynum, W. L.; *Normal structure coefficients for Banach spaces*, Pacific J. Math. **86** (1980), 427–436.

63. C. Byrne, *A unified treatment of some iterative algorithms in signal processing and image restoration,* Institute of Physics Publishing; Inverse Problems **20** (2004) 103–120.

64. Caccioppoli, R.; *Un teorenu generale sull'esistenza di elementi uniti in una trasformazione funzionale*, Rend. Accad. Naz. Lincei **13** (1931), 498–502.

65. Caristi, J.; *The fixed point theory for mappings satisfying inwardness conditions*, Ph.D. Thesis, The University of Iowa, Iowa City, 1975.

66. Caristi, J.; *Fixed point theorems for mappings satisfying inwardness condition,* Trans. Amer. Math. Soc. **215** (1976), 241-251.

67. Chang, S. S.; *On Chidume's open questions and approximation solutions of multi-valued strongly accretive mapping equations in Banach spaces*, J. Math. Anal. Appl. **216** (1997), 94–111.

68. Chang, S. S.; *The Mann and Ishikawa iterative approximation of solutions to variational inclusions with accretive type mappings*, Computers Math. Appl. **37**(9) (1999), 18–25.

69. Chang, S. S.; *Variational inequality and complementarity problem: theory and applications*, Shanghai Scientific and Technological Literature Publishing House, (1991).

70. Chang, S. S.; *Set-valued variational inclusions in Banach spaces*, J. Math. Anal. Appl. **248**(2) (2000), 438–454.

71. Chang, S. S.; *Some results for asymptotically pseudocontractive mappings and asymptotically nonexpansive mappings*, Proc. Amer. Math. Soc. **129**(2000), 845–853.

72. Chang, S. S.; *On the convergence problems of Ishikawa and Mann iteration process with errors for ψ-pseudo contractive type mappings*, Appl. Math. Mechanics, **21** (2000), 1–10.

73. Chang, S. S., Agarwal, R. P. and Cho, Y. J.; *Strong convergence of iterative sequences for asymptotically nonexpansive mappinga in Banach spaces,* Dynam. Systems Appl. **14** (2005), no. 3-4, 447–454.

74. Chang, S. S., Cho, Y.J. and Lee, B. S.; *Iterative approximations of fixed points and solutions for strongly accretive and strongly pseudocontractive mappings in Banach spaces,* J. Math. Anal. Appl. **224**(1998) 149–165.

75. Chang, S. S., Cho, Y. J. and Zhou, H.; *Demi-closed principle and weak convergence problems for asymptotically nonexpansive mappings,* J. Korean Math. Soc., **38** (2001), no. 6, 1245–1260.

76. Chang, S. S., Cho, Y. J. and Zhou, H.; *Iterative methods for nonlinear operator equations in Banach spaces,* Nova Science Publishers, Inc., Huntington, NY(2002).

77. Chang, S. S., Kim, J. K. and Cho, Y. J.; *Approximations of solutons of set-valued ϕ-strongly accretive equations,* Dynam. Systems Appl. **14** (2005), no. 3-4, 515–524.

78. Chang, S. S., Kim, J. K. and Kim, K. H.; *On the existence of iterative approximation problems of solutions for set-valued variational inclusions in Banach spaces,* J. Math. Anal. Appl. **268** (2002) 89–108.

79. Chang, S. S., Kim, J. K. and Cho, Y. J.; *On the equivalence for the convergence of Mann iteration and Ishikawa iteration with mixed errors for Lipschitz strongly pseudo-contractive mappings,* Comm. Appl. Nonlinear Anal. **12** (2005), no. 4, 79–88.

80. Chang, S. S., Lee, H. W. J., Yang, L and Liu, J. A.; *On the strong and weak convergence for implicit iteration process,* Comm. Appl. Nonlnear Anal. **12** (2005), no. 3, 47–60.

81. Chang, S. S., Park, J. Y. and Cho, Y. J.; *Iterative approximations of fixed points for asymptotically nonexpansive mappings in Banach spaces,* Bull. Korean Math. Soc. **37**(2000), 1, 109–119.

82. Chang, S. S., Park, J. K., Cho, Y. J. and Hung, I. H.; *Convergence theorems of the Ishikawa iterative scheme for asymptotically pseudo-contractive mappings in Banach spaces,* J. Comput. Anal. Appl. **6** (2004), no. 4, 313–322.

83. Chang, S. S. and Tan, K. K.; *Iteration process of fixed point for operators of monotone type in Banach spaces,* Bull. Austral. Math. Soc., **57** (1998), 433–445.

84. Chang, S. S., Tan, K. K., Lee, H. W. J., and Chi, K. C.; *On the convergence of implicit iteration process with error for a finite family of asymptotically nonexpansive mappings,* J. Math. Anal. Appl. **313** (2006), 273–283.

85. Chen, R., Song, Y. and Zhou, H.; *Convergence theorems for implicit iteration process for a finite family of continuous pseudocontractive mappings,* J. Math. Anal. Appl. **314** (2006), no. 2, 701–709.

86. Chepanovich, R. Sh.; *Nonlinear Hammerstein equations and fixed points,* Publ. Inst. Math. (Beograd) N. S.**35**(49) (1984), 119–123.

87. Chidume, C. E.; *On the approximation of fixed points of nonexpansive mappings,* Houston J. Math. **7**, (1981), 345-355.

88. Chidume, C. E.; *Iterative methods and nonlinear functional equations*, Ph.D. Dissertation, June (1984), The Ohio State University, Columbus, Ohio, U.S.A.

89. Chidume, C. E.; *The solution by iteration of nonlinear equations in certain Banach spaces*, J. Nigerian Math. Soc. **3** (1984), 57–63.

90. Chidume, C. E.; *On the Ishikawa fixed point iterations for quasi-contractive mappings*, J. Nigerian Math. Soc. **4** (1985), 1–11.

91. Chidume, C. E.; *The iterative solution of the equation $f \in x + Tx$ for a monotone operator T in L^p spaces*, J. Math. Anal. Appl. Vol. **116** (1986), no. 2, 531–537.

92. Chidume, C. E.; *An approximation method for monotone Lipschitzian operators in Hilbert spaces*, J. Austral. Math. Soc. **41** (series A) (1986), no. 1, 59–63.

93. Chidume, C. E.; *Quasi-nonexpansive mappings and uniform asymptotic regularity*, Kobe J. Math. **3** (1986), no. 1, 29–35.

94. Chidume, C. E.; *Iterative construction of fixed points for multi-valued operators of the monotone type*, Appl. Anal. **23** (1986), no. 3, 209–218.

95. Chidume, C. E.; *Iterative approximation of fixed points of Lipschitzian strictly pseudo-contractive mappings*, Proc. Amer. Math. Soc.**98** (1987), no. 4, 283–288.

96. Chidume, C. E.; *On the Ishikawa and Mann iteration methods for nonlinear quasi-contractive mappings*, J. Nigerian Math. Soc. **7** (1988), 1–9.

97. Chidume, C. E.; *Fixed point iterations for certain classes of nonlinear mappings*, Appl. Anal. **27** (1988), no. 1-3, 31–45.

98. Chidume, C. E.; *Fixed point iterations for nonlinear Hammerstein equations involving nonexpansive and accretive mappings*, Indian J. Pure Appl. Math. **120**, (1989) no. 2, 129–135.

99. Chidume, C. E.; *Iterative solution of nonlinear operator equation of the monotone and dissipative type*, Appl. Anal. **33** (1989), no. 1-2, 79–86.

100. Chidume, C. E.; *Approximation of fixed points of Lipschitz pseudo-contractive mappings in Banach spaces*, Discovery and Innovation **1** (1989), 65–69.

101. Chidume, C. E.; *Fixed point iterations for certain classes of nonlinear mappings*, II, J. Nigerian Math. Soc. **8** (1989), 11–23.

102. Chidume, C. E.; *An iterative process for nonlinear Lipschitzian strongly accretive mapping in L^p spaces*, J. Math. Anal. Appl. **152** (1990), no. 2, 453–461.

103. Chidume, C. E.; *Iterative solution of nonlinear equations of the monotone type in Banach spaces*, Bull. Austral. Math. Soc. **42** (1990), no. 1, 35–45.

104. Chidume, C. E.; *Iterative method for nonlinear set-valued operators of the monotone type with applications to operator equations*, J. Nigerian Math. Soc. **9** (1990), 7–20.

105. Chidume, C. E.; *Approximation of fixed points of quasi-contractive mappings in L^p spaces*, Indian J. Pure Appl. Math. **22** (1991), no. 4, 273–286.

106. Chidume, C. E.; *Approximation of fixed points of strongly pseudo-contractive mappings*, Proc. Amer. Math. Soc. **120** (1994), no. 2, 545–551.

107. Chidume, C. E.; *An iterative method for nonlinear demiclosed-type operators*, Dynam. Systems Appl. **3** (1994), 349–356.

108. Chidume, C. E.; *Iterative solutions of nonlinear equations with strongly accretive operators*, J. Math. Anal. Appl. **192** (1995), no. 2, 502–518.

109. Chidume, C. E.; *Steepest descent approximations for accretive operator equations*, Nonlinear Anal. **26** (1996), no. 2, 299–311.

110. Chidume, C. E.; *Iterative solution of nonlinear equations in smooth Banach spaces*, Nonlinear Anal. **26** (1996), no. 11, 1823–1834.

111. Chidume, C. E.; *Steepest descent method for locally accretive mappings*, J. Korean Math. Soc. **33** (1996), 1–14.

112. Chidume, C. E.; *Global iteration schemes for strongly pseudo-contractive maps*, Proc. Amer. Math. Soc. **126** (1998), no. 9, 2641–2649.

113. Chidume, C. E.; *Iterative solution of nonlinear equations of strongly accretive type*, Math. Nachr. **189** (1998), 49–60.

114. Chidume, C. E.; *Convergence theorems for strongly pseudocontractive and strongly accretive nonlinear maps*, J. Math. Anal. Appl. **228** (1998), no. 1, 254–264.

115. Chidume, C. E.; *Iterative methods for nonlinear Lipschitz pseudocontractive opera-tors,* J. Math. Anal. Appl. **251** (2000), no. 1, 84–92.

116. Chidume, C. E.; *Iterative approximation of fixed points of Lipschitz pseudocontractive maps,* Proc. Amer. Math. Soc. **129** (2001), no. 8, 2245–2251.

117. Chidume, C. E.; *Convergence theorems for asymptotically pseudocontractive map-pings,* Nonlinear Anal. **49** (2002) no. 1, 1–11.

118. Chidume, C. E.; *Nonexpansive mappings, generalisations and iterative algorithms,* Nonlinear analysis and applications: to V. Lakshmikantham on his 80th birthday.Vol. 1,2, 383–421, Kluwer Acad. publ., Dordrecht, 2003.

119. Chidume, C. E.; *Strong convergence theorems for fixed points of asymptotically pseu-docontractive semigroups,* J. Math. Anal. Appl. **296** (2004), no. 2, 410–421.

120. Chidume, C. E., Abbas, M., and Ali, B.; *Convergence of the Mann algorithm for a class of pseudo-contractive mappings,* Appl. Math. Comput. **194** (2007), no. 1, 1–6.

121. Chidume, C. E. and Ali, B.; *Convergence theorems for common fixed points for finite families of asymptotically nonexpansive mappings in reflexive Banach spaces,* PanAmer. Math. J. **16** (2006), no. 4, 81–95.

122. Chidume, C. E. and Ali, B.; *Approximation of common fixed points for finite families of non-self asymptotically nonexpansve mappings in Banach spaces,* J. Math. Anal. Appl. **326** (2007), no. 2, 960–973.

123. Chidume, C. E. and Ali, B.; *Weak and strong convergence theorems for finite families of asymptotically nonexpansive mappings in Banach spaces,* J. Math. Anal. Appl. **330** (2007), 377–387.

124. Chidume, C. E. and Ali, B.; *Convergence theorems for finite families of asymptot-ically quasi-nonexpansive mappings,* J. Inequal. Appl. **2007** (2007), Art. ID 68616, 10pp.

125. Chidume, C. E. and Ali, B.; *Convergence theorems for common fixed points for finite families of nonexpansive mappings in reflexive Banach spaces,* Nonlinear Anal. **68** (2008) 3410–3418.

126. Chidume, C. E. and Ali, B.; *Convergence of path and an iterative method for famlies of nonexpansive mappings,* Appl. Anal. **67** (2008), no. 1 117–129.

127. Chidume, C. E. and Aneke, S. J.; *Existence, uniqueness and approximation of a solution of a k−positive definite operator equation,* Appl. Anal. **50** (1993), no. 3-4, 285–294.

128. Chidume, C. E. and Aneke, S. J.; *Local approximation methods for the solution of k−positive definite operator equations,* Bull. Korean Math. Soc. **40** (2003), no. 4, 603–611.

129. Chidume, C. E. and Chidume, C. O.; *Convergence theorems for zeros of general-ized Lipschitz generalized phi-quasi-accretive operators,* Proc. Amer. Math. Soc. **134** (2006), no. 1, 243–251.

130. Chidume, C. E. and Chidume, C. O.; *Iterative methods for common fixed points for a countable family of nonexpansive mappings in uniformly convex spaces,* (submitted).

131. Chidume, C. E. and Chidume, C. O.; *Convergence theorems for fixed points of uni-formly contnuous generalized Phi-hemicntractive mappings,* J. Math. Anal. Appl. **303** (2005), no. 2, 545–554.

132. Chidume, C. E. and Chidume, C. O.; Iterative approximation of fixed points of non-expansive mapings, J. Math. Anal. Appl. **318** (2006), no. 1, 288–295.

133. Chidume, C. E., Chidume, C. O., and Ali, B.; *Approximation of fixed points of non-expansive mappings and solutions of variational inequalities,* J. Ineq. Appl. **2008**, Art. ID 284345, 12 pages.

134. Chidume, C. E., Chidume, C. O., and Ali, B.; *Convergence of Hybrid steepest de-scent method for variational inequalities in Banach spaces,* (accepted: Appl. Math. Comput.).

135. Chidume, C. E., Chidume, C. O. and Nwogbaga, A. P.; *Approximation methods for common fixed points for countable families of nonexpansive mappings,* (accepted: Nonlinear Anal.).

136. Chidume, C. E., Chidume, C. O. and Shehu, Y.; *Approximation of fixed points of nonexpansive mappings and solutions of variational inequalities,* PanAmer. Math. J. **18**(2008), no. 2, 73–82.

137. Chidume, C.E. and Djitte, N.; *Iterative approximation of solutions of Nonlinear equations of Hammerstein type,* (accepted: Nonlinear Anal.).

138. Chidume, C. E. and Djitte, N.; *Approximation of solutions of Hammerstein equations with bounded strongly accretive nonlinear operators,* (accepted: Nonlinear Anal.).

139. Chidume, C. E., Kazmi, K. R. and Zegeye, H.; *General auxiliary problem and algorithm for a general multivalued variational-like inequality problem in reflexive Banach spaces,* Appl. Anal. **82** (2003), no. 12, 1099–1109.

140. Chidume, C. E., Kazmi, K. R. and Zegeye, H.; *Iterative approximation of a solution of a general variational-like inclusion in Banach spaces,* Intern. J. Math. Math. Sci. (2004), no. 21-24, 1159–1168.

141. Chidume, C. E., Khumalo, M. and Zegeye, H.; *Generalized projection and approximation of fixed points of non-self maps,* J. Aprox. Theory, **120** (2003), no. 2, 242–252.

142. Chidume, C. E. and Li, J.; *Projection methods for approximating fixed points of Lipschitz suppressive operators.,* PanAmer. Math. J. **15** (2005), no. 1, 29–40.

143. Chidume, C. E., Li, J. and Udomene, A.; *Convergence of paths and approximation of fixed points of asymptotically nonexpansive mappings,* Proc. Amer. Math. Soc. **133**(2005), no. 2, 473–480.

144. Chidume, C. E. and Lubuma, M. S.; *Solution of the Stokes system by boundary integral equations and fixed point schemes,* J. Nigerian Math. Soc. **11** (1992), no. 3, 1–7.

145. Chidume, C. E. and Moore, C.; *Fixed point iterations for pseudocontractive maps,* Proc. Amer. Math. Soc. **127**(1999), no. 4, 1163–1170.

146. Chidume, C. E. and Moore, C.; *Steepest descent method for equilibrium points of nonlinear systems with accretive operators,* J. Math. Anal. Appl. **245** (2000), no. 1, 142–160.

147. Chidume, C. E., Morales, C. H.; *Accretive operators which are always single-valued in normed spaces,* Nonlinear Anal. **67** (12) (2007), 3328–3334.

148. Chidume, C. E. and Mutangadura, S. A.; *An example on the Mann iteration method for Lipschitz pseudocontrations,* Proc. Amer. Math. Soc. **129** (2001), no. 8, 2359–2363.

149. Chidume, C. E. and Nnoli, B. V. C.; *A necessary and sufficient condition for convergence of the Mann sequence for a class of nonlinear operators,* Bull. Korean Math. Soc. **39** (2002), no. 2, 269–276.

150. Chidume, C. E., Ofoedu, E. U. and Zegeye, H.; *Strong and weak convergence theorems for asymptotically nonexpansive mappings,* J. Math. Anal. Appl. **280** (2003), no. 2, 364–374.

151. Chidume, C. E., Ofoedu, E. U. and Zegeye, H.; *Strong convergence theorems for uniformly L-Lipschitzian mappings in Banach spaces,* PanAmer. Math. J. **16** (2006), no. 4, 1–11.

152. Chidume, C. E., Ofoedu, E. U.; *A new iteration process for generalized Lipschitz pseudo-contractive generalized Lipschitz accretive mappings,* Nonlinear Anal. **67** (2007), 307–315.

153. Chidume, C. E., Ofoedu, E. U.; *Approximation of common fixed points for finite families of total asymptotically nonexpansive mappings,* J. Math. Anal. Appl. (333) (2007) 128–141.

154. Chidume, C. E. and Osilike, M. O.; *Iterative solution of nonlinear integral equations of the Hammerstein type,* J. Nigerian Math. Soc. **11** (1992), 9–19.

155. Chidume, C. E. and Osilike, M. O.; *Fixed point iterations for quasi-contractive maps in uniformly smooth Banach spaces,* Bull. Korean Math. Soc. **30** (1993), no. 2, 201–212.

156. Chidume, C. E. and Osilike, M. O.; *Fixed point iterations for strictly hemicontractive maps in uniformly smooth Banach spaces,* Numer. Funct. Anal. Optimz. **15** (1994), (7&8), 779–790.

157. Chidume, C. E. and Osilike, M. O.; *Approximation methods for nonlinear operator equations of the m-accretive type,* J. Math. Anal. Appl. **189** (1995), no. 1, 25–239.

158. Chidume, C. E. and Osilike, M. O.; *Ishikawa iteration process for nonlinear Lipschitz strongly accretive mappings,* J. Math. Anal. Appl. **192** (1995), no. 3, 727–741.

159. Chidume, C. E. and Osilike, M. O.; *Approximation of a solution for a k-positive definite operator equation,* J. Math. Anal. Appl. **210** (1997), no. 1, 1–7.

160. Chidume, C. E. and Osilike, M. O.; *Nonlinear accretive and pseudo-contractive operator equations in Banach spaces,* Nonlinear Anal. **31** (1998), no. 7, 79–789.

161. Chidume, C. E. and Osilike, M. O.; *Iterative solution of nonlinear accretive operator equations in arbitrary Banach spaces,* Nonlinear Anal. **36** (1999), no. 7, 863–872.

162. Chidume, C. E. and Osilike, M. O.; *Equilibrium points for a system involving m−accretive operators,* Proc. Edinburgh Math. Soc. **44** (2001), no. 1, 187–199.

163. Chidume, C. E. and Shahzad, N. *Strong convergence of an implicit iteration process for finite family of nonexpansive mappings,* Nonlinear Anal. **62**(2005), no. 6, 1149–1156.

164. Chidume, C. E., Shahzad, N. and Zegeye, H.; *Convergence theorems for mappings which are asymptotically nonexpansive in the intermediate sense,* Numer. Funct. Anal. Optim. **25** (2004), nos. 3-4, 39–257.

165. Chidume, C. E. and Udomene, A.; *Convergence theorems for uniformly continuous pseudocontractions,* J. Math. Anal. Appl. **323** (2006), no. 1, 89–99.

166. Chidume, C. E. and Zegeye, H.; *Approximation of the zeros of nonlinear m−Accretive operators,* Nonlinear Anal. **37** (1999), no. 1, 81–96.

167. Chidume, C. E. and Zegeye, H.; *Global iterative schemes for accretive operators,* J. Math. Anal. Appl. **257**(2001), no. 2, 364–377.

168. Chidume, C. E. and Zegeye, H.; *Iterative solution of 0 ∈ Ax for an m−accretive operator A in certain Banach spaces,* J. Math. Anal. Appl. **269** (2002), no. 2, 421–430.

169. Chidume, C. E. and Zegeye, H.; *Approximation methods for nonlinear operator equations,* Proc. Amer. Math. Soc. **131** (2003), no. 8, 2467–2478.

170. Chidume, C. E. and Zegeye, H.; *Iterative approximation of solutions of nonlinear equations of Hammerstein type,* Abstr. Appl. Anal. **6** (2003), 353–367.

171. Chidume, C. E. and Zegeye, H.; *Approximate fixed point sequences and convergence theorems for asymptotically pseudocontractive mappings,* J. Math. Anal. Appl. **278** (2003), no. 2, 354–366.

172. Chidume, C. E. and Zegeye, H.; *Iterative solution of nonlinear equations of accretive and pseudocontractive types,* J. Math. Anal. Appl. **282** (2003), no. 2, 756–765.

173. Chidume, C. E. and Zegeye, H.; *Approximation of solutions of nonlinear equations of monotone and Hammerstein type,* Appl. Anal. **82** (2003), no. 8, 747–758.

174. Chidume, C. E. and Zegeye, H.; *Approximate fixed point sequences and convergence theorems for Lipschitz pseudocontractive maps,* Proc. Amer. Math. Soc. **132** (2004), no. 3, 831–840.

175. Chidume, C. E. and Zegeye, H.; *Approximation of solutions of nonlinear equations of Hammerstein type in Hilbert space,* Proc. Amer. Math. Soc. **133** (2005), no. 3, 851–858.

176. Chidume, C. E. and Zegeye, H.; *Convergence theorems for fixed points of demicontinuous pseudocontractive mappings,* Fixed Point Theory Appl. **1** (2005), no. 1, 67–77.

177. Chidume, C. E. and Zegeye, H.; *Strong convergence theorems for asymptotically quasi-nonexpansive mappings,* Comm. Appl. Nonlinear Anal. **12**(2005), no. 1, 43–50.

178. Chidume, C. E., Zegeye, H.; *Strong convergence theorems for common fixed points of uniformly L-Lipschitzian pseudocontractive semi-groups,* Appl. Anal. **86** (2007), no. 3, 353–366.

179. Chidume, C. E., Zegeye, H. and Aneke, S. J.; *Approximation of fixed points of weakly contractive non-self maps in Banach spaces,* J. Math. Anal. Appl. **270** (2002), no. 1, 189–199.

180. Chidume, C. E., Zegeye, H. and Aneke, S. J.; *Iterative methods for fixed points of asymptotically weakly contractive maps,* Appl. Anal. **82** (2003), no. 7, 701–712.

181. Chidume, C. E., Zegeye, H. and Kazmi, K. R.; *Existence and convergence theorems for a class of multi-valued variational inclusions in Banach spaces,* Nonlinear Anal. **59** (2004), no. 5, 649–656.

182. Chidume, C. E., Zegeye, H. and Ntatin, B.; *A generalized steepest descent approximation for the zeros of m-accretive operators,* J. Math. Anal. Appl. **236** (1999), no. 1, 48–73.

183. Chidume, C. E., Zegeye, H. and Prempeh, E.; *Strong convergence theorems for a finite family of nonexpansive mappings,* Comm. Appl. Nonlinear Anal., **11**(2004), no. 2, 25–32.

184. Chidume, C. E., Zegeye, H. and Shahzad, N.; *Convergence theorems for a common fixed point of finite family of nonself nonexpansive mappings,* Fixed Point Theory Appl. (2005), no. 2, 233–241.

185. Chidume, C. O. and De Souza, G.; *Convergence theorems for nonexpansive mappings and variational inequality problem in certain Banach spaces,* (submitted).

186. Chidume, C. O. and De Souza, G.; *A strong convergence theorem for fixed points of asymptotically nonexpansive mappings in Banach spaces,* J. Nigerian Math. Soc. **25** (2006), 69–78.

187. Cho, Y. J., Kang, J. I. and Zhou, H.; *Approximating common fixed points of asymptotically nonexpansive mappings,* Bull. Korean Math. Soc. **42** (2005), no. 4, 661–670.

188. Cho, Y. J., Zhou, H. and Kim, J. K.; *Iterative approximatons of zeros for accretive operators in Banach spaces,* Commun. Korean Math. Soc. **21** (2006), no. 2, 237–251.

189. Ciorenescu, I.; *Geometry of Banach spaces, Duality Mapping and Nonlinear Problems,* Kluwer Academic Publishers, 1990.

190. Ciric, Lj. B.; *A generalization of Banach's contraction principle,* Proc. Amer. Math. Soc. **45** (1974), 267–273.

191. Clarkson, J. A.; *Uniformly convex spaces,* Trans. Amer. Math. Soc. **40** (1936), 396–414.

192. Colao, V., Marino, G. and Xu, H. K.; *An iterative method for finding common solutions of equilibrium and fixed point problems,* J. Math. Anal. Appl. **344**(2008), 340–352.

193. Ceng, L. C., Cubiotti, P. and Yao, J. C.; *Strong convergence theorems for finitely many nonexpansive mappings and applications,* Nonlinear Anal. **67**(2007), 1464–1473.

194. Crandall, M. G. and Pazy, A.; *Semigroups of nonlinear contractions and dissipative sets,* J. Funct. Anal. **3** (1969), 376–418.

195. Cudia, D. F.; *The theory of Banach spaces: Smoothness,* Trans. Amer. Math. Soc. **110** (1964), 284–314.

196. Day, M. M.; *Uniform Convexity in Factor and Conjugate Spaces,* Ann. Math. **45**, no. 2, (1944).

197. De Figueiredo, D. G.; *Topics in nonlinear functional analysis,* University of Maryland, (1967).

198. Deimling, K.; *Zeros of accretive operators,* Manuscripta Math. **13** (1974), 365–374.

199. Deimling, K.; *Nonlinear Functional Analysis,* Springer-Verlag, 1985.

200. De Marr, R.; *Common fixed points for commuting contraction mappings,* Pacific J. Math. **13** (1963), 1139–1141.

201. Deng, L.; *On Chidume's open questions,* J. Math. Anal. Appl. **174**(1993), no. 2, 441–449.

202. Deng, L.; *Iteration processes for nonlinear Lipschitzian strongly accretive mappings in L_p spaces,* J. Math. Anal. Appl. **188** (1994), no. 1, 128–140.

203. Deng, L.; *An iterative process for nonlinear Lipschitz strongly accretive mappings in uniformly convex and uniformly smooth Banach spaces*, Acta Appl. Math. **32** (1993), no. 2, 183–196.

204. Ding, X. P.; *Iterative approximation of Lipschitz strictly pseudocontractive mappings in uniformly smooth Banach spaces*, Nonlinear Anal. **24**(1995), no. 7, 981–987.

205. Diaz, J. B. and Metcalf, F. T.; *On the set of subsequential limit points of successive approximations*, Trans. Amer. Math. Soc. **135** (1969), 459–485.

206. Diestel, J.; *Geometry of Banach spaces − selected topics*, Lecture Notes in Math. **485**, Springer-Verlag, Berlin, Heidelberg, New York, 1976.

207. Ding, X. P.; *Iteration method of constructing fixed point of nonlinear mappings*, Comput. Math. **3** (1981), 285–295.

208. Ding, X. P.; *Generalized strongly nonlinear quasi-variational inequalities*, J. Math. Anal. Appl. **173** (1993), 577–587.

209. Ding, X. P.; *Perturbed proximal point algorithms for generalized quasi-variational inclusions*, J. Math. Anal. Appl. **210** (1997), 88–101.

210. Dolezale, V.; *Monotone operators and applications in automation and network theory*, Studies in automation and control, **3** Elsevier Science Publishers, New York, 1979.

211. Dotson, W. G.; *On the Mann iterative process*, Trans. Amer. Math. Soc.**149** (1970), 65–73.

212. Dotson, W. G.; *Fixed points of quasi-nonexpansive mappings*, J. Austral. Math. Soc. **13** (1972), 167–170.

213. Dotson, W. G.; *An iterative process for nonlinear monotone nonexpansive operators in Hilbert spaces*, Math. Comput. **32** (1978), no. 4, 223–225.

214. Downing, D.; *Fixed point theorems and surjectivity results for nonlinear mappings in Banach spaces*, Ph.D. Thesis, The university of Iowa, Iowa City, 1977.

215. Dunn, J. C.; *Iterative construction of fixed points for multivalued operators of the monotone type*, J. Funct. Anal. **27** (1978), 38–50.

216. Edelstein, M.; *On nonexpansive mappings*, Proc. Amer. Math. Soc. **15** (1964), 689–695.

217. Edelstein, M.; *A remark on a theorem of Krasnoselskii*, Amer. Math. Monthly **13** (1966), 507–510.

218. Edelstein, M. and O'Brian, R. C.; *Nonexpansive mappings, asymptotic regularity and successive approximations*, J. London Math. Soc. **17** (1978), no. 3, 547–554.

219. Elton, J., Lin, P.K., Odell, E., and Szarek, S.; *remarks on the fixed point problem for nonexpansive maps*, Preprint.

220. Eshita, K., Miyake, H. and Takahashi, W.; *Strong convergence theorems for asymptotically nonexpansive semigroups in general Banach spaces*, Dyn. Contin. Discrete Impuls. Syst. Ser. A, Math. Anal. **13** (2006), no. 5, 621–640.

221. Eshita, K. and Takahashi, W.; *Strong convergence theorems for commutative semigroups of continuous linear operators on Banach spaces*, Taiwanese J. Math. **9** (2005), no. 4, 531–550.

222. Falset, J.G., Kaczor, W., Kuczumow, T. and Reich, S.; *Weak convergence theorems for asymptotically nonexpansive mappings and semigroups*, Nonlinear Anal. **43** (2001), no. 3, 377–401.

223. Figiel, T.; *On the moduli of convexity and smoothness*, Studia Math., **56** (1976), 121–155.

224. Fitzpatrick, P. M., Hess, P. and Kato, T.; *Local boundedness of monotone type operators*, Proc. Japan Acad. **48** (1972), 275–277.

225. Franks, R. L. and Marzec, R. P.; *A theorem on mean-value iterations*, Proc. Amer. Math. Soc. **30** (1971), 324–326.

226. Gao, G. L., Zhou, H. and Yang, J. F.; *Convergence theorems for Ishikawa iteration processes in Banach spaces*, (Chinese) J. Hebei Univ. Nat. Sci. **26** (2006), no. 4, 345–347.

227. Garcia-Falset, J. and Morales, C. H.; *Existence theorems for m-accretive operators in Banach spaces*, J. Math. Anal. Appl. **309** (2005), 453–461.

228. Genel, A. and Lindenstrauss, J.; *An example concerning fixed points*, Israel J. Math. **22** (1975), 81–86.

229. Ghosh, M. K. and Debnath, L.; *Convergence of Ishikawa iterates of quasi-nonexpansive mappings*, J. Math. Anal. Appl. **207** (1997), no. 1, 96–103.

230. Goebel, K. and Kirk, W. A.; *Topics in metric fixed point theory, Cambridge Studies in Advanced Mathematics* **28**, CUP, Cambridge, New York, Port Chester, Melbourne, Sydney, 1990.

231. Goebel, K. and Kirk, W. A.; *A fixed point theorem for asymptotically nonexpansive mappings*, Proc. Amer. Math. Soc. **35** (1972), 171–174.

232. Goebel, K and Kirk, W. A.; *A fixed point theorem for transformations whose iterates have uniform Lipschitz constant*, Studia Math. **47** (1973), 137–140.

233. Goebel, K., Kirk, W. A. and Shimi, T. N.; *A fixed point theorem in uniformly smooth spaces*, Boll. U.M.I. **7** (1973), no. 4, 67–75.

234. Goebel, K. and Reich, S.; *Uniform convexity, hyperbolic geometry, and nonexpansive mappings*. Monographs and Textbooks in Pure and Applied Mathematics, **83**. Marcel Dekker, Inc., New York, 1984. ix+170 pp. ISBN: 0-8247-7223-7

235. Goebel, K. and Koter, M.; *Fixed points of rotative Lipschitzian mappings,* Rend. Sem. Mat. Fis. Milano **51** (1981), 145–156 (1983).

236. Goebel, K. and Koter, M.; *A remark on nonexpansive mappings,* Canad. Math. Bull. **24** (1981), no. 1, 113–115.

237. Goebel, K. and Koter, M.; *Regularly nonexpansive mappings,* An. Ştiinţ. Univ. "Al. I. Cuza" Iaşi Secţ. I a Mat. (N.S.) **24** (1978), no. 2, 265–269.

238. Göhde, D.; *Zum prinzip der kontraktiven abbidungen*, Math. Nachr. **30** (1965), 251–258.

239. Gornicki, J.; *Weak convergence theorems for asymptotically nonexpansive mappings in uniformly convex spaces*, Comment. Math. Univ. Carolinae **30** (1989), no. 2, 249–252.

240. Gornicki, J.; *Nonlinear ergodic theorems for asymptotically nonexpansive mappings in Banach spaces satisfying Opial's condition*, J. Math. Anal. Appl. **161** (1991), no. 2, 440–446.

241. Gossez, J. P. *and* Lami-Dozo, E.; *Some geometric properties related to the fixed point theory for nonexpansive mappings*, Pacific J. Math. **40** (1972), 365–573.

242. Groetsch, G. W.; *A note on segmenting Mann iterates*, J. Math. Anal. Appl. **40** (1972), 369-372.

243. Gu, G. H., Zhou, H. and Gao, G. L., *Iterative construction of fixed points for a class of asymptotically nonexpansive mappings,* (Chinese) J. Hebei Norm. Univ. Nat. Sci. Ed. **29** (2005), no. 5, 441–444.

244. Gu, F.; *Convergence theorems for Φ-pseudo contractive type mappings in normed lnear spaces*, Northeast Math. J. **17**(2001) no. 3, 340–346.

245. Halpern, B.; *Fixed points of nonexpansive maps*, Bull. Amer. Math. Soc. **3** (1967), 957–961.

246. Hammerstein, A.; *Nichtlineare integralgleichungen nebst anwendungcn*, Acta Math. **54** (1930), 117–176.

247. Hanner, O.; *On uniform convexity of L_p and ℓ_p spaces*, Ark. Mat. **3**(19) (1956), 239–244.

248. Hardy, G. and Rogers, T.; *A generalization of a fixed point theorem of Reich*, Canad. Math. Bull. **16** (1973), no. 2, 201–206.

249. Hassouni, A. and Moudafi, A.; *A perturbed algorithm for variational inclusions*, J. Math. Anal. Appl. **185** (1994), no. 2, 706–721.

250. He, X.; *On ψ- strongly accretive mapping and some set-valued variational problems*, J. Math. Anal. Appl. **227** (2003), 504–511.

251. Hicks, T. L. and Kubicek, J. R.; *On the Mann iteration process in a Hilbert space*, J. Math. Anal. Appl. **59** (1979), 498–504.

252. Hirano, H. and Huang, Z.; *Convergence theorems for multi-valued Φ-hemicontractive operators and Φ-strongly accretive operators,* Comput. Math. Appl. **46** (2003), no. 10-11, 1461–1471.

253. Holmes, R. H. and Lau, A. T.; *Nonexpansive actions of topological semigroups and fixed points,* J. London Math. Soc. **5** (1972), 330–336.

254. Huang, J.; *Convergence theorems of the sequence of iterates for a finite family asymptotically nonexpansive mappings,* Int. J. Math. Math. Sci. **27** (2001), no. 11, 653–662.

255. Iiduka, H. and Takahashi, W.; *Weak convergence theorems by Cesro means for non-expansive mappings and inverse-strongly-monotone mappings,* J. Nonlinear Convex Anal. **7** (2006), no. 1, 105–113.

256. Iiduka, H. and Takahashi, W.; *Strong convergence theorem by a hybrid method for nonlinear mappings of nonexpansive and monotone type and applications,* Adv. Nonlinear Var. Inequal. **9** (2006), no. 1, 1–10.

257. Isac, G.; *Complementarity problem,* Lecture Notes in Mathematics, No. 1528, Springer- Verlag, Berlin, 1992.

258. Isac, G. and Li, J.; *The convergence property of Ishikawa iteration schemes in non-compact subsets of Hilbert spaces and its applications to complementarity theory,* Comput. Math. Appl. **47** (2004), no. 10-11, 1745–1751.

259. Ishikawa, S.; *Fixed points and iteration of nonexpansive mapping in a Banach space,* Proc. Amer. Math. Soc. **73** (1976), 61–71.

260. Ishikawa, S.; *Fixed points by a new iteration Method,* Proc. Amer. Math. Soc. **44**(1974), no. 1, 147–150.

261. James, R. C.: *Orthogonality and linear functionals in normed linear spaces,* Trans. Amer. Math. Soc. **61** (1947), 265–292.

262. Johnson, G. A.; *Nonconvex set which has the unique nearest point property,* J. Approx. Theory **51** (1987), 289–332.

263. Johnson, G. G.; *Fixed points by mean value iterations,* Proc. Amer. Math. Soc. **34** (1972), 193–194.

264. Jung, J. S.; *Iterative approaches to common fixed points of nonexpansive mappings in Banach spaces,* J. Math. Anal. Appl. **302** (2005), no. 2, 509–520.

265. Jung, J. S., Cho, Y. J. and Agarwal, R. P.; *Iterative shemes with some control conditions for family of finite nonexpansive mappings in Banach spaces,* Fixed Point Theory Appl. **2** (2005), 125–135.

266. Jung J. S., and Kim, S. S.; *Strong convergence theorems for nonexpansive nonself mappings in Banach spaces,* Proc. Amer. Math. Soc. **73** (1998), no. 3, 321–329.

267. Kaczor, W.; *Weak convergence of almost orbits of asymptotically nonexpansive commutative semigroups,* J. Math. Anal. Appl. **272** (2002), no. 2, 565–574.

268. Kaczor, W., Kuczumow, T. and Reich, S.; *A mean ergodic theorem for mappings which are asymptotically nonexpansive in the intermediate sense,* Nonlinear Anal. **47** (2001), no. 4, 2731–2742.

269. Kang, J. I., Cho, Y. J. and Zhou, H.; *Approximation of common fixed points for a class of finite nonexpansive mappings in Banach spaces,* J. Comput. Anal. Appl. **8** (2006), no. 1, 25–38.

270. Kakutani, S.; *Two fixed point theorems concerning bicompact convex sets,* Proc. Imp. Acad. Tokyo **14** (1938), 242–245.

271. Kannan, R.; *Some results on fixed points II,* Amer. Math. Monthly **76** (1969), 405–408.

272. Karlovitz, L. A.; *On nonexpansive mappings,* Proc. Amer. Math. Soc. **55** (1976), 91–103.

273. Karlovitz, L. A.; *Some fixed point results for nonexpansive mappings,* in Fixed Point Theory and Its Applications, Academic Press, New York, (1976), 91–103.

274. Karlovitz, L. A.; *Existence of fixed points of nonexpansive mappings in a space without normal structure,* Pacific J. Math. **66** (1976), 153–159.

275. Kato, T.; *Nonlinear semigroups and evolution equations,* J. Math. Soc. Japan **19** (1967), 508–520.

276. Kay, D. C.; *A parallelogam law for certain* L_p *spaces*, Amer. Math. Monthly **74** (1967), 140–147.

277. Kazmi, K. R.; *Mann and Ishikawa type perturbed iterative algorithms for generalized quasi-variational inclusions*, J. Math. Anal. Appl. **209** (1997), 572–584.

278. Kim, G. E. and Kim, T. H.; *Mann and Ishikawa iterations with errors for non-Lipschitian mappings in Banach spaces*, Comput. Math. Appl., **42** (2001), 1565–1570.

279. Kim, H. K. and Xu, H. K.; *Strong convergence of modified Mann iteration for asymptotically nonexpansive mappings and semigroups*, Nonlinear Anal., **64** (2006), 1140–1152.

280. Kim, G. E. and Takahashi, W.; *Approximating common fixed points of nonexpansive semigroups in Banach spaces*, Sci. Math. Japan. **63** (2006), no. 1, 31–36.

281. Kinderlehrer, D., and Stampacchia, G.; *An introduction to variational inequalities and their applications*, Academic press, Inc. 1980.

282. Kirk, W. A.; *Fixed point theorems for non-Lipschitzian mappings of asymptotically nonexpansive type*, Israel J. Math. **17** (1974), 339–346.

283. Kirk, W. A.; *A fixed point theorem for mappings which do not increase distance*, Amer. Math. Soc. **72** (1965), 1004–1006.

284. Kirk, W. A.; *Locally nonexpansive mappings in Banach spaces*, pp. 178–198, Lecture Notes in Math., **886**, Springer-Verlag, Berlin, 1981.

285. Kirk, W. A.; *On local expansions and accretive mappings*, Internat. J. Math. Math. Sci. **6** (1983), 419–429.

286. Kirk, W. A. and Morales, C. H.; *On the approximation of fixed points of locally nonexpansive mappings*, Canad. Math. Bull. **24**(1981), no. 4, 441–445

287. Kirk, W. A. and Schöneberg, R.; *Mapping theorems for local expansion in metric and Banach spaces*, J. Math. Anal. Appl. **72** (1979), 114–121.

288. Kirk, W. A. and Schöneberg, R.; *Zeros of m-accretive operators in Banach spaces*, Israel J. Math. **35** (1980), no. 1-2, 1–8.

289. Kobayashi, Y.; *Difference approximation of Cauchy problems for quasi-dissipative operators and generation of nonlinear semigroups*, J. Math. Soc. Japan **27** (1975), 640–655.

290. Kohsaka, F. and Takahashi, W.; *Proximal point algorithms with Bregman functions in Banach spaces*, J. Nonlinear Convex Anal. **6** (2005), no. 3, 505–523.

291. Krasnosel'skiĭ, M. A.; *Two observations about the method of successive approximations*, Uspehi Math. Nauk **10** (1955), 123–127.

292. Krasnosel'skiĭ, M. A. and Rutickiĭ, Y. B.; *Convex Functions and Orlicz Spaces*, P. Noordhoff Ltd. Groningen, The Netherlands, 1961.

293. Lan, H. Y.; Huang, N. J. and Cho, Y. J.; *New iterative approximation for a system of generalized nonlinear variational inclusions with set-valued mappings in Banach spaces*, Math. Inequal. Appl. **9** (2006), no. 1, 175–187.

294. Lau, A. T.; *Semigroups of nonexpansive mappings on a Hilbert space*, J. Math. Anal. Appl. **105** (1985), 514–522.

295. Lau, A. T. and Takahashi, W.; *Weak convergence and nonlinear ergodic theorems for reversible semigroups of nonexpansive mappings*, Pacific J. Math. **126** (1987), 277–294.

296. Li, J.; *On the existence of solutions of variational inequalities in Banach spaces*, J. Math. Anal. Appl. **295** (2004), no. 1, 115–126.

297. Li, J.; *The metric projection and its applications to solving variational inequalities in Banach spaces*, Fixed Point Theory **5** (2004), no. 2, 285–298.

298. Li, J.; *The generalized projection operator on reflexive Banach spaces and its applications*, J. Math. Anal. Appl. **306** (2005), no. 1, 55–71.

299. Li, J., Whitaker, John; *Exceptional family of elements and solvability of variational inequalities for mappings defined only on closed convex cones in Banach spaces*, J. Math. Anal. Appl. **310** (2005), no. 1, 254–261.

300. Li, J., Rhoades, B. E.; *An approximation of solutions of variational inequalities,* Fixed Point Theory Appl. (2005), no. 3, 377–388.

301. Li, J., Park, Sehie; *On solutions of generalized complementarity and eigenvector problems,* Nonlinear Anal. **65** (2006), no. 1, 12–24.

302. Liang, Z.; *Iterative solution of nonlinear equations involving m-accretive operator equations in Banach spaces,* J. Math. Anal. Appl. **188** (1994), no. 2, 410–416.

303. Lim, T. C.; *Characterization of normal structure,* Proc. Amer. Math. Soc. **43** (1974), 313–319.

304. Lim, T. C.; *A fixed point theorem for families of non expansive mappings,* Pacific J. Math. **53** (1974), 487–493.

305. Lim, T. C.; *Fixed point theorems for uniformly Lipschitzian mappings in L_p spaces,* Nonlinear Anal. **7**(1983), no. 5, 555–563.

306. Lim, T. C.; *Some L_p inequalities and their applications to fixed point theorems of uniformly Lipschizian mappings,* in *Proc. of Symposia in Pure Math.* **45** Part 2, (1986), 119–125. Amer. Math. Soc. Providence, RI.

307. Lim, T. C. and Xu, H. K.; *Fixed point thoerems for asymptotically nonexpansive mappings,* Nonlinear Anal. **22**(1994), 1345–1355.

308. Lin, P. K. and Tan, K. K.; *Demiclosedness principle and asymptotic behaviour for asymptotically nonexpansive mappings,* Nonlinear Anal. **24**(1995), 929–946.

309. Lindenstrauss, J.; *On the modulus of smoothness and divergent series in Banach spaces,* Michigan Math. J. **10** (1963), 241–252.

310. Lindenstrauss, J.; *On nonlinear projections in Banach spaces,* Michigan Math. J. **11** (1964), 262–287.

311. Lindenstrauss, J. and Tzafriri, L.; *On the complemented subspace problem,* Israel J. Math. **9** (1971), 263–269.

312. Lindenstrauss, J. and Tzafriri, L.; *Classical Banach spaces II: Function Spaces,* Ergebnisse Math. Grenzgebiete Bd. **97**, Springer-Verlag, Berlin, 1979.

313. Lions, P. L.; *Approximation de points fixed de contractions,* C. R. Acad. Sci. Paris Ser. A **284** (1977), 1357–1359.

314. Liu, L.; *Ishikawa and Mann iterative process with errors for nonlinear strongly accretive mappings in Banach spaces,* J. Math. Anal. Appl. **194**(1995), no. 1, 114–125.

315. Liu, L.; *Approximation of fixed points of a strictly pseudocontractive mapping,* Proc. Amer. Math. Soc. **125** (1997), no. 5, 1363–1366.

316. Liu, Z. and Kang, S. M.; *Convergence theorems for $\phi-$strongly accretive and $\phi-$hemicontractive operators,* J. Math. Anal. Appl. **253** (2001), 35–49.

317. Liu , L. W. and Li, Y. Q.; *On generalized set-valued variational inclusions,* J. Math. Anal. Appl. **261** (2001), 231–240.

318. Maingé, P.; *Approximation methods for common fixed points of nonexpansive mappings in Hilbert space,* J. Math. Anal. Appl. 325 (2007), 469–479.

319. Mann, W. R.; *Mean value methods in iteration,* Proc. Amer. Math. Soc. **4** (1953), 506–510.

320. Markov, A.; *Quelques theoremes sur les ensembles Abeliens,* Doklady Akad. Nauk SSSR (N.S.) **10** (1936), 311–314.

321. Marino, G. and Xu, H. K.; *A general iterative method for nonexpansive mappings in Hilbert spaces,* J. Math. Anal. Appl. **318** (2006), no. 1, 43–52.

322. Martin, R. H.; *Nonlinear operators and differential equations in Banach spaces,* Interscience, New York, 1976.

323. Martin, R. H.; *A global existence theorem for autonomous differential equations in Banach spaces,* Proc. Amer. Math. Soc. **26** (1970), 307–314.

324. Martines-Yanes, C. and Xu, H. K.; *Strong convergence of CQ method for fixed point iteration process,* Nonlinear Anal., **64** (2006), 2400–2411.

325. Maruster, S.; *The solution by iteration of nonlinear equations,* Proc. Amer. Math. Soc. **66** (1977), 69–73.

326. Matsuhita, S. Y. and Kuroiwa, D.; *Strong convergence of averaging iteration of nonexpansive nonself mappings,* J. Math. Anal. Appl. **294** (2004), 206-204.

327. Matsushita, S. Y. and Takahashi, W.; *On the existence of zeros of monotone operators in reflexive Banach spaces*, J. Math. Anal. Appl. **323** (2006), no. 2, 1354–1364.

328. Maurey, B.; *points fixes des contractons sur un convexe forme de L_1*, Seminaire d'Analyse fontionnelle , Ecole Polytechnique, Palaiseau.

329. Minty, G. J.; *Monotone (nonlinear) operators in Hilbert spaces*, Duke Math. J. **29** (1962), 541–546.

330. Minty, G. J.; *On a "monotonicity" method for the soluton of nonlinear equations in Banach spaces*, Proc. Nat. Acad. Sci. USA **50** (1963), 1038–1041.

331. Miyake, H. and Takahashi, W.; *Ergodic theorems for almost expansive curves in Hilbert spaces*, Dyn. Contin. Discrete Impuls. Syst. Ser. A Math. Anal. **12** (2005), no. 6, 825–835.

332. Miyake, H. and Takahashi, W.; *Nonlinear ergodic theorems for nonexpansive mappings in general Banach spaces*, J. Nonlinear Convex Anal. **7** (2006), no. 2, 199–209.

333. Moore, C.; *Picard iterations for solution of nonlinear equations in certain Banach spaces*, J. Math. Anal. Appl. **245**(2000), no. 2, 317–325

334. Moore, C.; *The solution by iteration of nonlinear equations involving psi-strongly accretive operators*, Nonlinear Anal. **37**(1999), no. 1, 125–138.

335. Moore, C. and Nnoli, B. V. C.; *Strong convergence of averaged approximants for Lipschitz pseudocontractive maps*, J. Math. Anal. Appl. **260** (2001), 269–278.

336. Moore, C. and Nnoli, B. V. C.; *Iterative solution of nonlinear equations involving set-valued uniformly accretive operators,*, Comput. Math. Appl. **42** (2001), 131–140.

337. Morales, C. H.; *Zeros for strongly accretive set-valued mappings*, Comm. Math. Univ. Carolin. **27** (1986), 455–469.

338. Morales, C. H.; *Locally accretive mappings in Banach spaces*, London Math. Soc. **28** (1995), 627–633.

339. Morales, C.H.; *Strong convergence theorems for pseudocontractive mappings in Banach spaces*, Houston J. Math. **16**(1990), 549–557.

340. Morales, C. H. and Chidume, C. E.; *Convergence of the steepest descent method for accretive operators*, Proc. Amer. Math. Soc. **127**(1999), 3677–2683.

341. Morales C. H. and Jung J. S.; *Convergence of paths for pseudo-contractive mappings in Banach spaces*, Proc. Amer. Math. Soc. **128** (2000), no. 2, 3411–3419.

342. Müller, G. and Reinermann, J.; *Fixed point theorems for pseudocontractive mappings and a counter-example for compact maps*, Comment. Math. Univ. Carolinae **18** (1977), no. 2, 281–298.

343. Nadezhkina, N. and Takahashi, W.; *Weak convergence theorem by Cesro means for nonexpansive mappings and monotone mappings*, Fixed Point Theory **6** (2005), no. 2, 311–321.

344. Nadezhkina, N. and Takahashi, W.; *Strong convergence theorem by the hybrid and extragradient methods for monotone mappings and countable families of nonexpansive mappings*, Sci. Math. Jpn. **63** (2006), no. 2, 217–227.

345. Nadezhkina, N. and Takahashi, W.; *Strong convergence theorem by a hybrid method for nonexpansive mappings and Lipschitz-continuous monotone mappings*, SIAM J. Optim. **16** (2006), no. 4, 1230–1241 (electronic).

346. Nadezhkina, N. and Takahashi, W.; *Weak convergence theorem by an extragradient method for nonexpansive mappings and monotone mappings*, J. Optim. Theory Appl. **128** (2006), no. 1, 191–201.

347. Nadler, S. B.; *Multivalued contraction mappings*, Pacific J. Math. **30** (1969), 475–488.

348. Nakajo, K., Shimoji, K., and Takahashi, W.; *On weak convergence by products of mappings in Hilbert spaces*, Comm. Appl. Nonlinear Anal. **13** (2006), no. 1, 27–50.

349. Nakajo, K., Shimoji, K., and Takahashi, W.; *Strong convergence theorems by the hybrid method for families of nonexpansive mappings in Hilbert spaces*, Taiwanese J. Math. **10** (2006), no. 2, 339–360.

350. Nakajo, K. and Takahashi, W.; *Strong convergence theorems for nonexpansive mappings*, J. Math. Anal. Appl. **279** (2003), 372–379.

351. Nevanlinna, O.; *Global iteration schemes for monotone operators*, Nonlinear Anal. **3**(1979), no. 4, 505–514.

352. Nevanlinna, O. and Reich, S.; *Strong convergence of contraction semigroups and of iterative methods for accretive operators in Banach spaces*, Israel J. Math. **32** (1979), 44–58.

353. Nilsrakoo, W. and Saejung, S.; Weak and strong convergence theorems for countable Lipschitzian mappings and its applications, to appear in Nonlinear Anal.

354. Nishiura, K., Shioji, N. and Takahashi, W.; *Nonlinear ergodic theorems for asymptotically nonexpansive semigroups in Banach spaces*, Dyn. Contin. Discrete Impuls. Sys. Ser. A Math. Anal. **10** (2003), no. 4, 563–578.

355. Noor, M. A.; *General variational inequalities*, J. Math. Lett. **1** (1988), 119–122.

356. Noor, M. A.; *An iterative algorithm for variational inequalities*, J. Math. Anal. Appl. **158** (1991), 446–455.

357. Noor, M. A.; *Equivalence of variational inclusions with resolvent equations*, Nonlinear Anal. **41**(2000), 963–970.

358. Noor, M. A.; *General variational inequalities and nonexpansive mappings*, J. Math. Anal. Appl. **331** (2007), 810–822.

359. Noor, M. A., Noor, K. I., and Rassias, T. M.; *Set-valued resolvent equations and mixed variational inequalities*, J. Math. Anal. Appl. **220** (1998), 741–756.

360. Nördlander, G.; *The modulus of convexity in normed linear spaces*, Ark. Mat. **4** (1960), 15–17.

361. O'Hara, J.G., Pillay, P., and Xu, H. K., *Iterative approaches to finding nearest common fixed points of nonexpansive mappings in Hilbert spaces*, Nonlinear Anal. **54** (2003), 1417–1426.

362. Oka, H.; *Nonlinear ergodic theorems for commutative semigroups of asymptotically nonexpansive mappings*, Nonlinear Anal. **18** (1992), 619–635.

363. Ofoedu, E. U.; *Strong convergence theorem for uniformly L-Lipschitzian asymptotically pseudocontractive mapping in real Banach space,* J. Math. Anal. Appl. **321** (2006), no. 2, 722–728.

364. Ofoedu, E. U. and Shehu, Y.; *Iterative construction of a common fixed point of finite family of nonlinear mappings,* to appear in Appl. Anal.

365. Oka, H.; *An ergodic theorem for asymptotically nonexpansive mappings in the intermediate sense,* Proc. Amer. Math. Soc. **125** (1997), 1693–1703.

366. Opial, Z.; *Weak convergence of the sequence of successive approximations for nonexpansive mappings*, Bull. Amer. Math. Soc. **73** (1967), 591–597.

367. Opial, Z.; *Nonexpansive and monotone mappings in Banach spaces*, Lecture Notes **67-1**, Lefschetz Centre for Dynamical Systems, Division of Applied Mathematics, Brown University, Providence, RI 02912, 1967.

368. Osilike, M. O.; *Ishikawa and Mann iteration methods for nonlinear strongly accretive mappings*, Bull. Austral. Math. Soc. **46** (1992), 411–422.

369. Osilike, M. O.; *Iterative solution of nonlinear equations of the ϕ-strongly accretive type*, J. Math. Anal. Appl. **200**(2) (1996), 259–271.

370. Osilike, M. O.; *Ishikawa and Mann iteration methods with errors for nonlinear equations of the accretive type*, J. Math. Anal. Appl. **213** (1997), 91–105.

371. Osilike, M. O.; *Approximation methods for nonlinear m-accretive operator equations*, J. Math. Anal. Appl. **20** (1997), no. 1, 20–24.

372. Osilike, M. O.; *Iterative solutions of nonlinear ϕ-strongly accretive operator equations in arbitrary Banach spaces*, Nonlinear Anal. **36**(1999), no. 1, 1–9.

373. Osilike, M. O. and Aniagbosor, S. C.; *Weak and strong convergence theorems for fixed points of asymptotically nonexpansive mappings,* Math. Comput. Modelling, **32** (2000), 1181–1191.

374. Osilike, M. O. and Udomene, A.; *Demiclosedness principle and convergence theorems for strictly pseudocontractive mappings of Browder-Petryshyn type*, J. Math. Anal. Appl. **256** (2001), no. 2, 431–445.

375. Osilike, M. O. and Udomene, A.; *A note on approximation of solutions of a K-positive definite operator equation*, Bull. Korean Math. Soc. **38** (2001), no. 2, 231–236.

376. Osilike, M. O., Udomene, A., Igbokwe, D. I. and Akuchu, B. G.; *Demiclosedness principle and convergence theorems for k-strictly asymptotically pseudocontractive maps*, J. Math. Anal. Appl. **326** (2007), no. 2, 1334–1345.

377. Outlaw, C. L.; *Mean Value iterations of nonexpansive mappings in Banach spaces*, Pacific J. Math. **30** (1969), 747–750.

378. Outlaw, C. L. and Groetsch, C. W.; *Averaging iterations in a Banach space*, Bull. Amer. Math. Soc. **75** (1969), 430–432.

379. Pascali, D. and Sburlan, S.; *Nonlinear mappings of monotone type*, Editura Academia Bucaresti, Romania, 1978.

380. Passty, G. B.; *Construction of fixed points for asymptotically nonexpansive mappings*, Proc. Amer. Math. Soc. **84** (1982), 212–216.

381. Petryshyn, W. V.; *Construction of fixed points of demicompact mappings in Hilbert Spaces*, J. Math. Anal. Appl. **14** (1966), 276–284.

382. Petryshyn, W. V. and Williamson, W. E.; *Strong and weak convergence of the sequence of successive approximations for quasi-nonexpansive mappings*, J. Math. Anal. Appl. **43** (1973), 459–497.

383. Phelps, R. R.; *Convex Functions, Monotone Operators and Differentiability*, Lecture Notes in Math. **1364**, Springer-Verlag, Berlin, 1989.

384. Picard, E.; *Sur l'application des methodes d'approximations successive a l'etude de certaines equations differentielles ordinaires*, Journ. de Math. **9** (1893), 217–271.

385. Plubtieng, S. and Punpaeng, R.; *A general iterative method for equilibrium problems and fixed point problems in Hilbert spaces*, J. Math. Anal. Appl. **336**(2007), 455–469.

386. Prus, S.; *Banach spaces with uniform Opial property*, Nonlinear Anal. **18** (1992), 697–704.

387. Prus, S. and Smarzewski, R.; *Strongly unique best approximation and centers in uniformly convex spaces*, J. Math. Anal. Appl. **121** (1987), 10–12.

388. Prüss, J.; *A characterization of uniform convexity and applications of accretive operators*, Hiroshima Math. J. **11** (1981), 229–234.

389. Qihou, L.; *On Naimpally and Singh's open question*, J. Math. Anal. Appl. **124** (1987), 157–164.

390. Qihou, L.; *The convergence theorems of the sequence of Ishikawa iterates for hemicontractive mappings*, J. Math. Anal. Appl. **148** (1990), 55–62.

391. Qihou, L.; *A convergence theorem for Ishikawa iterates of continuous generalized nonexpansive maps*, J. Math. Anal. Appl. **165** (1992), 305–309.

392. Qihou, L.; *Convergence theorems of the sequence of iterates for asymptotically demicontractive and hemicontractive mappings*, Nonlinear Anal. **26**(1996), no. 11, 1835–1842.

393. Qihou, L.; *Iterative sequences for asymptotically quasi-nonexpansive mappings*, J. Math. Anal. Appl. **259** (2001), 1–7.

394. Qihou, L.; *Iterative sequences for asymptotically quasi-nonexpansive mappings with error terms*, J. Math. Anal. Appl. **259** (2001), 18–24.

395. Qihou, L.; *Iterative sequences for asymptotically quasi-nonexpansive mapping with an error member of Uniformly Convex Banach spaces*, J. Math. Anal. Appl. **266** (2002), 468–471.

396. Quan, J., Chang, S. S., and Long, J.; *Approximation of asymptotically quasi-nonexpansive-type mappings by finite steps iterative sequences*, Fixed Point Theory Appl. 2006(**2006**), article ID 70830.

397. Ray, W. O.; *An elementary proof of surjective for a class of accretive operators*, Proc. Amer. Math. Soc. **75** (1979), 255–258.

398. Ray, B. K.; *A fixed point theorem in Banach spaces*, Indian J. Pure Appl. Math. **8** (1977), 903–907.

399. Reich, S.; *Kannan's fixed point theorem*, Boll. U. M. I. **4** (1971), 1–11.

400. Reich, S.; *Some remarks concerning contraction mappings*, Canad. Math. Bull. **14**(1971), no. 1, 121–124.

401. Reich, S.; *Fixed points of contractive functions*, Boll. U. M. I. **5** (1972), no. 4, 26–42.

402. Reich, S.; *Remarks on fixed points*, Accad. Naz. Lincei **52** (1972), 689–697.

403. Reich, S.; *Asymptotic behavior of contractions in Banach spaces*, J. Math. Anal. Appl. **44** (1973), 57–70.

404. Reich, S.; *An iterative procedure for constructing zeros of accretive sets in Banach spaces*, Nonlinear Anal. **2** (1978), 85–92.

405. Reich, S.; *Iterative methods for accretive sets,* in "Nonlinear Equations in Abstract Spaces", pp. 317–326, Academic Press, New York, 1978.

406. Reich, S.; *Weak convergence theorems for nonexpansive mappings in Banach spaces*, J. Math. Anal. Appl. **67** (1979), 274–276.

407. Reich, S.; *Constructing zeros of accretive operators I & II*, Appl. Anal. **9** (1979), 159–163.

408. Reich, S.; *Constructive techniques for accretive and monotone operators* in "Applied Nonlinear Analysis", Academic Press, New York (1979), 335–345.

409. Reich, S.; *Strong convergent theorems for resolvents of accretive operators in Banach spaces*, J. Math. Anal. Appl. **75** (1980), 287–292.

410. Reich, S.; *Product formulas, nonlinear semigroups and accretive operators,* J. Func. Anal. **36** (1980), 147–168.

411. Reich, S.; *On the asymptotic behavior of nonlinear semi groups and the range of accretive Operators*, J. Math. Anal. Appl. **79** (1981), 113–126.

412. Reich, S.; *Some problems and results in fixed point theory,* Contemp. Math. **21** (1983), 179–187

413. Reich, S.; *Approximating fixed points of nonexpansive mappings*, PanAmer. Math. J. **4** (1994), 23–28.

414. Reinermann, J.; *Uber Fipunkte kontrahierender Abbildungen und schwach konvergente Toeplitz- Verfahren,* Arch. Math. **20**(1969), 59–64.

415. Rhoades, B. E.; *Fixed point iterations using infinite matrices*, Trans. Amer. Math. Soc. **196** (1974), 161–176.

416. Rhoades, B. E.; *Fixed point iterations using matrices II, Constructive and computational methodes for differential and integral equations*, Springer- Verlag Lecture Notes Series **430**, Springer Verlag, New York, (1974) 390–395.

417. Rhoades, B. E.; *Comments on two fixed point iteration methods*, J. Math. Anal. Appl. **56**(3) (1976), 741–750.

418. Rhoades, B. E.; *Fixed point iterations using infinite matrices III*, Fixed points; Algorithms and Applications, (R. Karamordian ed.), Academic press (1977), 337–347.

419. Rhoades, B. E.; *A comparison of various definitions of contractive mappings*, Trans. Amer. Math. Soc. **226** (1977), 257–290.

420. Rhoades, B. E.; *Extensions of some fixed point theorems of Ciric, Maiti and Pal*, Math. Sem. Notes **6** (1978), 41–46.

421. Rhoades, B. E.; *Fixed point iterations for certain nonlinear mappings*, J. Math. Anal. Appl. **183** (1994), 118–120.

422. Rhoades, B. E. and Saliga L.; *Some fixed point iteration procedures, II*, Nonlinear Anal. Forum **6**(2001), 193–217.

423. Rhoades, B. E. and Soltuz, S. M.; *The convergence of mean value iteration for a family of maps,* Int. J. Math. Math. Sci. (2005), no. 21, 3479–3485.

424. Rhoades, B. E. and Soltuz, S. M.; *The Mann-Ishikawa iterations and the Mann-Ishikawa iterations with errors are equivalent models dealing with a non-Lipschitzian map,* Rev. Anal. Numer. Theot. Approx. **34** (2005), no. 2, 181–193.

425. Rhoades, B. E. and Soltuz, S. M.; *The equivalence of Mann and Ishikawa iteratons dealing with ψuniformly pseudocontractive maps without bounded range,* Tamkang J. Math. **37** (2006), no. 3. 285–295.

426. Rhoades, B. E. and Soltuz, S. M.; *The convergence of an implicit mean value theorem,* Int. J. Math. Math. Sci. (2006), Art. ID 68369, 7pp.

427. Rhoades, B. E. and Soltuz, S. M.; *The equivalence between the Mann and Ishikawa iteratons dealing with generalized contractions* Int. J. Math. Math. Sci. (2006), Art. ID 54653, 5pp.

428. Rhoades, B. E. and Soltuz, S. M.; *The equivalence between the T stabilities of Mann and Ishikawa iteratons* J. Math. Anal. Appl. **318** (2006), no. 2, 472–475.

429. Rhoades, B. E. and Soltuz, S. M.; *The class of asymptotically demicontractive maps a proper subclass of asymptotically pseudocontractive maps,* PanAmer. Math. J. **16** (2006), , no. 2, 93–97.

430. Rockafellar, R. T.; *Local boundedness of nonlinear monotone operators,* Michigan Math. J. **16** (1969), 397-407.

431. Schaefer, H.; *ber die Methode sukzessiver Approximationen,* (German) Jber. Deutsch. Math. Verein. **59** (1957), Abt. 1, 131–140.

432. Schöneberg, R.; *On the structure of fixed point sets of pseudocontractive mappings II,* Comment. Math. Univ. Carolin. **18**(1977), no. 2, 299–310.

433. Schöneberg, R.; *On the domain invariance theorem for accretive mappings,* J. London Math. Soc. **24**(1981), 548–554.

434. Schu, J.; *Approximating fixed points of Lipschitzian pseudocontractive mappings,* Lehrstuhl C für Mathematik, Preprint No. **17** (1989).

435. Schu, J.; *Approximation of fixed points of asymptotically nonexpansive mappings,* Proc. Amer. Math. Soc. **112** (1991), no. 1, 143–151.

436. Schu, J.; *Iterative constraction of fixed points of strictly pseudocontractive mappings,* Appl. Anal. **40** (1991), 67–72.

437. Schu, J.; *On a theorem of Chidume, C. E. concerning the iterative approximation of fixed points,* Math. Nachr. **153** (1991), 313–319.

438. Schu, J.; *Iterative construction of fixed points of asymptotically nonexpansive mappings,* J. Math. Anal. Appl. **158** (1991), 407–413.

439. Schu, J., *Weak and strong convergence of fixed points of asymptotically nonexpansive mappings,* Bull. Austral. Math. Soc. **43** (1991), 153–159.

440. Schu, J., *Approximating fixed points of Lipschitzian pseudocontractive mappings,* Houston J. Math. **19** (1993), 107–115.

441. Senter, H. F. and Dotson, W. G.; *Approximating fixed points of nonexpansive mappings,* Proc. Amer. Math. Soc. **44** (1974), 375–380.

442. Shahzad, N.; *Approximating fixed points of nonself nonexpansive mappings in Banach spaces,* Nonlinear Anal. **61** (2005), 1031–1039.

443. Shahzad, N. and Al-Dubiban, R.; *Approximating common fixed points of nonexpansive mappings in Banach spaces,* Georgian Math. J. **13** (2006), no. 3, 529–537.

444. Shahzad, N. and Udomene, A.; *Approximating common fixed points of two asymptotically quasi-nonexpansive mappings in Banach spaces,* Fixed Point Theory Appl. **2006** Art. ID 18909, 10 pp.

445. Shahzad, N. and Udomene, A.; *Fixed point solutions of variational inequalities for asymptotically nonexpansive mappings in Banach spaces,* Nonlinear Anal. **64** (2006), no. 3, 558–567.

446. Shahzad, N. and Zegeye, H.; *On stability for ϕ-strongly pseudocontractive mappings,* Nonlinear Anal. **64** (2006), no. 12, 2619–2630.

447. Shahzad, N. and Zegeye, H.; *Strong convergence of an implicit iteration process for a finite family of generalized asymptotically quasi-nonexpansive maps,* Appl. Math. Comput. **189**, no. 2, 1058–1065.

448. Shimi, T. N.; *Approximation of fixed points of certain nonlinear mappings,* J. Math. Anal. Appl. **65** (1978), 565–571.

449. Shimizu, T. and Takahashi, W.; *Strong convergence to common fixed points of families of nonexpansive mappings,* J. Math. Anal. Appl. **211** (1997), 71–83.

450. Shioji, S. and Takahashi, W.; *Strong convergence of approximated sequences for nonexpansive mappings in Banach spaces,* Proc. Amer. Math. Soc. **125** (1997), 3641–3645.

451. Shioji, N. and Takahashi, W.; *Strong convergence theorems for asymptotically non-expansive semigroups in Banach spaces,* J. Nonlinear Convex Anal. **1** (2000), no. 1, 73–87.

452. Shioji, N. and Takahashi, W.; *Strong convergence theorems for asymptotically nonexpansive mappings in Banach spaces,* Arch. Math. (Basel) **72** (1999), no. 5, 354–359.

453. Shioji, N. and Takahashi, W.; *Strong convergence of averaged approximants for asymptotically nonexpansive mappings in Banach spaces,* J. Approx. Theory **97** (1999), no. 1, 53–64.

454. Shioji, N. and Tahakashi, W.; *Strong convergence theorems for asymptotically non-expasive semigroups in Hilbert spaces,* Nonlinear Anal. **34** (1998), 87–99.

455. Siddiqi, A. H. and Ansari, Q. H.; *General strongly nonlinear variational inequalities,* J. . Math. Anal. Appl. **166** (1992), 386–392.

456. Siddiqi, A. H., Ansari, Q. H.; and Kazmi, K. R.; *On nonlinear variational inequalities,* Indian J. Pure Appl. Math. **25** (1994), 969–973.

457. Soardi, R.; *Su un problema di punto unito di S. Reich,* Boll. U. M. I. **4** (1971), no. 4, 841–845.

458. Sun, Z.H.; *Strong convergence of an implicit iteration process for a finite family of asymptotically quasi-nonexpansive mappings,* J. Math. Anal. Appl. **286** (2003), 351–358.

459. Suzuki, T.; *Strong convergence of Krasnoselskii and Mann's type sequences for one-parameter nonexpansive semigroups without Bochner integrals,* J. Math. Anal. Appl. **305** (2005), 227–239.

460. Suzuki, T.; *On strong convergence to common fixed points of nonexpansive semi-groups in Banach spaces* Proc. Amer. Math. Soc. **131** (2003) 2133–2136.

461. Suzuki, T.; *The set of common fixed points of a n-parameter continiuous semigroup of mappings,* Nonlinear Anal. **63** (2005), no. 38, 1180–1190.

462. Suzuki, T.; *An example for a one-parameter nonexpansive semigroup,* Abstr. Appl. Anal. (2005), no. 2, 173–183.

463. Suzuki, T.; *Strong convergence theorems for infinite families of nonexpansive map-pings in general Banach spaces,* Fixed Point Theory Appl. (2005), no. 1, 103–123.

464. Suzuki, T.; *Strong convergence theorems of Browder's type sequences for infinite families of nonexpansive mappings in Hilbert spaces,* Bull. Kyushu Inst. Technol. Pure Appl. Math., (2005), no. 52, 21–28.

465. Suzuki, T.; *Common fixed point of one-parameter nonexpansive semi-group,* Bull. London Math. Soc. **38** (2006) 1009–1018.

466. Suzuki, T.; *Characterizations of common fixed points of one-parameter nonexpansive semigroups, and convergence theorem to common fixed points,* J. Math. Anal. Appl. **324** (2006), no. 2, 1006–1019.

467. Suzuki, T.; *Browder's type strong convergence theorems for infinite families of nonex-pansive mappings in Banach spaces,* Fixed Point Theory Appl. **2006**, Art. ID 59692, 16pp.

468. Suzuki, T.; *Common fixed points of one-parameter nonexpansive semigroups in strictly convex Banach spaces,* Abstr. Appl. Anal. (2006), Art. ID 58684, 10pp.

469. Suzuki, T.; *The set of common fixed points of a one-parameter continuous semigroup of nonexpansive mappings is $F(\frac{1}{2}T(1) + \frac{1}{2}T\sqrt{2})$ in strictly convex Banach spaces,* Taiwanese J. Math. **10** (2006), no. 2, 381–397.

470. Suzuki, T.; *The set of common fixed points of a one-parameter continuous semigroup of mappings is $F(T(1)) \cap F(T\sqrt{2})$,* Proc. Amer. Math. Soc. **134** (2006), no. 3, 673–681.

471. Suzuki, T.; *Browder's type convergence theorems for one parameter semi-groups of nonexpansive mappings in Banach spaces,* Israel J. Math. **157** (2007) 239–257.

472. Suzuki, T.; *Moudafi's viscosity approximations with Meir-Keeler contractions,* J. Math. Anal. Appl. **325** (2007), no. 1, 342–352.

473. Suzuki, T. and Takahashi, W.; *Strong convergence of Mann's sequences for one-parameter nonexpansive semigroups in general Banach spaces,* J. Nonlinear Convex Anal. **5** (2004), 209–216.

474. Takahashi, W.; *Fixed point theorems for families of nonexpansive mappings on unbounded sets,* J. Math. Soc. Japan **36** (1984), 543–553.

475. Takahashi, W.; Nonlinear Functional Analysis, *Yokohama Publishers, Yokohama* 2000.

476. Takahashi, W. and Kim, G.E.; *Strong convergwence of approximatants to fixed points of nonexpansive nonself mappings in Banach spaces,* Nonlinear Anal. (1998), 447–454.

477. Takahashi, W. and Tamura, T.; *Convergence theorems for pair of nonexpansive mappings,* J. Conv. Anal. **5** (1998), 45–56.

478. Takahashi, W. and Ueda, Y.; *On Reich's strong convergence theorems for resolvents of accretive operators,* J. Math. Anal. Appl. **104** (1984), no. 2, 546–553.

479. Takahashi, W. and Zembayashi, K.; *Nonlinear strong ergodic theorems for asymptotically nonexpansive semigroups with compact domains,* Yokohama Math. J. **52** (2006), no. 2, 131–149.

480. Takahashi, W. and Zembayashi, K.; *Fixed point theorems for one-parameter asymptotically nonexpansive semigroups in general Banach spaces,* Nonlinear Anal. **65** (2006), no. 2, 433–441.

481. Takahashi, W., Tamura, T. and Toyoda, M.; *Approximation of common fixed points of a family of finite nonexpansive mappings in Banach spaces,* Sci. Math. Jpn. **56** (2002), no.3, 475–480.

482. Takahashi, S. and Takahashi, W.; *Viscosity approximation methods for equilibrium problems and fixed point problems in Hilbert spaces,* J. Math. Anal. Appl. **331** (2007), 506–515.

483. Tan, B. and Zhou, H.; *An iteration method for fixed points of nonexpansive nonself-mappings in Hilbert spaces,* (Chinese) Acta Anal. Funct. Appl. **8** (2006), no. 3, 272–275.

484. Tan, K. K. and Xu, H. K.; *A nonlinear ergodic theorem for asymptotically nonexpansive mappings,* Bull. Austral. Math. Soc. **45** (1992), 25–36.

485. Tan, K. K. and Xu, H. K.; *The nonlinear ergodic theorem for asymptotically nonexpansive mapping in Banach spaces,* Proc. Amer. Math. Soc., **114** (1992), 399–404.

486. Tan, K. K. and Xu, H. K.; *Iterative solutions to nonlinear equations of strongly accretive operators in Banach spaces,* J. Math. Anal. Appl. **178** (1993), no. 1, 9–12.

487. Tan, K. K. and Xu, H. K.; *Approximating fixed points of nonexpensive mappings by the Ishikawa iteration process,* J. Math. Anal. Appl. **178**(1993), no. 2, 301–308.

488. Tan, K. K. and Xu, H. K.; *Fixed point iteration processes for asymptotically nonexpansive mappings,* Proc. Amer. Math. Soc., **122** (1994), 733–739.

489. Tingly, D.; *Noncontractive uniformly Lipschitzian semigroups in Hilbert space,* Proc. Amer. Math. Soc. **92** (1984), 255–261.

490. Tricomi, F.; *Un teorema sulla convergenza delle successioni formate dalle successive iterate di una funzione di una variable reale,* Giorn. Math. Battaglini **54** (1916), 1–9.

491. Trubnikov, Yu. V.; *The Hanner inequality and the convergence of iterative processes,* Soviet Math. Izvestiya **31** (1987), no. 7, 74–83.

492. Turett, B.; *A dual view of a theorem of Baillon, Nonlinear Analysis and applications,* (S.P. Singh and J. H. Burry, eds), Marcel Dekker, New York, 1982, pp. 279–286.

493. Udomene, A.; *Construction of zeros of accretive mappings,* J. Math. Anal. Appl. **262** (2001), no. 2, 623–632.

494. Udomene, A.; *Fixed point variational solutions for uniformly continuous pseudocontractions in Banach spaces,* Fixed Point Theory Appl. 2006, Art. ID 69758, 12 pp.

495. Vainberg, M. M.; *On the convergence of the method of steepest descent for nonlinear equations,* Sov. Math., Dokl. **1** (1960), 1–4.

496. Vidossich, G.; *Applications of topology to analysis: On the topological properties of theset of fixed ponts of nonlinear operators,* Confer. Sem. Mat. Univ. Bari **70** (1964), 781–787.

497. Vijayaraju, P.; *Fixed point Theorems for asymptotically nonexpansive mappings*, Bull. Calcutta Math. Soc. **80** (1988), 133–136.

498. Wang, L., *Strong and weak convergence theorems for common fixed points of nonself asymptotically nonexpansive mappings*, J. Math. Anal. Appl. **323** (2006), 550–557.

499. Wang, L.; *An iteration method for nonexpansive mappings in Hilbert spaces*, Fixed Point Theory Appl. **2007**, Art. ID 28619, 8 pages.

500. Weng, X. L.; *Fixed point iteration for local strictly pseudocontractive mappings*, Proc. Amer. Math. Soc. **113** (1991), no. 3, 727–731.

501. Weng, X. L.; *Iterative construction of fixed points of a dissipative type operator*, Tamkang J. Math. **23** (1992), 205–215.

502. Wei, L. and Zhou, H.; *An iterative convergence theorem of zero points for maximal monotone operators in Banach spaces and its application*, (Chinese) Math. Practice Theory **36** (2006), no. 5, 235–242.

503. Wei, L. and Zhou, H., *A new iterative scheme with errors for the zero point of maximal monotone operators in Banach spaces*, (Chinese) Math. Appl. (Wuhan) **19** (2006), no. 1, 101–105.

504. Wen, S. and Cao Z., *The generalized decomposition theorem in Banach spaces and its applications*, J. Approx. Theory **129** (2004), no. 2, 167–181.

505. Wittmann, R.; *Approximation of fixed points of nonexpansive mappings*, Arch. Math. (Basel), **58**(1992), 486–491.

506. Wojtaszczyk, P.; *Banach spaces for analysts*, Cambridge University Press, Cambridge, 1991.

507. Xiao, R.; *Chidume's open problems and fixed point theorems*, Xichuan Daxue Xuebao **35** (1998), no. 4, 505–508.

508. Xu, H. K.; *Strong asymptotic behavior of almost orbits of nonlinear smigroups*, Nonlinear Anal. **46** (2001),135–151.

509. Xu, H. K.; *Inequalities in Banach spaces with applications*, Nonlinear Anal. **16** (1991), no. 12, 1127–1138.

510. Xu, H. K.; *Another control condition in an iterative method for nonexpansive mappings*, Bull. Austral. Math. Soc. **65** (2002), 109–113.

511. Xu, H. K.; *Iterative algorithms for nonlinear operators*, J. London Math. Soc. **66** (2002), no. 2, 240–256.

512. Xu, H. K.; *Viscosity approximaton methods for nonexansive mappings*, J. Math. Anal. Appl. **298** (2004), no. 1, 279–291.

513. Xu, H. K.; *A strong convergence theorem for contraction semigroups in Banach spaces*, Bull. Austral. Math. Soc. **72** (2005), no. 3, 371–379.

514. Xu, H. K.; *A variable Krasnoselskii-Mann algorithm and the multiple-set split feasibility problem*, Inverse Problems **22** (2006), no. 6, 2021–2034.

515. Xu, H. K.; *A regularization method for the proximal point algorithm,*. J. Global Optim. **36** (2006), no. 1, 115–125.

516. Xu, H. K.; *Strong convergence of approximating fixed point sequences for nonexpansive mappings*, Bull. Austral. Math. Soc. **74** (2006), no. 1, 143–151.

517. Xu, H. K.; *Strong convergence of an iterative method for nonexpansive and accretive operators*, J. Math. Anal. Appl. **314** (2006), no. 2, 631–643.

518. Xu, H. K. and Kim, T. H.; *Convergence of hybrid steepest-decent methods for variational inequalities*, J. Optim. Theory Appl. 119 (2003), 185–201.

519. Xu, H. K. and Yamada, I.; *Asymptotic regularity of linear power bounded operators*, Taiwanese J. Math. **10**(2006), no. 2, 417–429.

520. Xu, H. K. and Yin, X. M.; *Strong convergence theorems for nonexpansive nonself mappings*, Nonlinear Anal. **24** (1995), no. 2, 223–228.

521. Xu, Y.; *Existence and convergence for fixed points of mappings of the asymptotically nonexpansive type*, Nonlinear Anal. **16** (1991), 1139–1146.

522. Xu, Y.; *Ishikawa and Mann iterative processes with errors for nonlinear strongly accretive operator equations*, J. Math. Anal. Appl. **224** (1998), 91–101.

523. Xu, Z. B.; *Characterstic inequalities of L_p spaces and their applications*, Acta. Math. Sinica **32**(2) (1989), 209–218.

524. Xu, Z. B.; *A note on the Ishikawa iteration schemes*, J. Math. Anal. Appl. **167** (1992), 582–587.

525. Xu, Z. B. and Roach, G. F.; *Characteristic inequalities for uniformly convex and uniformly smooth Banach spaces*, J. Math. Anal. Appl. **157** (1991), 189–210.

526. Xu, Z. B. and Roach, G. F.; *An alternating procedure for operators on uniformly convex and uniformly smooth Banach spaces*, Proc. Amer. Math. Soc. **111** (1991), no. 4, 1057–1074.

527. Xu, Z. B. and Roach, G. F.; *A necessary and sufficient condition for convergence of a steepest descent approximation to accretive operator equations*, J. Math. Anal. Appl. **167** (1992), 340–354.

528. Xu, Z. B. and Roach, G. F.; *On the uniform continuity of metric projections in Banach spaces*, Approx. Theory Appl. **8** (1992), no. 3, 11–20.

529. Xu, Z. B., Jiang, Y. L. and Roach, G. F.; *A further necessary and sufficient condition for strong convergence of nonlinear contraction semigroups and of iteration methods for accretive operators in Banach spaces*, Proc. Edinburgh Math. Soc.**38** (1995), no. 2, 1–12.

530. Xue, Z.Q., Zhou, H. and Cho, Y. J.; *Iterative solutions of nonlinear equations for m-accretive operators in Banach spaces*, J. Nonlinear Convex Anal. **3** (2003), no. 1, 313–320.

531. Yamada, I.; *The Hybrid Steepest-Descent Method for Variational Inequality Problems over the Intersection of the Fixed-Point Sets of Nonexpansive Mappings*, Inherently Parallel Algorithms in Feasibility and Optimization and Their Applications, Edited by D. Butnariu, Y. Censor, and S. Reich, North-Holland, Amsterdam, Holland, pp. 473–504, 2001.

532. Yao, Y.; *A general iterative method for a finite family of nonexpansive mappings*, Nonlinear Analysis **66**(2007),

533. Yao, Y., Liou Y. C., and Chen, Q.; *A general iterative method for any infinite family of nonexpansive mappings*, Nonlinear Anal. **69** (2008), 1644–1654.

534. Yao, Y., Chen, R., and Zhou, H.; *Iterative process for certain nonlinear mappings in uniformly smooth Banach spaces*, Nonlinear Funct. Anal. Appl. **10** (2005), no. 4, 651–664. 2676-2687.

535. Yao, Y., Chen, R., and Zhou, H.; *Strong convergence to common fixed points of non-expansive mappings without commutativity assumption*, Fixed Point Theory Appl. **2006,** Art. ID 89470, 8 pp.

536. Yao, Y., Chen, R., and Zhou, H.; *Strong convergence and control condition of modified Halpern iterations in Banach spaces*, Int. J. Math. Math. Sci. **2006,** Art. ID 29728, 10 pp.

537. Zalinescu, C.; *On uniformly convex functions*, J. Math. Anal. Appl. **95** (1983), 344–374.

538. Zarantonello, E. H.; *Solving functional equations by constructive averaging*, Technical Report, #**160** U. S. Army Math. Research Center, Madison, Wisconsin 1960.

539. Zarantonello, E. II.; *The closure of the numerical range contains the spectrum*, Bull. Amer. Math. Soc. **70** (1964), 781–787.

540. Zegeye, H. and Prempeh, E; *Strong convergence of approximants to fixed points of Lipschitzian pseudocontractive maps.*, Comput. Math. Appl. **44** (2002), no. 3-4, 339–346.

541. Zegeye, H. and Shahzad, N.; *Strong convergence theorems for a common zero of a finite family of m-accretive mappings*, Nonlinear Anal. **66** (2007), no. 5, 1161–1169.

542. Zegeye, H. and Shahzad, N.; *Convergence theorems for ψ-expansive and accretive mappings*, Nonlinear Anal. **66** (2007), no. 1, 73–82.

543. Zegeye, H. and Shahzad, N.; *Viscosity approximation methods for a common fixed point of finite family of nonexpansive mappings*, Appl. Math. Comput. **191** (2007), no. 1, 155–163.

544. Zegeye, H. and Shahzad, N.; *Approximation methods for a common fixed point of a finite family of nonexpansive mappings*, Numer. Funct. Anal. Optim., **28** (2007), no. 11-12, 1405–1419.

545. Zegeye, H. and Shahzad, N.; *Strong convergence theorems for a common zero point of a finite family of α-inverse strongly accretive mappings*, J. Nonlinear Convex Anal. **9** (2008), no. 1, 95–104.

546. Zegeye, H. and Shahzad, N.; *Viscosity methods of approximation for a common fixed point of a family of quasi-nonexpansive mappings*, Nonlinear Anal. **68** (2008), no. 7, 2005–2012.

547. Zegeye, H. and Shahzad, N.; *Strong convergence theorems for a finite family of asymptotically nonexpansive mappings and semigroups*, to appear in Nonlinear Anal.

548. Zegeye, H. and Shahzad, N.; *Strong convergence theorems for monotone mappings and relatively weak nonexpansive mappings*, to appear in Nonlinear Anal.

549. Zeidler, E.; *Nonlinear Functional Analysis and its Applications Part II: Monotone Operators*, Springer-Verlag, Berlin, 1985.

550. Zeng, L. C.; *Iterative algorithms for finding approximate solutions for general strongly nonlinear variational inequalities*, J. Math. Anal. Appl. **187** (1994), 352–360.

551. Zeng, L. C.; *Error bounds for approximation solutions to nonlinear equations of strongly accretive operators in uniformly smooth Banach spaces*, J. Math. Anal. Appl. **209** (1997), 67–80.

552. Zhang, S; *On the convergence problems of Ishikawa and Mann iteration process with errors for ψ-pseudocontractive type mappings*, App. Math. Mechanics, **21** (2000), 1–10.

553. Zheng, L. C. and Yao, J. C.; *Implicit iteration scheme with perturbed mapping for common fixed points of finite family of nonexpansive mappings*, Nonlinear Anal. **64** (2006), 2507–2515.

554. Zhou, H.; *Iterative solutions of nonlinear equations involving strongly accretive operators without the Lipschitz assumption*, J. Math. Anal. Appl. **213**(1997), no. 1, 296–307.

555. Zhou, H., *Iterative approximation of zeros for α-strongly accretive operators*, (Chinese) Chinese Ann. Math. Ser. **A 27** (2006), no. 3, 383–388.

556. Zhou, H. and Cheng, D. Q.; *Iterative approximations of fixed points for nonlinear mappings of φ-hemicontractive type in normed linear spaces*, Math. Appl. (Wuhan) **11** (1998), no. 3, 118–121.

557. Zhou, H. and Jia, Y.; *Approximating the zeros of accretive operators by the Ishikawa iteration process*, Abstr. Appl. Anal. **1** (1996), no. 2, 153–167.

558. Zhou, H., Jia Y.; *Approximation of fixed points of strongly pseudocontractive maps without Lipschitz assumption*, Proc. Amer. Math. Soc. **125** (1997), no. 6, 1705–1709.

559. Zhou, H., Wei, L. and Cho, Y. J.; *Strong convergence theorems on an iterative method for family of finite nonexpansive mappings in reflexive Banach spaces*, Appl. Math. Comput. **173** (2006), 196–212.

560. Zhou, H. Y., Zhang, M. H., and Zhou, H.; *A convergence theorem on Mann iteration for strictly pseudo-contraction mappings in Hilbert spaces*, (Chinese) J. Hebei Univ. Nat. Sci. **26** (2006), no. 4, 348–349.

561. Zhu, L.; *Iterative solution of nonlinear equations involving m-accretive operators in Banach spaces*, J. Math. Anal. Appl. **188** (1994), 410–415.

Index

Lecture Notes in Mathematics

For information about earlier volumes
please contact your bookseller or Springer
LNM Online archive: springerlink.com

Vol. 1825: J. H. Bramble, A. Cohen, W. Dahmen, Multiscale Problems and Methods in Numerical Simulations, Martina Franca, Italy 2001. Editor: C. Canuto (2003)

Vol. 1826: K. Dohmen, Improved Bonferroni Inequalities via Abstract Tubes. Inequalities and Identities of Inclusion-Exclusion Type. VIII, 113 p, 2003.

Vol. 1827: K. M. Pilgrim, Combinations of Complex Dynamical Systems. IX, 118 p, 2003.

Vol. 1828: D. J. Green, Gröbner Bases and the Computation of Group Cohomology. XII, 138 p, 2003.

Vol. 1829: E. Altman, B. Gaujal, A. Hordijk, Discrete-Event Control of Stochastic Networks: Multimodularity and Regularity. XIV, 313 p, 2003.

Vol. 1830: M. I. Gil', Operator Functions and Localization of Spectra. XIV, 256 p, 2003.

Vol. 1831: A. Connes, J. Cuntz, E. Guentner, N. Higson, J. E. Kaminker, Noncommutative Geometry, Martina Franca, Italy 2002. Editors: S. Doplicher, L. Longo (2004)

Vol. 1832: J. Azéma, M. Émery, M. Ledoux, M. Yor (Eds.), Séminaire de Probabilités XXXVII (2003)

Vol. 1833: D.-Q. Jiang, M. Qian, M.-P. Qian, Mathematical Theory of Nonequilibrium Steady States. On the Frontier of Probability and Dynamical Systems. IX, 280 p, 2004.

Vol. 1834: Yo. Yomdin, G. Comte, Tame Geometry with Application in Smooth Analysis. VIII, 186 p, 2004.

Vol. 1835: O.T. Izhboldin, B. Kahn, N.A. Karpenko, A. Vishik, Geometric Methods in the Algebraic Theory of Quadratic Forms. Summer School, Lens, 2000. Editor: J.-P. Tignol (2004)

Vol. 1836: C. Năstăsescu, F. Van Oystaeyen, Methods of Graded Rings. XIII, 304 p, 2004.

Vol. 1837: S. Tavaré, O. Zeitouni, Lectures on Probability Theory and Statistics. Ecole d'Eté de Probabilités de Saint-Flour XXXI-2001. Editor: J. Picard (2004)

Vol. 1838: A.J. Ganesh, N.W. O'Connell, D.J. Wischik, Big Queues. XII, 254 p, 2004.

Vol. 1839: R. Gohm, Noncommutative Stationary Processes. VIII, 170 p, 2004.

Vol. 1840: B. Tsirelson, W. Werner, Lectures on Probability Theory and Statistics. Ecole d'Eté de Probabilités de Saint-Flour XXXII-2002. Editor: J. Picard (2004)

Vol. 1841: W. Reichel, Uniqueness Theorems for Variational Problems by the Method of Transformation Groups (2004)

Vol. 1842: T. Johnsen, A. L. Knutsen, K_3 Projective Models in Scrolls (2004)

Vol. 1843: B. Jefferies, Spectral Properties of Noncommuting Operators (2004)

Vol. 1844: K.F. Siburg, The Principle of Least Action in Geometry and Dynamics (2004)

Vol. 1845: Min Ho Lee, Mixed Automorphic Forms, Torus Bundles, and Jacobi Forms (2004)

Vol. 1846: H. Ammari, H. Kang, Reconstruction of Small Inhomogeneities from Boundary Measurements (2004)

Vol. 1847: T.R. Bielecki, T. Björk, M. Jeanblanc, M. Rutkowski, J.A. Scheinkman, W. Xiong, Paris-Princeton Lectures on Mathematical Finance 2003 (2004)

Vol. 1848: M. Abate, J. E. Fornaess, X. Huang, J. P. Rosay, A. Tumanov, Real Methods in Complex and CR Geometry, Martina Franca, Italy 2002. Editors: D. Zaitsev, G. Zampieri (2004)

Vol. 1849: Martin L. Brown, Heegner Modules and Elliptic Curves (2004)

Vol. 1850: V. D. Milman, G. Schechtman (Eds.), Geometric Aspects of Functional Analysis. Israel Seminar 2002-2003 (2004)

Vol. 1851: O. Catoni, Statistical Learning Theory and Stochastic Optimization (2004)

Vol. 1852: A.S. Kechris, B.D. Miller, Topics in Orbit Equivalence (2004)

Vol. 1853: Ch. Favre, M. Jonsson, The Valuative Tree (2004)

Vol. 1854: O. Saeki, Topology of Singular Fibers of Differential Maps (2004)

Vol. 1855: G. Da Prato, P.C. Kunstmann, I. Lasiecka, A. Lunardi, R. Schnaubelt, L. Weis, Functional Analytic Methods for Evolution Equations. Editors: M. Iannelli, R. Nagel, S. Piazzera (2004)

Vol. 1856: K. Back, T.R. Bielecki, C. Hipp, S. Peng, W. Schachermayer, Stochastic Methods in Finance, Bressanone/Brixen, Italy, 2003. Editors: M. Fritelli, W. Runggaldier (2004)

Vol. 1857: M. Émery, M. Ledoux, M. Yor (Eds.), Séminaire de Probabilités XXXVIII (2005)

Vol. 1858: A.S. Cherny, H.-J. Engelbert, Singular Stochastic Differential Equations (2005)

Vol. 1859: E. Letellier, Fourier Transforms of Invariant Functions on Finite Reductive Lie Algebras (2005)

Vol. 1860: A. Borisyuk, G.B. Ermentrout, A. Friedman, D. Terman, Tutorials in Mathematical Biosciences I. Mathematical Neurosciences (2005)

Vol. 1861: G. Benettin, J. Henrard, S. Kuksin, Hamiltonian Dynamics – Theory and Applications, Cetraro, Italy, 1999. Editor: A. Giorgilli (2005)

Vol. 1862: B. Helffer, F. Nier, Hypoelliptic Estimates and Spectral Theory for Fokker-Planck Operators and Witten Laplacians (2005)

Vol. 1863: H. Führ, Abstract Harmonic Analysis of Continuous Wavelet Transforms (2005)

Vol. 1864: K. Efstathiou, Metamorphoses of Hamiltonian Systems with Symmetries (2005)

Vol. 1865: D. Applebaum, B.V. R. Bhat, J. Kustermans, J. M. Lindsay, Quantum Independent Increment Processes I. From Classical Probability to Quantum Stochastic Calculus. Editors: M. Schürmann, U. Franz (2005)

Vol. 1866: O.E. Barndorff-Nielsen, U. Franz, R. Gohm, B. Kümmerer, S. Thorbjønsen, Quantum Independent Increment Processes II. Structure of Quantum Lévy Processes, Classical Probability, and Physics. Editors: M. Schürmann, U. Franz, (2005)

Vol. 1867: J. Sneyd (Ed.), Tutorials in Mathematical Biosciences II. Mathematical Modeling of Calcium Dynamics and Signal Transduction. (2005)

Vol. 1868: J. Jorgenson, S. Lang, $Pos_n(R)$ and Eisenstein Series. (2005)

Vol. 1869: A. Dembo, T. Funaki, Lectures on Probability Theory and Statistics. Ecole d'Eté de Probabilités de Saint-Flour XXXIII-2003. Editor: J. Picard (2005)

Vol. 1870: V.I. Gurariy, W. Lusky, Geometry of Müntz Spaces and Related Questions. (2005)

Vol. 1871: P. Constantin, G. Gallavotti, A.V. Kazhikhov, Y. Meyer, S. Ukai, Mathematical Foundation of Turbulent Viscous Flows, Martina Franca, Italy, 2003. Editors: M. Cannone, T. Miyakawa (2006)

Vol. 1872: A. Friedman (Ed.), Tutorials in Mathematical Biosciences III. Cell Cycle, Proliferation, and Cancer (2006)

Vol. 1873: R. Mansuy, M. Yor, Random Times and Enlargements of Filtrations in a Brownian Setting (2006)

Recent Reprints and New Editions

LECTURE NOTES IN MATHEMATICS   🐎 **Springer**

Edited by J.-M. Morel, F. Takens, B. Teissier, P.K. Maini

Editorial Policy (for the publication of monographs)

1. Lecture Notes aim to report new developments in all areas of mathematics and their applications - quickly, informally and at a high level. Mathematical texts analysing new developments in modelling and numerical simulation are welcome.

    Monograph manuscripts should be reasonably self-contained and rounded off. Thus they may, and often will, present not only results of the author but also related work by other people. They may be based on specialised lecture courses. Furthermore, the manuscripts should provide sufficient motivation, examples and applications. This clearly distinguishes Lecture Notes from journal articles or technical reports which normally are very concise. Articles intended for a journal but too long to be accepted by most journals, usually do not have this "lecture notes" character. For similar reasons it is unusual for doctoral theses to be accepted for the Lecture Notes series, though habilitation theses may be appropriate.

2. Manuscripts should be submitted either to Springer's mathematics editorial in Heidelberg, or to one of the series editors. In general, manuscripts will be sent out to 2 external referees for evaluation. If a decision cannot yet be reached on the basis of the first 2 reports, further referees may be contacted: The author will be informed of this. A final decision to publish can be made only on the basis of the complete manuscript, however a refereeing process leading to a preliminary decision can be based on a pre-final or incomplete manuscript. The strict minimum amount of material that will be considered should include a detailed outline describing the planned contents of each chapter, a bibliography and several sample chapters.

    Authors should be aware that incomplete or insufficiently close to final manuscripts almost always result in longer refereeing times and nevertheless unclear referees' recommendations, making further refereeing of a final draft necessary.

    Authors should also be aware that parallel submission of their manuscript to another publisher while under consideration for LNM will in general lead to immediate rejection.

3. Manuscripts should in general be submitted in English. Final manuscripts should contain at least 100 pages of mathematical text and should always include

    - a table of contents;
    - an informative introduction, with adequate motivation and perhaps some historical remarks: it should be accessible to a reader not intimately familiar with the topic treated;
    - a subject index: as a rule this is genuinely helpful for the reader.

    For evaluation purposes, manuscripts may be submitted in print or electronic form, in the latter case preferably as pdf- or zipped ps-files. Lecture Notes volumes are, as a rule, printed digitally from the authors' files. To ensure best results, authors are asked to use the LaTeX2e style files available from Springer's web-server at:

    ftp://ftp.springer.de/pub/tex/latex/svmonot1/ (for monographs).

Additional technical instructions, if necessary, are available on request from: lnm@springer.com.

4. Careful preparation of the manuscripts will help keep production time short besides ensuring satisfactory appearance of the finished book in print and online. After acceptance of the manuscript authors will be asked to prepare the final LaTeX source files (and also the corresponding dvi-, pdf- or zipped ps-file) together with the final printout made from these files. The LaTeX source files are essential for producing the full-text online version of the book (see www.springerlink.com/content/110312 for the existing online volumes of LNM).

   The actual production of a Lecture Notes volume takes approximately 12 weeks.

5. Authors receive a total of 50 free copies of their volume, but no royalties. They are entitled to a discount of 33.3% on the price of Springer books purchased for their personal use, if ordering directly from Springer.

6. Commitment to publish is made by letter of intent rather than by signing a formal contract. Springer-Verlag secures the copyright for each volume. Authors are free to reuse material contained in their LNM volumes in later publications: a brief written (or e-mail) request for formal permission is sufficient.

Addresses:
Professor J.-M. Morel, CMLA,
École Normale Supérieure de Cachan,
61 Avenue du Président Wilson, 94235 Cachan Cedex, France
E-mail: Jean-Michel.Morel@cmla.ens-cachan.fr

Professor F. Takens, Mathematisch Instituut,
Rijksuniversiteit Groningen, Postbus 800,
9700 AV Groningen, The Netherlands
E-mail: F.Takens@math.rug.nl

Professor B. Teissier, Institut Mathématique de Jussieu,
UMR 7586 du CNRS, Équipe "Géométrie et Dynamique",
175 rue du Chevaleret
75013 Paris, France
E-mail: teissier@math.jussieu.fr

For the "Mathematical Biosciences Subseries" of LNM:

Professor P.K. Maini, Center for Mathematical Biology,
Mathematical Institute, 24-29 St Giles,
Oxford OX1 3LP, UK
E-mail: maini@maths.ox.ac.uk

Springer, Mathematics Editorial I, Tiergartenstr. 17
69121 Heidelberg, Germany,
Tel.: +49 (6221) 487-8259
Fax: +49 (6221) 4876-8259
E-mail: lnm@springer.com